Pushing the Limits
iOS 7 Programming

Developing Extraordinary Mobile Apps for
Apple iPhone®, iPad®, and iPod touch®

Rob Napier and Mugunth Kumar

D1418202

A John Wiley and Sons, Ltd, Publication

This edition first published 2014

Registered office

John Wiley & Sons Ltd, The Atrium, Southern Gate, Chichester, West Sussex, PO19 8SQ, United Kingdom

For details of our global editorial offices, for customer services and for information about how to apply for permission to reuse the copyrigh
material in this book please see our website at www.wiley.com.

A catalogue record for this book is available from the British Library.

ISBN 978-1-118-81834-3 (pbk); ISBN 978-1-118-81832-9 (ebk); ISBN 978-1-118-81833-6 (ebk)

Set in 9.5/12 MyriadPro-Regular by TCS/SPS, Chennai, India

Printed in United States by Bind-Rite

Dedication

To the Neverwood Five. We're getting the band back together.
Rob

To my mother who shaped the first twenty years of my life
Mugunth

Publisher's Acknowledgements

Some of the people who helped bring this book to market include the following:

Editorial and Production

VP Consumer and Technology Publishing Director: Michelle Leete

Associate Director–Book Content Management: Martin Tribe

Associate Publisher: Chris Webb

Executive Commissioning Editor: Craig Smith

Project Editor: Tom Dinse

Copy Editor: Chuck Hutchinson

Technical Editor: Jay Thrash

Editorial Manager: Jodi Jensen

Senior Project Editor: Sara Shlaer

Editorial Assistant: Annie Sullivan

Marketing

Marketing Manager: Lorna Mein

Marketing Assistant: Polly Thomas

About the Authors

Rob Napier is a builder of tree houses, hiker, and proud father. He began developing for the Mac in 2005, and picked up iPhone development when the first SDK was released, working on products such as The Daily, PandoraBoy, and Cisco Mobile. He is a major contributor to Stack Overflow and maintains the *Cocoaphony* blog (cocoaphony.com).

Mugunth Kumar is an independent iOS developer based in Singapore. He graduated in 2009 and holds a Masters degree from Nanyang Technological University, Singapore, majoring in Information Systems. He writes about mobile development, software usability, and iOS-related tutorials on his blog (blog.mugunthkumar.com). Prior to iOS development he worked for Fortune 500 companies GE and Honeywell as a software consultant on Windows and .NET platforms. His core areas of interest include programming methodologies (Object Oriented and Functional), mobile development and usability engineering. If he were not coding, he would probably be found at some exotic place capturing scenic photos of Mother Nature.

About the Technical Editor

Jay Thrash is a veteran software developer who has spent the past three years designing and developing iOS applications. During his career, he developed a keen interest in the areas of user interaction and interface design.

Prior to settling down as an iOS developer, Jay has worked on a variety of platforms and applications, including flight simulators and web application development. He has also spent over six years in the PC and console gaming industry.

Acknowledgments

One more time, Rob thanks his family for their patience. He also thanks the Triangle Cocoaheads for their great ideas, inspiration, and support. Mugunth thanks his parents and friends for their support while writing this book. Thanks to Wiley, especially Craig Smith, for the continued support, encouragement, and nudging that it takes to get a book out the door. Thanks to Jay Thrash, Tom Dinse, and Chuck Hutchinson for keeping our words intelligible, and for keeping everything on schedule.

Contents

Introduction

In some ways, iOS 7 is the most radical change to iOS since the software development kit (SDK) was released in iPhone OS 2. The press and blogosphere have discussed every aspect of the new "flat" user interface and what it means for app developers and users. Suffice to say, no iOS upgrade has required so many developers to redesign their UI so much.

But in other ways, iOS 7 is a nearly seamless transition from iOS 6. Compared to the multitasking changes in iOS 4, iOS 7 may require very little change to your apps, particularly if you have either a very standard UI or a completely custom UI. For those on the extremes, the UI changes are either nearly free or irrelevant.

For all developers, though, iOS brings changes. There are more ways to manage background operations, but the rules for running in the background are even stricter than before. UIKit Dynamics means even more dynamic animations, but they can be challenging to implement well. TextKit brings incredible features to text layout, coupled with maddening limitations and bugs. iOS 7 is a mixed bag, both wonderful and frustrating. But you need to learn it. Users are upgrading quickly.

If you're ready to take on the newest Apple release and push your application to the limits, this book will get you there.

Who This Book Is For

This is not an introductory book. Many other books out there cover Objective-C and walk you step by step through Interface Builder. However, this book assumes that you have a little experience with iOS. Maybe you're self-taught, or maybe you've taken a class. Perhaps you've written at least most of an application, even if you haven't submitted it yet. If you're ready to move beyond the basics, to learn the best practices and the secrets that the authors have gleaned from practical experience writing real applications, this is the book for you.

This book also is not just a list of recipes. It contains plenty of sample code, but the focus is on discovering how to design, code, and maintain great iOS apps. Much of this book is about *why* rather than just *how*. You find out as much about design patterns and writing reusable code as about syntax and new frameworks.

What This Book Covers

The iOS platforms always move forward, and so does this book. Most of the examples here require at least iOS 6, and many require iOS 7. All examples use Automatic Reference Counting (ARC), automatic property synthesis, and object literals. Except in a very few places, this book does not cover backward compatibility. If you've been shipping code long enough to need backward compatibility, you probably know how to deal with it. This book is about writing the best possible apps using the best features available.

This book focuses on iPhone 5, iPad 3, and newer models. Most topics here are applicable to other iOS devices. Chapter 15 is devoted to dealing with the differences between the platforms.

What Is New in This Edition

This edition covers most of the newest additions to iOS 7, including the new background operations (Chapter 11), Core Bluetooth (Chapter 13), UIKit Dynamics (Chapter 19), and TextKit (Chapter 21). We provide guidance on how to best deal with the new flat UI (Chapter 2) and have added a chapter of "tips and tricks" you may not be aware of (Chapter 3).

We wanted to keep this book focused on the most valuable information you need for iOS 7. Some chapters from earlier editions have been moved to our website (iosptl.com). There, you can find chapters on common Objective-C practices, Location Services, error handling, and more.

How This Book Is Structured

iOS has a very rich set of tools, from high-level frameworks such as UIKit to very low-level tools such as Core Text. Often, you can achieve a goal in several ways. How do you, as a developer, pick the right tool for the job?

This book separates the everyday from the special purpose, helping you pick the right solution to each problem. You discover why each framework exists, how the frameworks relate to each other, and when to choose one over another. Then you learn how to make the most of each framework for solving its type of problem.

There are four parts to this book, moving from the most common tools to the most powerful. Chapters that are new in this edition or have been extensively updated are indicated.

Part I: What's New?

If you're familiar with iOS 6, this part quickly introduces you to the new features of iOS 7.

- **(Updated) Chapter 1: "The Brand New Stuff"**—iOS 7 adds a lot of new features, and here you get a quick overview of what's available.

- **(New) Chapter 2: "The World Is Flat: New UI Paradigms"**—iOS 7 dramatically changes what it means to look and act like an iOS app. In this chapter, you learn the new patterns and design language you need to make the transition.

Part II: Getting the Most Out of Everyday Tools

As an iOS developer, you have encountered a wide variety of common tools, from notifications to table views to animation layers. But are you using these tools to their full potential? In this part, you find the best practices from seasoned developers in Cocoa development.

- **(New) Chapter 3: "You May Not Know…"**—Even if you're an experienced developer, you may not be familiar with many small parts of Cocoa. This chapter introduces you to best practices refined over years of experience, along with some lesser-known parts of Cocoa.

- **(Updated) Chapter 4: "Storyboards and Custom Transitions"**—Storyboards can still be confusing and a bit intimidating for developers familiar with nib files. In this chapter, you learn how to use storyboards to your advantage and how to push them beyond the basics.

- **(Updated) Chapter 5: "Get a Handle on Collection Views"**—Collection views are steadily replacing table views as the preferred layout controller. Even for table-like layouts, collection views offer incredible flexibility that you need to understand to make the most engaging apps. This chapter shows you how to master this important tool.

- **(New) Chapter 6: "Stay in Bounds with Auto Layout"**—If there was one consistent message from WWDC 2013, it was this: use Auto Layout. There was hardly a UIKit session during the conference that didn't stress this point repeatedly. You may have shied away from Auto Layout due to its many Interface Builder problems in Xcode 4. Xcode 5 has dramatically better support for Auto Layout. Whether you're a constraints convert or longing to return to springs and struts, you should check out what's new in Auto Layout.

- **Chapter 7: "Better Custom Drawing"**—Custom drawing is intimidating to many new developers, but it's a key part of building beautiful and fast user interfaces. Here, you discover the available drawing options from UIKit to Core Graphics and how to optimize them to look their best while keeping them fast.

- **Chapter 8: "Layers Like an Onion: Core Animation"**—iOS devices have incredible facilities for animation. With a powerful GPU and the highly optimized Core Animation, you can build engaging, exciting, and intuitive interfaces. In this chapter, you go beyond the basics and discover the secrets of animation.

- **(Updated) Chapter 9: "Two Things at Once: Multitasking"**—Multitasking is an important part of many applications, and you discover how to do multiple things at once with operations and Grand Central Dispatch.

Part III: The Right Tool for the Job

Some tools are part of nearly every application, and some tools you need only from time to time. In this part, you learn about the tools and techniques that are a little more specialized.

- **Chapter 10: "Building a (Core) Foundation"**—When you want the most powerful frameworks available on iOS, you're going to want the Core frameworks such as Core Graphics, Core Animation, and Core Text. All of them rely on Core Foundation. In this chapter, you discover how to work Core Foundation data types so you can leverage everything iOS has to offer.

- **(Updated) Chapter 11: "Behind the Scenes: Background Processing"**—iOS 7 adds a lot more flexibility for background processing, but there are new rules you need to follow to get the most out of the changes. In this chapter, you discover the new NSURLSession and learn how to best implement state restoration.

- **Chapter 12: "REST for the Weary"**—REST-based services are a mainstay of modern applications, and you learn how to best implement them in iOS.

- **(New) Chapter 13: "Getting More Out of Your Bluetooth Devices"**—Apple keeps expanding iOS's capability to form ad hoc networks with other devices. This makes it possible to develop entirely new kinds of applications, from better games to micro-location services to easier file sharing. Get up to speed on what's new and jump into whole new markets.

- **Chapter 14: "Batten the Hatches with Security Services"**—User security and privacy are paramount today, and you find out how to protect your application and user data from attackers with the keychain, certificates, and encryption.

- **(Updated) Chapter 15: "Running on Multiple iPlatforms, iDevices, and 64-bit Architectures"**—The iOS landscape gets more complex every year with iPod touch, iPhone, iPad, Apple TV, and a steady stream of new editions. It's not enough just to write once, run everywhere. You need your applications to be their *best* everywhere. You learn how to adapt your apps to the hardware and get the most out of every platform.

■ **Chapter 16: "Reach the World: Internationalization and Localization"**—Although you may want to focus on a single market today, you can do small things to ease the transition to a global market tomorrow. Save money and headaches later, without interrupting today's development.

■ **Chapter 17: "Those Pesky Bugs:Debugging"**—If only every application were perfect the first time. Luckily, Xcode and LLDB provide many tools to help you track down even the trickiest of bugs. You go beyond the basics and find out how to deal with errors in development and in the field.

■ **(Updated) Chapter 18: "Performance Tuning Until It Flies"**—Performance separates the "okay" app from the exceptional app. It's critical to optimizing CPU and memory performance, but you also need to optimize battery and network usage. Apple provides an incredible tool for this task in Instruments. You discover how to use Instruments to find the bottlenecks and then how to improve performance after you find the problems.

Part IV: Pushing the Limits

This part is what this book is all about. You've learned the basics. You've learned the everyday. Now push the limits with the most advanced tools available. You discover the ins and outs of deep iOS.

■ **(New) Chapter 19: "Almost Physics: UIKit Dynamics"**—Apple constantly tries to make it easier to build dynamic, animated interfaces. UIKit Dynamics is its latest offering, giving a "physics-like" engine to UIKit. This tool is powerful but also can be challenging to use well. Learn to make the most of it.

■ **(New) Chapter 20: "Fantastic Custom Transitions"**—One of the favorite demos at WWDC 2013 was about custom transitions. Move beyond the "push" and learn how to make transitions dynamic and interactive.

■ **(Updated) Chapter 21: "Fancy Text Layout"**—The text-centric UI in iOS 7 demands incredible attention to detail in your font handling and text layout. TextKit brings many new features, from dynamic type to exclusion paths. It also brings bugs and frustrating limitations. No matter how you handle text, you need to master attributed strings. In this chapter, you learn the ins-and-outs of these powerful data structures and discover how to best use them with TextKit.

■ **Chapter 22: "Cocoa's Biggest Trick: Key-Value Coding and Observing"**—Many of Apple's most powerful frameworks rely on KVO for their performance and flexibility. You find out how to leverage the flexibility and speed of KVO, as well as the trick that makes it so transparent.

■ **(New) Chapter 23: "Beyond Queues: Advanced GCD"**—Dispatch queues are incredibly powerful tools that have become a key part of many applications. But there is more to Grand Central Dispatch than just queues. Learn about the tools such as semaphores, dispatch groups, and the incredibly powerful dispatch data and dispatch I/O.

■ **Chapter 24: "Deep Objective-C"**—When you're ready to pull back the curtain on how Objective-C really works, this is the chapter for you. You find out how to use the Objective-C runtime directly to dynamically modify classes and methods. You also learn how Objective-C method calls are dispatched to C function calls and how you can take control of the system to extend your programs in incredible ways.

You can skip around in this book to focus on the topics you need most. Each chapter stands alone, except for those that require Core Foundation data objects (particularly Core Graphics and Core Animation). Those chapters direct you to Chapter 10, "Building a (Core) Foundation," when you need that information.

What You Need to Use This Book

All examples in this book were developed with Xcode 5 on Mac OS X 10.8 and iOS 7. You need an Apple developer account to access most of the tools and documentation, and you need a developer license to run applications on your iOS device. Visit `developer.apple.com/programs/ios` to sign up.

Most of the examples in this book will run in the iOS Simulator that comes with Xcode 5. You can use the iOS Simulator without an Apple developer license.

Finding Apple Documentation

Apple provides extensive documentation at its website and within Xcode. The URLs change frequently and are often very long. This book refers to Apple documents by title rather than by URL. To find documents in Xcode, press Cmd-Option-? or click Help⇨Documentation and API Reference. In the Documentation Organizer, click the Search icon, type the name of the document, and then select the document from the search results. The following Figure shows an example of how to search for the *Coding Guidelines for Cocoa*.

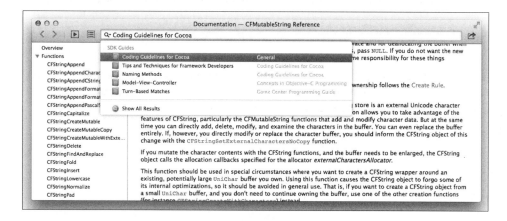

To find documents at the Apple developer site, visit `developer.apple.com`, click Member Center, and log in. Select the iOS Dev Center and enter the document title in the Search Developer search box.

The online documentation is generally identical to the Xcode documentation. You may receive results for both iOS and Mac. Be sure to choose the iOS version. Many iOS documents are copies of their Mac counterparts and occasionally include function calls or constants that are not available on iOS. This book tells you which features are available on iOS.

Source Code

As you work through the examples in this book, you may choose either to type in all the code manually or to use the source code files that accompany the book. All the source code used in this book is available for download at either `http://iosptl.com/code` or `www.wiley.com/go/ptl.ios7programming`.

For example, you can find the following sample code online in the Chapter 21 folder, in the SimpleLayout project, and the `CoreTextLabel.m` file:

CoreTextLabel.m (SimpleLayout)

```
- (id)initWithFrame:(CGRect)frame {
  if ((self = [super initWithFrame:frame])) {
    CGAffineTransform
    transform = CGAffineTransformMakeScale(1, -1);
    CGAffineTransformTranslate(transform,
                               0, -self.bounds.size.height);
    self.transform = transform;
    self.backgroundColor = [UIColor whiteColor];
  }
  return self;
}
```

Some source code snippets shown in the book are not comprehensive and are meant to help you understand the chapter. For those instances, you can refer to the files available on the website for the complete source code.

Errata

We try to get everything right, but sometimes things change, sometimes we mistype, and sometimes we're just mistaken. See `http://iosptl.com/code` for the latest updates, as well as blog posts on topics that haven't made the book yet. Send any mistakes you find to `robnapier@gmail.com` and `contact@mk.sg`.

Part I

What's New?

## Chapter 1

The Brand New Stuff

iOS 7 gives your Apple devices a fresh user interface, but the beauty of iOS 7's user interface changes are not just skin deep. Like this book, iOS 7 is a major update that benefits both users and developers. We start the chapter discussing the new UI and how that new UI will impact your app. Moving on, we discuss the various SDK additions and the new IDE and Compiler, Xcode 5/LLVM 5.

iOS 7 also brings a wealth of new core features to the SDK that can help your app stand out. iOS 7 adds two new technologies, UIKit Dynamics and UIMotionEffects, that can help you add animations that were not easy to do with Quartz Core. Custom transitions between views that were possible only when using Storyboards are now extended to support transitions between any two view controllers. The built-in Calendar and Photos apps are good examples that use custom transitions to seamlessly provide a sense of where the user is within the app. TextKit, a much improved (easy-to-use) version of Core Text that is Objective-C based, is arguably the most important and interesting addition. Sprite Kit, Dynamic Text, tighter integration with MapKit, true multitasking for all applications, and much better Bluetooth LE support are other SDK additions, to name a few.

The new features don't stop there. Along with these new SDK features, iOS 7 has a new IDE, Xcode 5, which is rewritten completely using the Automatic Reference Counting (ARC) compiler. Xcode 5 is much faster and less prone to crashes. The new IDE also brings in an updated compiler, LLVM 5.

> **Apple has again made huge changes to the platform that are disruptive enough to make most of the apps on the already over crowded App Store obsolete. This means that it's a niche market once again for a variety of apps and it's as good to be an iOS developer as it was in 2008 when iOS SDK was launched.**

Now let's delve deeper to gain an understanding of these new features, starting with the new UI.

The New UI

The changes to iOS 7's UI are profound, not just "skinning." The visual realism of the UI elements is muted and subdued while real world physics like motion realism is accentuated.

iOS 7 focuses on the following main themes: deference, clarity, and depth. Content is more important than chrome. As much as possible, apps now use the full screen to display content, and what was previously "chrome" is now translucent and shows muted, blurred content behind. All toolbars, navigation bars, and status bars also are now translucent. In addition to being translucent, they also blur and saturate the colors of the UI behind them, giving users a sense of depth, as you can see in Figure 1-1.

Figure 1-1 The area around the Timer tab of the Clock app shows a blurred black clock face behind it

Spatial depth is represented not just through the translucence of toolbars, navigation bars, and status bars, but throughout the UI. The home button on the multitasking bar is a great example of how spatial depth is shown in the iOS 7 UI. When you double tap the home button, you see app screenshots along with app icons. But that's not what is important. Pay close attention to the animation used: The whole screen "zooms out" to show all other running apps. Tapping an app icon "zooms" to that app. The calendar app also uses animation to provide a sense of spatial depth. Navigating from the year view to month view to date view uses a subtle transition that enhances the perception of spatial depth. Tapping a month zooms into the month, and tapping a year (from the month's screen) zooms out. Similarly, tapping an app icon launches the app by zooming in, not from the center of the screen as in prior iOS versions, but from the app icon's center. The reverse happens when you close the app. The app closes to the icon and not to the center of the screen, giving users a sense of where they are.

The new notification center, control center, and even some modal view controllers such as iPod's album list view controller also use blur effects to show what is "behind" them. Showing what is "behind" gives users a sense of where they are within your app.

The most interesting part of this new change is, as a developer, in most cases, you don't have to do any extra work to get these features working in your app if you have built your app using UIKit. You learn more about the new features of UIKit and how to design your app for iOS 7 in Chapter 2.

UIKit Dynamics and Motion Effects

The iOS 7 SDK adds new classes that allow you to add realism in motion to your own user interface. You can add realistic motion to any object that conforms to the `UIDynamicItem` protocol. Starting with iOS 7, all `UIViews` conform to this protocol, which means that you can add realistic motion to any of the `UIView` subclasses (including `UIControl`) in your application.

Implementing physics like realistic motion was possible prior to iOS 7 using animation methods in the `QuartzCore.framework`. But the new APIs in iOS 7's UIKit makes this task much easier. What makes implementing realistic motion in iOS 7 easier is that you specify the UIKit's dynamic behavior in a declarative fashion while `QuartzCore.framework` is intrinsically imperative. You do this by "attaching" dynamic behaviors to your `UIViews` (or one of its subclasses).

You implement realism in motion by adding a `UIDynamicAnimator` to a view and attaching one of the following dynamic behaviors, namely, `UIAttachmentBehavior`, `UICollisionBehavior`, `UIGravityBehavior`, `UIPushBehavior`, `UISnapBehavior` to the subviews of the view. The properties exposed by these behaviors can be used to customize and tweak the behavior. The dynamic animator takes care of the rest. As I said previously, the coding is more declarative and less imperative. You tell the `UIDynamicAnimator` what you want, and it does the right thing for you.

In addition to `UIViews`, `UICollectionViewLayoutAttributes` also conform to the `UIDynamicItem` protocol. That means you also can attach dynamic behaviors to collection view items. A good example of this in practice is the built-in Messages app. As you scroll up and down the message thread, you see a realistic bouncy motion between the chat bubbles. In fact, the Messages app no longer uses table views and uses collection views and dynamic behavior attached to its items.

`UIMotionEffect` is another new class added in iOS 7 that helps you add motion realism to your user interface. Using UIKit Dynamics, you simulated motion realism based on physics that you defined programatically. Using `UIMotionEffect`, you can simulate motion realism based on the device's motion. Anything that can be animated using a `CAAnimation` can be animated using `UIMotionEffect`. Whereas `CAAnimation`'s animation is a function of time, `UIMotionEffect`'s animation is a function of device motion.

You learn more about UIKit Dynamics and UIMotionEffects in Chapter 19. Chapter 5 covers applying UIKit Dynamics to your collection views like the Messages app.

Custom Transitions

Prior to iOS 7, custom transitions were available only for storyboard-created user interfaces. You normally create a subclass of `UIStoryboardSegue` and override the `perform` method to perform a transition. iOS 7 takes this concept to the next level by allowing any view controller you push using the method `pushViewController:animated:` to use your own custom animation. You do this by implementing the protocol `UIViewControllerAnimatedTransitioning`.

In Chapter 20, you learn how to create a transition using `UIStoryboardSegue` (the classical way) and the new way using `UIViewControllerAnimatedTransitioning`.

New Multitasking Modes

From iOS 4 to iOS 6, only selected categories of apps were allowed to run in the background. Examples include a location-based app that needs a user's location whenever it changes or a music player app that needs to stream and play back music in the background. In iOS 7, Apple finally allows all apps to run in the background. But unlike the desktop equivalent, apps don't truly run in the background. Instead, an app can register itself to be launched periodically so that it can fetch content in the background. A new `UIBackgroundMode` key, `fetch`, is introduced in iOS 7 to enable this capability.

Another interesting addition is the `remote-notification` background mode that allows an app to download or fetch new data when a remote notification (push notification) is received.

New multitasking modes are covered extensively in Chapter 9.

In addition, the `bluetooth-central` and `bluetooth-peripheral` background modes now support a new technology called State Preservation and Restoration. This technology is similar to the UIKit's equivalent by the same name. Bluetooth-related multitasking modes are covered in Chapter 13.

Text Kit

Text Kit is a fast, modern Unicode text layout engine built on top of Core Text. Whereas Core Text heavily depends on `CoreFoundation.framework`, Text Kit is more modern and uses `Foundation.framework` classes. This also means that Text Kit is ARC (Automatic Reference Counting) friendly, unlike Core Text. Prior to iOS 7, most of the UIKit elements that render text—namely, `UILabel`, `UITextView`, and `UITextField`—used WebKit. Apple rewrote these classes and built WebKit on top of Text Kit, which means that Text Kit is a first-class citizen in UIKit. This means you can and you should use TextKit for all text rendering. Text Kit centers around three classes—`NSTextStorage`, `NSLayoutManager`, and `NSTextContainer`—in addition to the extensions to `NSAttributedString`. There is no separate framework for TextKit per se. The related classes and extention methods are added to `UIKit.framework` and `Foundation.framework`. You learn about Text Kit in Chapter 21.

Dynamic Type

With iOS 7, users can set how big or small the size of text should be. Almost all built-in apps respect this setting and dynamically change the size of the text based on the user's preference. Note that increasing text size doesn't always make the font's point size bigger. When the text size setting is set to values that might render smaller point size text illegible, iOS 7 automatically makes the text bolder. By incorporating the Dynamic Type feature into your app, you can also dynamically change the size of text based on the Text Size setting on the iOS Settings app. There is one downside to Dynamic Type in iOS 7: It doesn't support custom fonts as of iOS 7. You learn about Dynamic Type in Chapter 21.

MapKit Integration

iOS 7 adds even tighter integration with Apple Maps. You can now show 3D maps using the `MKMapView` object. If you are a mapping provider, you can use the same `MKMapView` control but provide your own map tiles to replace the built-in tiles. Additionally, you can use `MKDirections` to show directions right on the map without exiting your app.

SpriteKit

SpriteKit is Apple's answer to cocos2d and box2d. SpriteKit provides a clean wrapper around OpenGL, much like `QuartzCore.framework`. Whereas `QuartzCore.framework` is predominantly used for UI animation, SpriteKit is more geared toward making games that simulate 3D (aka 2.5D games). SpriteKit is not covered in this edition of the book.

LLVM 5

Another important change along with iOS 7 is Xcode 5 and the new LLVM 5 compiler. LLVM 5 has both user-facing features and performance features. LLVM 5 fully supports generating armv7s and arm64 instruction set; this means that your app will run faster on iPhone 5s and the latest iPad Air just by recompiling with the new LLVM 5 compiler.

On the C++ side, LLVM 5 now fully supports the C++ 11 standard and uses the new libc++ library instead of the old gnuc++ library.

On the Objective-C side, LLVM 5 adds new compiler warnings and enables some already-existing warnings by default. For example, the compiler notifies you of dead code—code that cannot be reached from anywhere else.

Arguably the most important addition is the treatment of enums as first-class citizens. Enums are proper types, and LLVM 5 shows a warning when you attempt a conversion from one enumeration type to another.

LLVM also detects whether a selector you use is declared in the scope. If it is not declared, LLVM 5 now warns you.

Enhancements to the Objective-C Language

The Objective-C language has two major enhancements this year. They are modules and some changes to Automatic Reference Counting.

Modules

Apple introduced modules, which are intended to replace `#import` statements. For more than 30 to 40 years, `#import` statements were used in C/C++, and they were great for including a third-party library or framework. Over the years, the size of C/C++ applications grew, and `#import` statements started having serious compilation complexities. Every single `.m` file can potentially have several thousand lines of code just because of a few import statements, and some of these imports are, at times, repeated over many files. For example, consider `#import <UIKit/UIKit.h>` or `#import <Foundation/Foundation.h>`.

The problem was that whenever you modify a file, the compiler has to compile that file and all the code included by that file. This clearly increased the compilation time to O(S × H), where S is the number of source code files and H is the number of included header files. To circumvent this problem, some seasoned developers used to write forward declarations in the header files and import statements in the implementation files and add often-used header files into the precompiled header files. This approach solved the problem to some extent, probably by reducing the compilation time. But sometimes, adding not-so-commonly used header files to precompiled header files causes namespace pollution, and a badly written framework (that defines a class or a macro without a prefix) can get overwritten by some other class in another badly written framework.

To solve all these issues, Apple introduced modules. Modules encapsulate a framework and import statements into your code semantically instead of copying and pasting the content of the framework into your code.

Modules can reduce the compilation complexity from O(S × H) to O(S + H), which means that no matter how many times you import a module, it gets compiled once. That means, S source files and H header files are compiled S + H times instead of S × H times. In fact, the H header files are precompiled into dynamic libraries (dylib) and will be directly linked automatically, therefore reducing the compile time from O(S × H) to just O(S).

Modules are supported only if you use the latest Xcode 5 with the latest LLVM 5 compiler. More importantly, modules are available only for the latest SDK, iOS 7. If your project still has to support iOS 6, you will probably have to wait. When you use modules to import built-in libraries, you see much-improved compilation speeds, especially on larger projects. All new projects created in Xcode 5 automatically use modules. When you open an existing project in Xcode 5, you can add module support by turning it on from the Build Settings tab. This setting is shown in Figure 1-2.

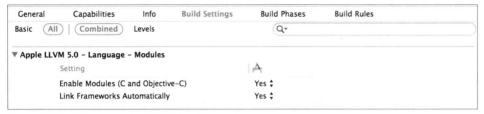

Figure 1-2 Xcode 5 showing the LLVM 5's Modules setting

Modules also eliminate the need to link the framework explicitly in your project by adding them from the Build Phases tab of the project's settings editor. However, modules on this version of Xcode/LLVM are supported only for built-in libraries. This might change in the near future (Read: iOS 8).

Improvements to ARC

LLVM 5 has better detection of retain cycles and also can predict whether a weak variable can become nil before being captured within a block. But what is more interesting is the ARC integration with Core Foundation classes. In prior versions of LLVM, the compiler mandated that you should bridge the cast even if there is no change in ownership. With LLVM 5, this case is implicit. That means, when there is no ownership change, you no longer need to "bridge" the cast between a Core Foundation object and an equivalent Foundation object.

Core Foundation and improvements to ARC are explained in detail in Chapter 20.

Xcode 5

Xcode 4 was an ambitious project at Apple. In 2011, Apple completely threw away the separate Xcode and Interface Builder in Xcode 3, something that had been around for almost a decade and integrated the capability to edit Interface Builder files right within Xcode. However, Xcode 4 was slow and often crashed a lot, including during some of the Worldwide Developers Conference (WWDC) sessions.

The main reason for Xcode 4's slow performance was that it was still using the old GCC compiler and used garbage collection (instead of ARC). Xcode 5 has few user-facing changes but is completely new from the ground up. It is compiled using LLVM 5 and rewritten using ARC, and that means it's more memory efficient and faster on your machine.

NIB File Format Changes

The first major change Xcode 5 brings up is a new nib file format that is easier to read. The format is still in XML but is more human friendly. What that means is, as a developer, you are able to fix merge conflicts in your XIB files much easier.

However, the new file format is not backward compatible with the older Xcode, but in most cases that shouldn't be a problem. You should ensure that all developers in your team use the latest Xcode. Xcode 4.6 cannot open the new file format.

Source Code Control Integration

Xcode 5 now has support for merging and switching Git branches right from Xcode. Xcode 5 also allows pushing and pulling commits from a remote repository right from the IDE. This means you will spend less time outside Xcode (in Terminal or a third-party Git client) and more time in Xcode.

Automatic Configuration

Xcode 4 had integration with the iOS developer program portal and allowed you to create and download provisioning profiles automatically. Xcode 5 takes this integration to the next level by allowing you to sign in to Xcode with your Apple ID. Xcode 5 automatically knows the various teams you work with and brings them all into Xcode. You can even set the team to a project, and Xcode automatically creates the necessary provisioning profiles on the developer portal for you.

Xcode 5 adds a new feature called capabilities that allows you to set high-level Apple features such as turning on In-App Purchases, iCloud, or Game Center right from Xcode. Xcode now automatically downloads the necessary provisioning profiles based on the new App ID and even updates your `Info.plist` file if required.

Refinements to the Debug Navigator

Xcode's debug navigator now shows real-time memory, CPU usage, and energy usage of your application right within Xcode. You no longer have to launch Instruments and profile your app as often. The debug navigator also lets you launch Instruments with the CPU/Memory profiling tool right from Xcode. Figures 1-3 and 1-4 show the CPU and Memory report features in Xcode 5.

Figure 1-3 Xcode 5's debug navigator showing CPU usage of the currently running application

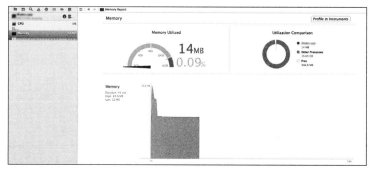

Figure 1-4 Xcode 5's debug navigator showing memory usage of the currently running application

Documentation Browser

Xcode 5 has an entirely revamped documentation browser complete with tabbed browsing. You can bookmark documentation pages (unfortunately no iCloud sync support for documentation bookmarks). The table of contents is now presented in a separate tab outside the documentation content.

> **Dash, a third-party application, is much better and faster than even Xcode 5's documentation browser. It's free from Mac App Store, and I highly recommend that you check it out. Dash works out of the box with the docsets installed with Xcode and also provides documentation for third-party libraries such as cocos2d.**

Asset Catalogs

Asset Catalogs provide a new way to group artwork you use in your application. An Asset Catalog contains image sets (the images and resources you use within your application), app icons, and launch images. These launch images, app icons, and image sets are grouped based on the device they're designed for.

Asset Catalogs are mostly an Xcode feature, so that means you can use it even if your app needs to be supported on iOS 6. On iOS 6, Xcode ensures that `UIImage`'s `imageNamed:` method will return the correct image within the catalog. On iOS 7, Xcode 5 compiles your Asset Catalog into a runtime binary (`.car` file) that reduces the time it takes for your app to download.

Asset Catalogs also provide an option to create stretchable images by allowing you to specify a resizable area within your image. All you have to do is select an image set within an Asset Catalog and click the Start Slicing button on top of your image. Adjust the slicing handle to specify the resizable area. You can choose to resize only vertically, only horizontally, or both.

Creating resizable areas (see Figure 1-5) within an Asset Catalog is, however, available only for projects with iOS 7 and higher as deployment target.

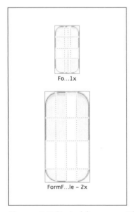

Figure 1-5 Resizable areas within an Image set in an Asset Catalog

Test Navigator

Xcode 5 makes writing unit test cases a first-class citizen. All new projects created in Xcode 5 automatically include a new testing framework called `XCTest.framework` that supposedly replaces OCUnit (`SenTestingKit.framework`). `SenTestingKit.framework` is still shipped with Xcode 5, and Xcode 5 has an option to automatically migrate your test cases from OCUnit to XCTest. The advantages are easy testing with command-line tools such as `xcodebuild`. Xcode also adds a new navigator called the test navigator along with the seven other navigators in the prior version of Xcode. The test navigator shows the results of running the test cases you have written.

> You can create a new type of breakpoint to stop the program execution when a test fails. It is illustrated in Figure 1-6. You can add this breakpoint from the breakpoints navigator.

Figure 1-6 Adding a test failure breakpoint

Continuous Integration

Xcode 5 supports continuous integration with Xcode services (running on OS X Mavericks server). Developers can now create "bots" that automate tasks such as "Run static analysis on every commit" or "Build the product every night at 12 a.m."

These bots are executed on the OS X Mavericks server, and you can view the results right in your client machine's test navigator.

> Xcode 5 doesn't natively support Jenkins, the pervasively used continuous integration tool. You still have to toy around with the command line, shell scripts, and post commit hooks (if any) manually. Xcode 5 also does not support continuous integration with OS X Mountain Lion server. There is, however, a slight advantage of using Xcode 5 + OS X Mavericks continuous integrations. Issues with your code, such as a breaking test, appear right in your client machine's test navigator.

Auto Layout Improvements

Xcode 5 improves developers' workflow with Auto Layout. The first and biggest change is that Xcode doesn't automatically add Auto Layout constraints. You have to manually ask Xcode to add these constraints for you. You can also delete Auto Layout constraints that were added by Xcode. If you delete a constraint and that causes layout ambiguities, Xcode warns you. Xcode also warns you if multiple constraints cannot be met simultaneously. Changes to Auto Layout and Xcode 5 cannot be explained in just a single section here, so we have added a whole new chapter, Chapter 6.

iOS Simulator

iOS simulator now supports iCloud, and the apps running inside the iOS simulator can access iCloud data. iOS simulator, however, drops support for Bluetooth LE emulation. (See Chapter 13 for information about supporting Bluetooth.)

Others

In addition to these new additions—the LLVM compiler, the Xcode IDE—iOS 7 also enhances the UIActivityViewController to support sharing your data with other nearby devices using AirDrop. A similar technology, called Multipeer Connectivity, allows you to connect and send any arbitrary data to a nearby device without a need for an access point. Multipeer connectivity also allows you to connect with devices that are nearby even if they are connected to different WiFi access points using a peer-to-peer WiFi network.

Summary

Whew! That's a lot of changes. I told you that iOS 7 is the biggest update to iOS since its inception. That's big for users and big for developers.

The adoption rate for iOS has always been much higher than that of the competition. When this was written, based on statistics from various websites, iOS 7 is used by more than 70% of the iOS devices in the market, and iOS 6 is used by less than 20% of devices. In fact, iOS 7 was adopted by more than 30% of devices in the first week of availability. At this rate, closer to 90% of devices will be running iOS 7 by the end of 2013. If you are starting to develop a new app, you might be better off to go with iOS 7 only. By the time you complete and publish the app, it's going to be an iOS 7 world.

If you have an existing app on App Store, you might consider rewriting it as a whole new app targeted just for iOS 7 and sell it along with your older app. When a considerable number of users have migrated to iOS 7, you can pull the older app off the App Store. There are plenty of killer features for the end users such as a brand new user interface, iTunes Radio, Air Drop, and enhanced multitasking that further augments the adoption rates. So why wait? Let's start the iOS 7 journey beginning with the new UI paradigms in the next Chapter.

Further Reading

Apple Documentation

The following document is available in the iOS Developer Library at `developer.apple.com` or through the Xcode Documentation and API Reference.

What's New in iOS 7

What's New in Xcode

Xcode Continuous Integration Guide

iOS 7 UI Transition Guide (TP40013174 1.0.43)

Assets Catalog Help

WWDC Sessions

The following session videos are available at `developer.apple.com`.

WWDC 2013 "Session 400: What's new in Xcode 5"

WWDC 2013 "Session 402: What is new in the LLVM compiler"

WWDC 2013 "Session 403: From Zero to App Store using Xcode 5"

WWDC 2013 "Session 406: Taking control of Auto Layout in Xcode 5"

WWDC 2013 "Session 409: Testing in Xcode 5"

WWDC 2013 "Session 412: Continuous integration with Xcode 5"

Chapter 2

The World Is Flat: New UI Paradigms

iOS 7 gives your Apple devices a fresh user interface (UI). The iOS 7 UI changes are the first major changes to happen to the iOS UI since its inception. iOS 7 focuses on three important characteristics: clarity, deference, and depth. It's important to understand these characteristics because they will help you design apps that look like native built-in system apps.

In this chapter, you learn about the various profound changes that iOS 7 introduces and how to enhance your app to use these new features. The first half of the chapter shows you every important UI paradigm that you should know and use to push the limits on your app. The latter half shows you how to transition your existing iOS 6 app to iOS 7 and (optionally) maintain backward compatibility.

Clarity, Deference, and Depth

Clarity simply means being clear to users. Most iOS users use your app in short bursts. Users open Facebook, look at your notifications or news feed, post a status update or a photo, and close the app. At least on the iPhone, very few apps are used for longer durations. In fact, statistics indicate that most apps are used for approximately 80 seconds per session. This means that you have limited time to convey the necessary information to users, so you should show what is most important to users clearly on the screen. If you are making a weather app, for example, temperature and weather conditions are the most important aspects that users are interested in. Make your UI clear enough that these two pieces of information are readily visible to users without their having to visually search for them.

Deference means the operating system doesn't provide UI that competes with your app's UI. Instead, it defers to your app and your content. This means every app will have its own unique look and feel. Prior to iOS 7, your app's UI is what your users see within the navigation bars and toolbars provided by the system. Visual chromes, bezels, and gradients are muted and replaced with a translucent bar that defers to the content behind it. The operating system's default apps defer to the content, and the UI, in most cases, play a supporting role.

Apps designed for iOS 7 should show a sense of "depth." Controls and content should be on different "layers." It all starts from the home screen. The parallax effect gives a sense that the icons are floating on top on a distinct layer. You can see a similar effect with modal alert views. The notification center and control center also show a sense of depth. Instead of using parallax effects, the notification center and control center use blur and translucency.

Animations Animations Animations

The first thing you will notice when you use iOS 7 is that visual realism (skeuomorphism) in apps has been muted considerably. Drop shadows are so subtle that they are barely visible. Glossy buttons are gone completely. Real-world skeuomorphic elements such as faux leather in the Notes app (iPhone) and the Contacts app (iPad) are gone.

Most importantly, application windows are full screen by default, and the status bar is now translucent and overlaid on top of the application window. Note that making applications fullscreen and the status bar translucent was an option until iOS 6. In iOS 6, you have to set the navigation bar's `translucent` property to `YES` and the `wantsFullScreenLayout` of the view controller to `YES explicitly`. In iOS 7, all windows default to this setting.

While visual realism may be muted, iOS 7 augments motion realism. Augmented motion realism starts with the lock screen. Try "lifting" the lock screen by panning the camera button and let it "drop" halfway through the screen. You will see that the lock screen bounces as it hits the bottom of the screen. Now try to "smash" the lock screen by pushing it toward the bottom of the screen. You will see that the lock screen now bounces more, just like a *real-world object* bouncing on a surface. So skeuomorphism hasn't been removed completely from iOS. Whereas previous versions of iOS gave importance to visual skeuomorphism, iOS 7 places more importance on physical skeuomorphism.

> iOS 7 is still skeuomorphic. Instead of visual skeuomorphism, the emphasis is placed on physical skeuomorphism. Objects onscreen obey the laws of physics and behave like real-world objects.

UIKit Dynamics

Implementing physics in your app is easy. iOS 7 introduces the new class UIDynamicAnimator, which emulates real-world physics. You can create a `UIDynamicAnimator` object and add it to your view. This view is usually the root view of your view controller and is also called a reference view in a UIKit dynamics context. Subviews of this reference view now obey the laws of physics based on the behavior attached to them.

Behaviors are declaratively defined and added to subviews. In fact, you can add behaviors to any object that conforms to the `UIDynamicItem` protocol. `UIView` and its subclasses (including `UIControl`) conform to this protocol. That means you can attach a behavior to almost anything that is visible onscreen.

An interesting addition here is that UICollectionViewLayoutAttributes conforms to the `UIDynamicItem` protocol, and that allows collection view elements to have a dynamic behavior. The default Messages app is an example of this behavior.

UIMotionEffect

Physical skeuomorphism doesn't stop there. Parallax effects on your home screen, alert boxes that appear as if they are floating on top of your views, and Safari's new tab-changing UI that shows "more" content when you tilt your device are all imitations of real-world behavior. You can easily add these features to your app using `UIMotionEffect`. A new class in iOS 7, `UIInterpolatingMotionEffect`, enables you to add these effects effortlessly to your app. Think of `UIMotionEffect` as a class that is similar to `CAAnimation`. `CAAnimation` animates layers to which it is attached. `UIMotionEffect` animates views to which it is attached. Whereas the animation of `CAAnimation` is a function of time, the animation of `UIMotionEffect` is a function of device motion. UIKit Dynamics and UIMotionEffects are covered in detail in Chapter 19.

> **Whereas the animation of** `CAAnimation` **is a function of time, the animation of** `UIMotionEffect` **is a function of device motion.**

Tint Colors

`UIViews` have a new property called `tintColor` that enables you to set a tint color for your app. All UI elements that are subviews of a view use the `tintColor` of the parent if a `tintColor` is not specified for the subview. That means you can set a global tint color for your app by setting a `tintColor` for your application's window.

It's important to note that when a modal view is presented, iOS 7 dims all UI elements behind it. If you have a custom view and you use the `tintColor` of the parent to do some custom rendering, you should override the tintColorDidChange method to update changes.

> **You also can set the** `tintColor` **from the File inspector panel within Xcode's storyboard editor.**

Layering and Context through Translucency

UIKit Dynamics and UIMotionEffect help users understand the spatial depth within your app. One other feature of iOS 7 that enhances spatial depth is the consistent use of blur and translucency on most modal windows. The Control Center and Notification Center blur the background in a subtle way so you can still figure out what is beneath.

iOS 7 creates these effects by directly reading from the GPU memory. Unfortunately, iOS 7 doesn't have any SDKs that allow you to do this easily, probably for security reasons. For example, you can't do something like this:

```
self.view.blurRadius = 50.0f;
self.view.saturationDelta = 2.0f;
```

However, the sample code from one of the WWDC sessions has a UIImage category, `UIImage+ImageEffects`, (you can get this from this books' sample code) that lets you do it, although using this approach is difficult and not straightforward. But you want to push the limits, and that's why you are reading this book, right? So, now you can re-create the effect in your own app.

Now I'll show you how to create a layer that looks like your notification center's or control center's background. Start by creating a single view application and add a beautiful background image of your choice and a button. You are going to write a button action handler that shows a pop-up blurring the background just behind the pop-up.

A shortcut (and kind of a cheat) is to create a hidden `UIToolbar` and use its layer instead of creating a new one. But this hack is fragile, and I don't recommend it. A better way is to take a screenshot using iOS 7's new UIScreenshotting APIs, crop the screenshot image, apply the blur effect on the screenshot image, and use the blurred image as the background for your pop-up.

Now look at the code:

Using UIImage+ImageEffects to Create a Blurred Popup Layer (SCTViewController.m)

```
// create the layer
self.layer = [CALayer layer];
self.layer.frame = CGRectMake(80, 100, 160, 160);
[self.view.layer addSublayer:self.layer];

// Take the screenshot
float scale = [UIScreen mainScreen].scale;
UIGraphicsBeginImageContextWithOptions(self.view.frame.size, YES, scale);
[self.view drawViewHierarchyInRect:self.view.frame afterScreenUpdates:NO];
__block UIImage *image = UIGraphicsGetImageFromCurrentImageContext();
UIGraphicsEndImageContext();

// Crop the screenshot
CGImageRef imageRef = CGImageCreateWithImageInRect(image.CGImage,
    CGRectMake(self.layer.frame.origin.x * scale,
               self.layer.frame.origin.y * scale,
               self.layer.frame.size.width * scale,
               self.layer.frame.size.height * scale));
image = [UIImage imageWithCGImage:imageRef];

// Apply the effect
image = [image applyBlurWithRadius:50.0f
                        tintColor:
[UIColor colorWithRed:0 green:1 blue:0 alpha:0.1]
            saturationDeltaFactor:2
                        maskImage:nil];
// assign it to the new layer's contents
self.layer.contents = (__bridge id)(image.CGImage);
```

The method `applyBlurWithRadius:tintColor:saturationDeltaFactor:maskImage:` is in the category UIImage+ImageEffects. You can download the complete code, including the category, from the book's website.

Figure 2-1 shows the sample code in action.

Figure 2-1 Screenshot of the app running in iOS 7 before and after the popup is shown

Dynamic Type

In iOS 7, users can set how big or small the size of text should be. The built-in Mail, Calendar, and most other apps respect this setting and dynamically change the size of the text based on user preference. Note that increasing text size doesn't always make the font's point size bigger. When the text size setting is set to values that might render smaller-point-size text illegible, iOS 7 automatically makes the text bolder. By incorporating the Dynamic Type feature into your app, you can also dynamically change the size of text based on the Text Size setting in the iOS Settings app. There is one downside to Dynamic Type in iOS 7, however: It doesn't support custom fonts as of when this chapter was written. You learn about Dynamic Type in Chapter 21.

Custom Transitions

Another new feature of iOS 7 is that Apple has minimized the number of screen transitions that happen in most of its built-in apps. When you open the Calendar app and tap on a date, the Date view animates from the Month view with a custom transition. Along similar lines, when you create a new event, the process of changing the event's start and end time is animated using a custom transition. Note that this animation used to be the navigation controller's default push navigation animation in previous versions of iOS. Another example is the photo application's Photos tab. The uber-long camera roll is now replaced with a collection view that uses a custom transition to navigate between Years, Collections, Moments, and Single Photo views. iOS 7's new UI paradigm emphasizes letting users know where they are, instead of letting users lose their way among the myriad of pushed view controllers. In most cases, these animations are done using a custom transition, as shown in Figure 2-2.

When you design your apps for iOS 7, you should carefully consider whether using a custom transition will help users know where they are within your app. The iOS 7 SDK adds APIs that enable you to do these animations without much difficulty.

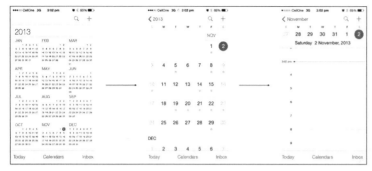

Figure 2-2 Calendar screen showing the screens that use custom transitions

Types of Custom Transitions

The iOS 7 SDK supports two kinds of custom transitions: Custom View Controller Transitions and Interactive View Controller Transitions. Custom View Controller Transitions were possible previously using storyboards and custom segues. Interactive View Controller Transitions enable users to use a gesture (usually a pan gesture) to control the amount of transition (from start to the end). So as a user pans or swipes his finger, transitions occur from one view controller to another.

When the transition is a function of time, it's usually a Custom View Controller Transition; and if it is a function of a parameter of a gesture recognizer or any such event, it is usually an Interactive View Controller Transition.

For example, you can think of the navigation controller's push transition (as of iOS 6) as a Custom View Controller Transition, whereas you can think of the `UIPageViewController` page transition as an Interactive View Controller Transition. When you use a `UIPageViewController` to page between views, the transition is not timed. Instead, the page transition follows your finger movement; hence, this is an Interactive View Controller Transition. The `UINavigationController` transition (in iOS 6) occurs over a period of time, so transitions like these are Custom View Controller Transitions.

The iOS 7 SDK allows you to customize almost any kind of transition: View Controller presentations and dismissals, the `UINavigationController` push and pop transition, the `UITabbarController` transitions (`UITabBarController` does not animate view controllers by default), or even a Collection view's layout change transitions.

Custom View Controller Transitions are easy to use compared to Interactive View Controller Transitions. I show you how to create a Custom View Controller Transition in this chapter. Interactive View Controller Transitions are relatively harder to use; that topic is covered in Chapter 20.

Transitioning (Migrating) Your App to iOS 7

So far in this chapter, you have learned the major UI paradigms introduced in iOS 7 and implemented one of them: the blur effect. In this section, I show you how to transition your app from iOS 6 to iOS 7.

As you already know, the iOS 7 UI is radically different from its predecessor. As such, you have to be aware of plenty of API changes. If possible, avoid supporting iOS 6 altogether as the adoption rate of iOS 7 is expected to be much higher than that of iOS 6. But if your business requires supporting iOS 6, continue reading the next section, where I show you how to support iOS 7 without sacrificing iOS 6.

UIKit Changes

In iOS 7, almost every UI element has changed. Buttons don't have borders, switches and segment controls are smaller, and sliders and progress indicators are thinner. If you have been using Auto Layout, most of these changes should not affect you much. If you haven't been using Auto Layout, now is the right time to update your nib files to use Auto Layout. If you don't use Auto Layout, supporting iOS 6 and iOS 7 on both 3.5- and 4-inch screens is hard without writing tons of layout code.

If you have been resisting Auto Layout, you have to start learning it now. Working with Auto Layout in Xcode 5 is a lot better than using Xcode 4.x. This book covers Auto Layout extensively in Chapter 6.

Custom Artwork

It's likely that most of your apps use custom artwork to "skin" the apps, and you need to think carefully about how you design that art. The reason is that iOS 7 design is very "flat." The visual skeuomorphism is muted down a lot. The reason for this is that prior to iPhone 4, almost all screens (phone/PC/Mac) had a pixel density of 70–160 ppi (pixels per inch). The iPhone 4's retina display doubled this density to 320 ppi. This pixel density is closer to what a healthy human eye can perceive and anything more is just extra data that cannot be perceived by a healthy human eye. This is also the reason why 300 ppi was used by the print industry.

The print industry does not mimic gloss, shiny buttons, drop shadows, bezels, gradients, or funky fonts. Have you ever seen a billboard or magazine cover that had artificial gloss or shine added to the artwork? No. Why? You don't need it. (Some magazine covers are glossy, but they are real gloss.) In fact, the gloss, shiny buttons, and so on were used as a crutch by software designers to make their design look nice on a (then) low-resolution (70–160 ppi) screen.

With iPhone 4's screen reaching pixel densities that are close to what you would see on a high quality printed magazine, adding gloss and shine to your UI felt redundant. This is why even before Apple, apps such as Twitterrific 5, LetterPress, and Clear (the todo app) heavily muted the interface. The Windows Phone 7 interface never had gloss, shine, or drop shadows from day 1.

Today, with the exception of the iPad mini, almost all iOS devices released by Apple in the two years from 2011-2012 use retina screens. "Flat" design is the future. Designing your UI with gloss will make your app look dated in iOS 7.

You will have to redesign your UI if it uses glossy or shiny buttons. Ask yourself, how would this UI look like if it were to be printed in a magazine? You should start thinking more like designers from the print industry than designers from the software industry.

Supporting iOS 6

Writing code that is backward compatible with iOS 6 is relatively harder than it used to be in the past. because iOS 7 is easily one of the major updates made to iOS since its inception. If your app requires iOS 6 support, start by skinning your iOS 6 app to look closer to the iOS 7 equivalent. That means, instead of using a flat button in iOS 7 and a glossy button in iOS 6, make the iOS 6 button flat. Start customizing the artwork for your iOS 6 version and try to make it look like the iOS 7 equivalent. When you are happy with the results, start adding iOS 7-specific features.

App Icon

iOS 7 uses a different app icon size than was used in iOS 6. It's 120×120 pixels for iPhone apps and 152×152 for iPad apps. You should update your app with these slightly bigger icons. Avoid using gloss or shine. The "Icon already includes gloss effects (UIPrerenderedIcon)" setting doesn't do anything in iOS 7. It might eventually be deprecated.

Launch Images

Your launch images should now be 480 points tall (or 568 on iPhone 5). Launch images in iOS 7 will be rendered below the status bar. If you provide a launch image that is shorter, you will see a black bar below, at the bottom of the screen.

Status Bar

The status bar in iOS 7 is translucent and blurs the content behind. In iOS 6 versions of your apps, you should make the status bar look flat using artwork that is less glossy (or has no gloss at all). In iOS 7, your view controllers will now extend to full screen, below the status bars. In the iOS 6 version of your app, change the status bar's style to UIStatusBarStyleBlackTranslucent to mimic this behavior.

Navigation Bar

The navigation bar in iOS 7 is also translucent by default and doesn't have a drop shadow. Instead, you see a one-pixel hairline at the bottom edge. Consider adding something like this for the iOS 6 version of your app instead of a drop shadow.

In iOS 6, setting the `tintColor` on your navigation bar changes the color of the bar. To get this effect in iOS 7, you should use `barTintColor`. The back button on your navigation bar is also different. You see a chevron and text instead of a button with a border. Remember, even the default buttons don't have borders, and the `UIButtonTypeRoundedRect` is deprecated. Customize your back button in the iOS 6 versions of your apps using the appearance proxy protocol of the navigation bar and set a back button image that looks similar to the iOS 7 equivalent.

The navigation bar's background images are usually 320 points by 44 points. In some cases you might have used a slightly taller navigation bar that adds a custom shadow. In iOS 7, bar background images are extended into the status bar instead of extending above your view controller as in iOS 6. Avoid using bar backgrounds that are larger than 44 points tall. You might have to redesign your bar backgrounds and think of different ways to add the shadow image. iOS 6 introduced a property called `shadowImage` to `UINavigationBar`. You should use this property to set the shadow to your navigation bar instead.

Toolbars

Toolbars also are translucent, but more important is that the buttons on toolbars don't have borders. If you have three or fewer buttons, consider using textual buttons instead of image-based buttons. For example, the Now Playing screen of iOS 7's Music app uses text-based buttons for Repeat, Shuffle, and Create (see Figure 2-3).

Figure 2-3 Screenshot showing the control buttons in iOS 7

View Controllers

In iOS 7, every view controller uses a full-screen layout. You can support this new layout in iOS 7 and preserve the old layout in iOS 6, but your app might look dated. Embrace the change and use the full-screen layout on your iOS 6 version. You can do this in iOS 6 by setting the `wantsFullscreenLayout` property of your view controller to `YES`. Note that this property is deprecated in iOS 7, and the behavior of the app in iOS 7 is undefined when this property is `NO`.

iOS 7 also enables you to use opaque bars. If you use opaque bars, you can control whether your view should appear below the translucent/opaque bars using the Interface Builder options. This is shown in Figure 2-4.

Figure 2-4 Interface Builder showing the option to extend your view below the translucent bars

You also see a Top layout guide and Bottom layout guide in your Interface Builder nib files (only when you turn on Auto Layout). You can use these guides to set constraints as well. So you can have a button that's always 50 points from the Top layout guide.

Table View Controllers

Table view controllers, especially the grouped table view style, no longer have the inset look. The separators are inset, start after the image, and run to the end of the screen. Cell separators are thinner, and the color also is lighter.

Section headers are lighter in color and are now a solid color instead of a gradient. In your iOS 7 version of the app, you can change this color by setting a new value to `sectionIndexBackgroundColor`. In iOS 6, you should instead return a view that looks similar to the iOS 7 equivalent in the tableView: viewForHeaderInSection: delegate.

Another important change is to the selection style. iOS 6 provided two styles: blue and gray. When you used the built-in style, the foreground color of the text was inverted from black to white. This no longer happens in iOS 7, and the highlight color is a lighter shade of gray. You can mimic this behavior by overriding the `setSelected:animated:` and `setHighlighted:animated:` methods in your `UITableViewCell` subclass.

The default swipe-to-delete gesture is changed from left-to-right to right-to-left. If you are using a right-to-left gesture in your table view cells to bring up a menu or perform another action, you have to consider other ways of doing that.

Pan Gestures

iOS 7 uses two pan gestures on all apps by default. The first is UIScreenEdgePanGestureRecognizer. In a navigation controller, swiping from the screen's edge lets the user go back to the previous view. This behavior is the default. If you are using a hamburger menu (colloquially called a side menu or side panels) as in Facebook's old version of the app, you should consider disabling the pan gesture that reveals the menu. In fact, with iOS 7's launch, Facebook completely got rid of the hamburger menu in favor of tab bars. The second gesture is a pan gesture from the bottom of the screen to reveal the control center. If your app uses a similar gesture to invoke a feature, you might have to redesign your interface.

Alert Views and Action Sheets

Alert views and Action sheets always use the system default style unless you create your own. If you are using your own custom alert view, consider adding `UIMotionEffect` to make the view appear as if it's floating on top of your view controller. You created the illusion of depth using drop shadows in iOS 6. You use motion effects instead in iOS 7. It's hard to mimic motion effects in iOS 6. Unless you are willing to spend the time and effort, consider leaving the drop shadows untouched in your iOS 6 version for this UI element.

Summary

This chapter introduced you to the new UI paradigm. You also learned the underlying reason behind the new design and learned in depth the different UI-related technologies that are new in iOS 7. Lastly, you learned about how to transition your apps to iOS 7 and (optionally) maintain backward compatibility. Writing code that works on both iOS 6 and iOS 7 and looks exceptionally good on both OSes is hard (at least relatively). If your business has the time and resources, do it. Otherwise, focus on using every possible feature of iOS 7 and make your app stand out.

iOS 7 provides developers another huge opportunity to be successful on the App Store by providing an iOS 7-only todo app, an iOS 7-only Twitter client, and an iOS 7-only calendar app, to name a few offerings. Now that you understand the iOS 7 UI paradigm, you can start on your next big thing.

Further Reading

Apple Documentation

The following document is available in the iOS Developer Library at `developer.apple.com` or through the Xcode Documentation and API Reference.

What's New in iOS 7

WWDC Sessions

The following session videos are available at developer.apple.com.

WWDC 2013, "Session 226: Implementing Engaging UI on iOS"

WWDC 2013, "Session 218: Custom Transitions Using View Controllers"

WWDC 2013, "Session 201: Building User Interfaces for iOS 7"

WWDC 2013, "Session 208: What's New in iOS User Interface Design"

WWDC 2013, "Session 225: Best Practices for Great iOS UI Design"

Part II

Getting the Most Out of Everyday Tools

Chapter 3
You May Not Know...

If you're reading this book, you likely already have a good grasp of iOS basics, but there are small features and practices that many developers are unfamiliar with, even after years of experience. In this chapter, you learn some of those tips and tricks that are important enough to know, but too small for their own chapter. You also learn some best practices that will keep your code more robust and maintainable.

Naming Best Practices

Throughout iOS, naming conventions are extremely important. In the following sections, you learn how to correctly name various items and why those names matter.

Automatic Variables

Cocoa is a dynamically typed language, and you can easily get confused about what type you are working with. Collections (arrays, dictionaries, and so on) don't have types associated with them, so it's very easy to code something accidentally like this:

```
NSArray *dates = @[@"1/1/2000"];
NSDate *firstDate = [dates firstObject];
```

This code compiles without a warning, but when you try to use `firstDate`, it will likely crash with an unknown selector exception. The mistake is calling an array of strings `dates`. This array should have been called `dateStrings`, or it should have contained `NSDate` objects. This kind of careful naming will save you a lot of headaches.

Methods

The names of methods should make it clear what types they accept and return. For instance, the name of this method is confusing:

```
- (void)add; // Confusing
```

It looks as though `add` should take a parameter, but it doesn't. Does it add some default object?

Names like these are much clearer:

```
- (void)addEmptyRecord;
- (void)addRecord:(Record *)record;
```

Now it's clear that `addRecord:` accepts a `Record`. The type of the object should match the name if there is any chance of confusion. For instance, this example shows a common mistake:

```
- (void)setURL:(NSString *)URL; // Incorrect
```

It's incorrect because something called `setURL:` should accept an `NSURL`, not an `NSString`. If you need a string, you need to add some kind of indicator to make this clear:

```
-   (void) setURLString: (NSString *) string;
-   (void) setURL: (NSURL *) URL;
```

This rule shouldn't be overused. Don't append type information to variables if the type is obvious. It's better to have a property called `name` than one called `nameString`, as long as there is no `Name` class in your system that might confuse the reader.

Methods also have specific rules related to memory management and key-value coding (KVC, discussed in detail in Chapter 22). Although Automatic Reference Counting (ARC) makes some of these rules less critical, incorrectly naming methods can lead to very challenging bugs when ARC and non-ARC code interact (including non-ARC code in Apple's frameworks).

Method names should always be "camelCase" with a leading lowercase letter.

If a method name begins with `alloc`, `new`, `copy`, or `mutableCopy`, the caller owns the returned object (that is, it has a net +1 retain count that the caller must balance). This rule can cause problems if you have a property with a name like `newRecord`. Rename this to `nextRecord` or something else.

Methods that begin with `get` should return a value by reference. For example:

```
-   (void) getPerson: (Person **) person;
```

Do not use the prefix `get` as part of a property accessor. The getter for the property `name` should be `-name`.

Property and Ivar Best Practices

Properties should represent the state of an object. Getters should have no externally visible side effects (they may have internal side effects such as caching, but those should be invisible to callers). Generally, they should be efficient to call and certainly should not block.

Avoid accessing ivars (instance variables) directly. Use accessors instead. I discuss a few exceptions in a moment, but first I want to discuss the reasons for using accessors.

Prior to ARC, one of the most common causes of bugs was direct ivar access. Developers would fail to retain and release their ivars correctly, and their programs would leak or crash. Because ARC automatically manages retains and releases, some developers may believe that this rule is no longer important, but there are other reasons to use accessors:

- **Key-value observing**—Perhaps the most critical reason to use accessors is that properties can be observed. If you don't use accessors, you need to make calls to `willChangeValueForKey:` and `didChangeValueForKey:` every time you modify a property's backing ivar. The accessor automatically makes these calls when they are needed.

- **Side effects**—You or one of your subclasses may include side effects in the setter. Notifications may be posted or events registered with `NSUndoManager`. You shouldn't bypass these side effects unless it's necessary. Similarly, you or a subclass may add caching to the getter that direct ivar access will bypass.

- **Lazy instantiation**—If a property is lazily instantiated, you must use the accessor to make sure it's correctly initialized.

- **Locking**—If you introduce locking to a property to manage multithreaded code, direct ivar access will violate your locks and likely crash your program.

- **Consistency**—One could argue that you should just use accessors when you know you need them for one of the preceding reasons, but this makes the code very hard to maintain. It's better that every direct ivar access be suspicious and explained instead of your having to constantly remember which ivars require accessors and which do not. This makes the code much easier to audit, review, and maintain. Accessors, particularly synthesized accessors, are highly optimized in Objective-C, and they are worth the overhead.

That said, you should not use accessors in a few places:

- **Inside accessors**—Obviously, you cannot use an accessor within itself. Generally, you also don't want to use the getter inside the setter either (this can create infinite loops in some patterns). An accessor should access its own ivar.

- **Dealloc**—ARC greatly reduces the need for `dealloc`, but it still comes up sometimes. It's best not to call external objects inside `dealloc`. The object may be in an inconsistent state, and it's likely confusing to the observer to receive several notifications that properties are changing when what's really meant is that the entire object is being destroyed.

- **Initialization**—Similar to `dealloc`, the object may be in an inconsistent state during initialization, and you generally shouldn't fire notifications or have other side effects during this time. This is also a common place to initialize read-only variables such as an `NSMutableArray`. This way, you avoid declaring a property `readwrite` just so you can initialize it.

Accessors are highly optimized in Objective-C and provide important features for maintainability and flexibility. As a general rule, you refer to all properties, even your own, using their accessors.

Categories

Categories allow you to add methods to an existing class at runtime. Any class, even Cocoa classes provided by Apple, can be extended with categories, and those new methods are available to all instances of the class. Declaration of a category is straightforward. It looks like a class interface declaration with the name of the category in parentheses:

```
@interface NSMutableString (PTLCapitalize)
- (void)ptl_capitalize;
@end
```

`PTLCapitalize` is the name of the category. Note that no ivars are declared here. Categories cannot declare ivars, nor can they synthesize properties (which is the same thing). You see how to add category data in the section "Associative References." Categories can *declare* properties because this is just another way of declaring methods. They just can't *synthesize* properties because that creates an ivar. The `PTLCapitalize` category doesn't require that `ptl_capitalize` actually be implemented anywhere. If `ptl_capitalize` isn't implemented and a caller attempts to invoke it, the system raises an exception. The compiler gives you no protection here. If you do implement `ptl_capitalize`, then *by convention* it looks like this:

```
@implementation NSMutableString (PTLCapitalize)
- (void)ptl_capitalize {
  [self setString:[self capitalizedString]];
}
@end
```

I say "by convention" because there is no requirement that this be defined in a category implementation or that the category implementation must have the same name as the category interface. However, if you provide an @implementation block named PTLCapitalize, it must implement all the methods from the @interface block named PTLCapitalize.

Technically, a category can override methods, but doing so is dangerous and not recommended. If two categories implement the same method, which one is used is undefined. If a class is later split into categories for maintenance reasons, your override could become undefined behavior, which is a maddening kind of bug to track down. Moreover, using this feature can make the code hard to understand. Category overrides also provide no way to call the original method. For debugging, I recommend swizzling, which is covered in Chapter 24.

> **Because of the possibility of collisions, you should add a prefix to your category methods, followed by an underscore, as shown in the** ptl_capitalize **example. Cocoa generally doesn't use embedded underscores like this, but in this case, it's clearer than the alternatives.**

A good use of categories is to provide utility methods to existing classes. When you do this, I recommend naming the header and implementation files using the name of the original class plus the name of the extension. For example, you might create a simple PTLExtensions category on NSDate:

NSDate+PTLExtensions.h

```
@interface NSDate (PTLExtensions)
- (NSTimeInterval)ptl_timeIntervalUntilNow;
@end
```

NSDate+PTLExtensions.m

```
@implementation NSDate (PTLExtensions)
- (NSTimeInterval)ptl_timeIntervalUntilNow {
  return -[self timeIntervalSinceNow];
}
@end
```

If you have only a few utility methods, it's convenient to put them together into a single category with a name such as PTLExtensions (or whatever prefix you use for your code). Doing so makes it easy to drop your favorite extensions into each project. Of course, this is also code bloat, so be careful about how much you throw into a "utility" category. Objective-C can't do dead-code stripping as effectively as C or C++.

+load

Categories are attached to classes at runtime. It's possible that the library that defines a category is dynamically loaded, so categories can be added quite late. (Although you can't write your own dynamic libraries in iOS, the system frameworks are dynamically loaded and include categories.) Objective-C provides a hook called +load that runs when the category is first attached. As with +initialize, you can use this to implement category-specific setup such as initializing static variables. You can't safely use +initialize in a category because the class may implement it already. If multiple categories implement +initialize, the one that runs is not defined.

I hope you're ready to ask the obvious question: "If categories can't use +initialize because they might collide with other categories, what if multiple categories implement +load?" This turns out to be one of the few really magical parts of the Objective-C runtime. The +load method is special-cased in the runtime so that every category may implement it and all the implementations run. There are no guarantees on order, and you shouldn't try to call +load by hand.

+load is called regardless of whether the category is statically or dynamically loaded. It's called when the category is added to the runtime, which often is at program launch, before main, but could be much later.

Classes can have their own +load method (not defined in a category), and they are called when the classes are added to the runtime. This approach is seldom useful unless you're dynamically adding classes.

You don't need to protect against +load running multiple times the way you do with +initialize. The +load message is sent only to classes that actually implement it, so you don't accidentally get calls from your subclasses the way you can in +initialize. Every +load is called exactly once. You shouldn't call [super load].

Associative References

Associative references allow you to attach key-value data to arbitrary objects. There are many uses for this capability, but one common use is to allow categories to add data for properties.

Consider the case of a Person class. Say you would like to use a category to add a new property called emailAddress. Maybe you use Person in other programs, and sometimes it makes sense to have an email address and sometimes it doesn't, so a category can be a good solution to avoid the overhead when you don't need it. Or maybe you don't own the Person class, and the maintainers won't add the property for you. In any case, how do you attack this problem? First, here is the basic Person class:

```
@interface Person : NSObject
@property (nonatomic, readwrite, copy) NSString *name;
@end

@implementation Person
@end
```

Now you can add a new property, emailAddress, in a category using an associative reference:

```
#import <objc/runtime.h>
@interface Person (EmailAddress)
@property (nonatomic, readwrite, copy) NSString *emailAddress;
@end

@implementation Person (EmailAddress)
```

```
static char emailAddressKey;

- (NSString *)emailAddress {
  return objc_getAssociatedObject(self, &emailAddressKey);
}

- (void)setEmailAddress:(NSString *)emailAddress {
  objc_setAssociatedObject(self, &emailAddressKey,
                           emailAddress,
                           OBJC_ASSOCIATION_COPY);
}
@end
```

Note that associative references are based on the key's memory address, not its value. It does not matter what is stored in `emailAddressKey`; it only needs to have a unique, unchanging address. That's why it's common to use an unassigned `static char` as the key.

Associative references have good memory management, correctly handling copy, assign, or retain semantics according to the parameter passed to `objc_setAssociatedObject`. They are released when the related object is deallocated. This fact means you can use associated objects to track when another object is destroyed. For example:

```
const char kWatcherKey;

@interface Watcher : NSObject
@end

#import <objc/runtime.h>

@implementation Watcher
- (void)dealloc {
  NSLog(@"HEY! The thing I was watching is going away!");
}
@end
...
NSObject *something = [NSObject new];
objc_setAssociatedObject(something, &kWatcherKey, [Watcher new],
                         OBJC_ASSOCIATION_RETAIN);
```

This technique is useful for debugging but can also be used for non-debugging tasks such as performing cleanup.

Using associative references is a great way of attaching a relevant object to an alert panel or control. For example, you can attach a "represented object" to an alert panel, as shown in the following code. This code is available in the sample code for this chapter.

ViewController.m (AssocRef)

```
id interestingObject = ...;
UIAlertView *alert = [[UIAlertView alloc]
                      initWithTitle:@"Alert" message:nil
                      delegate:self
                      cancelButtonTitle:@"OK"
                      otherButtonTitles:nil];
```

```
objc_setAssociatedObject(alert, &kRepresentedObject,
                         interestingObject,
                         OBJC_ASSOCIATION_RETAIN_NONATOMIC);
[alert show];
```

Now, when the alert panel is dismissed, you can figure out why you cared:

```
- (void)alertView:(UIAlertView *)alertView
clickedButtonAtIndex:(NSInteger)buttonIndex {
  UIButton *sender = objc_getAssociatedObject(alertView,
                                              &kRepresentedObject);
  self.buttonLabel.text = [[sender titleLabel] text];
}
```

Many programs handle this task with an ivar in the caller, but associative references are much cleaner and simpler. For those familiar with Mac development, this code is similar to `representedObject`, but more flexible.

One limitation of associative references (or any other approach to adding data through a category) is that they don't integrate with `encodeWithCoder:`, so they're difficult to serialize through a category.

Weak Collections

The most common Cocoa collections are `NSArray`, `NSSet`, and `NSDictionary`, which are excellent for most uses, but they are not appropriate in some cases. `NSArray` and `NSSet` retain the objects you store in them. `NSDictionary` retains values stored in it and also copies its keys. These behaviors are usually exactly what you want, but for some kinds of problems they work against you. Luckily, other collections have been available since iOS 6: `NSPointerArray`, `NSHashTable`, and `NSMapTable`. They are collectively known as *pointer collection classes* in Apple's documentation and are sometimes configured using the `NSPointerFunctions` class.

`NSPointerArray` is similar to `NSArray`, `NSHashTable` is similar to `NSSet`, and `NSMapTable` is similar to `NSDictionary`. Each of these new collection classes can be configured to hold weak references, pointers to nonobjects, or other unusual situations. `NSPointerArray` has the added benefit of being able to store `NULL` values, which is a common problem with `NSArray`.

> The Apple documentation for the pointer collection classes often refers to garbage collection because these classes were originally developed to work with garbage collection in 10.5. These classes are now compatible with ARC weak references. This is not always clear in the main class references but is documented in the `NSPointerFunctions` class reference.

The pointer collection classes can be extensively configured using an `NSPointerFunctions` object, but in most cases it is simpler to pass an `NSPointerFunctionsOptions` flag to `-initWithOptions:`. The most common cases, such as `+weakObjectsPointerArray`, have their own convenience constructors.

For more information, see the appropriate class reference documents, as well as Collections Programming Topics, and the article "NSHashTable & NSMapTable" by NSHipster (`nshipster.com`).

NSCache

One of the most common reasons to use a weak collection is to implement a cache. In many cases, though, you can use the Foundation caching object, NSCache, instead. For the most part, you use it like an NSDictionary, calling objectForKey:, setObject:forKey: and removeObjectForKey:.

NSCache has several underappreciated features, such as the fact that it is thread-safe. You may mutate an NSCache on any thread without locking. NSCache is also designed to integrate with objects that conform to <NSDiscardableContent>. The most common type of <NSDiscardableContent> is NSPurgeableData. By calling beginContentAccess and endContentAccess, you control when it is safe to discard this object. Not only does this provide automatic cache management while your app is running, it even helps when your app is suspended. Normally, when memory is tight and memory warnings have not freed up enough memory, iOS starts killing suspended background apps. In this event, your app doesn't get a delegate message; it's just killed. But if you use NSPurgeableData, iOS frees this memory for you, even while you are suspended.

For more information, see the references for NSCache, <NSDiscardableContent>, and NSPurgeableData in the Xcode documentation.

NSURLComponents

Sometimes Apple adds interesting classes with little fanfare. In iOS 7, Apple added NSURLComponents, which has no class reference document. It's mentioned in the "What's New in iOS 7" section of the iOS 7 release notes, but you have to read NSURL.h for its documentation.

NSURLComponents makes it easy to split an URL into its parts. For example:

```
NSString *URLString =
  @"http://en.wikipedia.org/wiki/Special:Search?search=ios";
NSURLComponents *components = [NSURLComponents
  componentsWithString:URLString];
NSString *host = components.host;
```

You can also use NSURLComponents to construct or modify URLs:

```
components.host = @"es.wikipedia.org";
NSURL *esURL = [components URL];
```

In iOS 7, NSURL.h adds several useful categories for dealing with URLs. For example, you can use [NSCharacterSet URLPathAllowedCharacterSet] to get the set of characters allowed in the path. NSURL.h also adds [NSString stringByAddingPercentEncodingWithAllowedCharacters:], which lets you control which characters are percent-encoded. Previously, you could do this only with Core Foundation, using CFURLCreateStringByReplacingPercentEscapes.

Search NSURL.h for 7_0 to find all the new methods and their documentation.

CFStringTransform

`CFStringTransform` is one of those functions that, after you discover it, you cannot believe you didn't know about it before. It transliterates strings in ways that simplify normalization, indexing, and searching. For example, it can remove accent marks using the option `kCFStringTransformStripCombiningMarks`:

```
CFMutableStringRef string = CFStringCreateMutableCopy(NULL, 0,
                            CFSTR("Schläger"));
CFStringTransform(string, NULL, kCFStringTransformStripCombiningMarks,
  false);
... => string is now "Schlager"
CFRelease(string);
```

`CFStringTransform` is even more powerful when you are dealing with non-Latin writing systems such as Arabic or Chinese. It can convert many writing systems to Latin script, making normalization much simpler. For example, you can convert Chinese script to Latin script like this:

```
CFMutableStringRef string = CFStringCreateMutableCopy(NULL, 0,
                            CFSTR("你好"));
CFStringTransform(string, NULL, kCFStringTransformToLatin, false);
... => string is now "nǐ hǎo"
CFStringTransform(string, NULL, kCFStringTransformStripCombiningMarks,
  false);
... => string is now "ni hao"
CFRelease(string);
```

Notice that the option is simply `kCFStringTransformToLatin`. The source language is not required. You can hand almost any string to this transform without having to know first what language it is in. `CFStringTransform` can also transliterate from Latin script to other writing systems such as Arabic, Hangul, Hebrew, and Thai.

Chinese Characters in Japanese

Chinese characters are always transliterated as Mandarin Chinese, even if they are being used in another writing system. This can be particularly tricky for Japanese because Chinese characters may be included in a Japanese string. In a three-character Japanese phase such as 白い月, the first and last characters are transliterated as Mandarin Chinese (*bái* and *yuè*), and the middle character is transliterated as Japanese (*i*), creating the nonsense string *báii yuè*.

Although `CFStringTransform` can handle Hiragana and Katakana, it cannot convert Kanji. If you need to transliterate complex Japanese text, see "`NSString-Japanese`" by 00StevenG at `https://github.com/00StevenG/NSString-Japanese`. It can handle Kanji and Romanji as well as Hiragana and Katakana.

`NSString-Japanese` is based on `CFStringTokenizer`, which offers more language-aware transforms at the cost of being somewhat more complicated to use.

For more information, see the CFMutableString Reference and the CFStringTokenizer Reference at `developer.apple.com`.

instancetype

Objective-C has long had some subtle subclassing problems. Consider the following situation:

```
@interface Foo : NSObject
+ (Foo *)fooWithInt:(int)x;
@end

@interface SpecialFoo : Foo
@end
...
SpecialFoo *sf = [SpecialFoo fooWithInt:1];
```

This code generates a warning, `Incompatible pointer types initializing 'SpecialFoo *'` `with an expression of type 'Foo *'`. The problem is that `fooWithInt` returns a `Foo` object. The compiler can't know that the returned type is really a more specific class (`SpecialFoo`). This is a fairly common situation. Consider `[NSMutableArray array]`. The compiler couldn't allow you to assign it to a subclass (`NSMutableArray`) without a warning if it returned an `NSArray`.

There are several possible solutions to this problem. First, you might try overloading `fooWithInt:` this way:

```
@interface SpecialFoo : Foo
+ (SpecialFoo *)fooWithInt:(int)x;
@end

@implementation SpecialFoo
+ (SpecialFoo *)fooWithInt:(int)x {
  return (SpecialFoo *)[super fooWithInt:x];
}
```

This approach works but is very inconvenient. You have to override many methods just to add the typecast. You could also perform the typecast in the caller:

```
SpecialFoo *sf = (SpecialFoo *)[SpecialFoo fooWithInt:1];
```

This approach is more convenient in `SpecialFoo` but is much less convenient for the caller. Adding lots of typecasts also eliminates type checking, so it is more error-prone.

The most common solution is to make the return type `id`:

```
@interface Foo : NSObject
+ (id)fooWithInt:(int)x;
@end

@interface SpecialFoo : Foo
@end
...
SpecialFoo *sf = [SpecialFoo fooWithInt:1];
```

This approach is reasonably convenient but also eliminates type checking. Of the available solutions, however, it has been the best choice until recently. This is why most constructors in Cocoa return `id`.

Cocoa has extremely consistent naming practices. Any method that begins with `init` is supposed to return an object of that type. Couldn't the compiler just enforce that? The answer is yes, and the latest versions of Clang do that. So now, if you have a method called `initWithFoo:` that returns `id`, the compiler assumes that the return type is really the class of the object, and it gives warnings if you have a type mismatch.

This automatic conversion is great for `init` methods, but this example is a convenience constructor, `+fooWithInt:`. Can the compiler help in that case? Yes, but not automatically. The naming convention for convenience constructors is not as strong as that for `init` methods. `SpecialFoo` might have a convenience constructor like `+fooWithInt:specialThing:`. The compiler has no good way to figure out that this is meant to return a `SpecialFoo` from its name, so it doesn't try. Instead, Clang adds a new type, `instancetype`. As a return type, `instancetype` represents "the current class." So now you can declare your methods this way:

```
@interface Foo : NSObject
+ (instancetype)fooWithInt:(int)x;
@end

@interface SpecialFoo : Foo
@end
...
SpecialFoo *sf = [SpecialFoo fooWithInt:1];
```

For consistency, it is best to use `instancetype` as the return type for both `init` methods and convenience constructors.

Base64 and Percent Encoding

Cocoa has long needed easy access to Base64 encoding and decoding. Base64 is the standard for many web protocols and is useful for many cases in which you need to store arbitrary data in a string.

In iOS 7, new `NSData` methods such as `initWithBase64EncodedString:options:` and `base64EncodedStringWithOptions:` are available for converting between Base64 and `NSData`.

Percent encoding is also important for web protocols, particularly URLs. You can now decode percent-encoded strings using `[NSString stringByRemovingPercentEncoding]`. Although it has always been possible to percent-encode strings with `stringByAddingPercentEscapesUsingEncoding:`, iOS 7 adds `stringByAddingPercentEncodingWithAllowedCharacters:`, which allows you to control which characters are percent-encoded.

-[NSArray firstObject]

It's a tiny change, but I have to mention it because we've waited so long for it: After years of developers implementing their own category to get the first object of an array, Apple has finally added the method `firstObject`. Like `lastObject`, if the array is empty, `firstObject` returns `nil` rather than crashing like `objectAtIndex:0`.

Summary

Cocoa has a long history, full of traditions and conventions. Cocoa is also an evolving, actively developed framework. In this chapter you have learned some of the best practices that have grown over decades of Objective-C development. You have learned how to choose the best names for classes, methods, and variables. You have also learned how to best use some lesser-known features like associative references, and new features like NSURLComponents. Even for experienced Objective-C developers, you hopefully have picked up a few more Cocoa tricks, and learned a few things you didn't know before.

Further Reading

Apple Documentation

The following documents are available in the iOS Developer Library at developer.apple.com or through the Xcode Documentation and API Reference.

CFMutableString Reference

CFStringTokenizer Reference

Collections Programming Topics

Collections Programming Topics, "Pointer Function Options"

Programming with Objective-C

Other Resources

Thompson, Matt. *NSHipster*. Matt Thompson's blog is a fantastic weekly update on little-known Cocoa features.

nshipster.com

00StevenG. *NSString-Japanese*. If you need to normalize Japanese text, this is a very useful category to deal with the complexities of its multiple writing systems.

https://github.com/00StevenG/NSString-Japanese

Chapter 4

Storyboards and Custom Transitions

Prior to iOS 5, interface elements and views were created using Interface Builder (IB) and saved in nib files. Storyboards are yet another way to create interfaces, and in addition to creating interface elements, you can specify the navigation (called *segues*) between those interfaces. This was something you could not do previously without writing code. You can think of storyboards as a graph of all your view controllers connected by segues that dictate the transition between them.

The benefits of storyboards don't stop there. They also make it easy for developers to create static table views without a data source. How many times have you wanted to create a table view that's not bound to a real data source—for example, a table that shows a list of options instead of data. A common use for this is your app's settings page. Storyboards also help co-developers and clients understand the complete workflow of the app.

Storyboards aren't all romantic, and in my opinion, they pose some significant drawbacks. Later in this chapter, you learn how to use storyboards without being hurt by those drawbacks.

> **Storyboards are the future of building your user interface. If you have been avoiding storyboards, it is high time that you learn them. In fact, Xcode 5 does not even provide a way to turn off storyboards when you create a project from the new project template.**

You begin by learning how to start using storyboards and how to do things using storyboards that you might now do with nib files, such as communicating between controllers. In the "Static Tables" section later in this chapter, you find out how to create a static table view without a data source. Finally, you discover the most interesting aspect of storyboards: writing your own custom transition animations. Although these cool transition animations sound complicated, the iOS SDK has made it really easy to write them.

Getting Started with Storyboards

You can use storyboards for new projects or add them to an existing project that doesn't have a storyboard yet. For existing projects, you can add storyboards in the same way you add a new file to a project. You learn how to instantiate view controllers in this storyboard in the upcoming section "Instantiating a Storyboard."

For new projects, you can create storyboards in Xcode 4.5 by using the new project template and selecting the default Use Storyboard option, as shown in Figure 4-1. In Xcode 5, you will not find this option. In fact, Xcode 5 defaults to the "Use Storyboard" option and makes it harder to remove storyboards from your app.

Figure 4-1 New project template in Xcode 4.5 showing the Use Storyboard option

When you create a new project using storyboards, the `info.plist` key of your app contains a key called `UIMainStoryboardFile`. This key supersedes `NSMainNibFile` that was used prior to iOS 5. You can continue to use `NSMainNibFile` if your app's main window is loaded from a nib file instead of a storyboard. However, you can't use both `UIMainStoryboardFile` and `NSMainNibFile` in the same app. `UIMainStoryboardFile` takes precedence, and your nib file specified in `NSMainNibFile` never gets loaded.

> Your application can store the complete storyboard in one file, and IB automatically builds it into separate files optimized for loading. In short, you don't have to be worried about loading time or performance when using storyboards.

Instantiating a Storyboard

When your `UIMainStoryboardFile` is set, the compiler automatically generates code for instantiating it and loads it as your application's startup window. If you're adding storyboards in an existing app, you do so programmatically. The methods for instantiating view controllers within a storyboard are defined in the `UIStoryboard` class.

When you want to display a view controller specified in your storyboard, you load the storyboard using this method:

```
+ storyboardWithName:bundle:
```

Loading View Controllers Within a Storyboard

Loading view controllers within a storyboard is similar to the nib loading method, and with the `UIStoryboard` object, you can instantiate view controllers using the following methods:

```
- instantiateInitialViewController
- instantiateViewControllerWithIdentifier:
```

Segues

Segues are transitions defined in your storyboard file. UIKit provides two default transition styles: Push and Modal. They behave similarly to the `pushViewController:animated:completion:` and `presentViewController:animated:completion` methods you use in iOS 5 and prior versions. In addition, you can create custom segues and create new kinds of transitions between view controllers. You look at these tasks later in this chapter in the section "Custom Transitions."

You create segues by connecting certain events on view controllers with other view controllers on your storyboard file. You can drag from a button to a view controller, from a gesture recognizer object to a view controller, and so on. IB creates a segue between them, and you can select the segue and use the attributes inspector panel to modify the transition styles.

The attributes inspector panel also allows you to set a custom class if you select a custom transition style. You can think of a segue as something that connects an action with a transition. Actions that can trigger segues can be button tap events, row selection events on static table views, a recognized gesture, or even audio events. The compiler automatically generates the necessary code to perform a segue when the event to which you connected the segue occurs.

When a segue is about to be performed, a `prepareForSegue:sender:` method is invoked on the source view controller, and an object of type `UIStoryboardSegue` is passed to it. You can override this method to pass data to the destination view controller. The next section explains how to complete this task.

When a view controller performs multiple segues, the same `prepareForSegue:sender:` method gets called for every segue. To identify the performed segue, use the segue identifier to check whether the performed segue is the intended one and pass data accordingly. As a defensive programming practice, I advise you to perform this check even if the view controller performs only one segue. This way, you ensure that later when you add a new segue, your app will continue to run without crashing.

Passing Data

When you use storyboards, view controllers are instantiated and presented to the user automatically. You're given a chance to fill in data by overriding the `prepareForSegue:sender:` method. By overriding this method, you can get the pointer to the destination view controller and set the initial values there.

The framework calls the same methods that you used before, such as `viewDidLoad`, `initWithCoder:`, and the `NSObject awakeFromNib` method, which means that you can continue writing your view controller's initialization code as you do when you don't use storyboards.

Returning Data

With storyboards, you communicate data back to the parent view controller exactly as you do with nib files or manually coded user interfaces. Data created or entered by the user on modal forms that you present can be returned to the parent via delegates or blocks. The only difference is that on your parent view controller, you have to set the delegate in the `prepareForSeque:sender:` method to `self`.

Instantiating Other View Controllers

`UIViewController` has a `storyboard` property that retains a pointer to the storyboard object
(`UIStoryBoard`) from which it was instantiated. This property is `nil` if your view controller is created
manually or from a nib file. With this back reference, you can instantiate other view controllers defined in your
storyboard from any other view controller. You do so by using the view controller's identifier. The following
method on `UIStoryBoard` allows you to instantiate a view controller using its identifier:

```
- instantiateViewControllerWithIdentifier:
```

As a result, you can still have view controllers on your storyboard that aren't connected with any other view
controllers through segues, and yet these view controllers can be initialized and used.

Performing Segues Manually

Although storyboards can automatically trigger segues based on actions, in some cases, you may need
to perform segues programmatically. You might do so to deal with actions that cannot be handled by the
storyboard file. To perform a segue, you call the `performSegueWithIdentifier:sender:` method of
the view controller. When you perform segues manually, you can pass the caller and the context objects in the
sender parameter. This sender parameter will be sent to the `prepareForSegue:sender:` method later.

Unwinding Segues

Storyboards originally allowed you to instantiate and navigate to view controllers. iOS 6 introduced methods
in `UIViewController` that allow unwinding segues. By unwinding segues, you can implement methods to
navigate "back" without creating additional view controllers.

You can add unwinding support to a segue by implementing an `IBAction` method in your view controller
that takes a `UIStoryboardSegue` as a parameter, as shown in the following example:

```
- (IBAction)unwindMethod:(UIStoryboardSegue*)sender {
}
```

You can now connect an event in a view controller to its `Exit` object. Xcode automatically enumerates
all possible unwind events (any method that is an `IBAction` and accepts a `UIStoryboardSegue` as a
parameter) in a storyboard and allows you to connect to them. This is shown in Figure 4-2.

Building Table Views with Storyboards

One important advantage of storyboards is the capability to create static tables from IB. With storyboards, you
can build two types of table views: a static table that doesn't need a special class for providing a data source
and a table view containing a prototype cell (similar to custom table view cells in iOS 4) that binds data from a
model.

Static Tables

You can create static tables in your storyboard by dragging a table, selecting it, and from the attributes
inspector, choosing Static Cells, as shown in Figure 4-3.

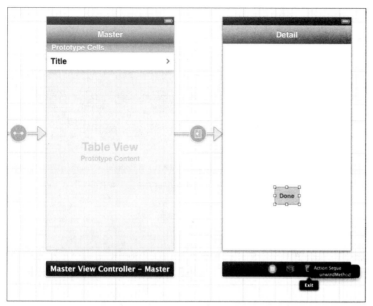

Figure 4-2 Connecting an IBAction for unwinding a segue

Figure 4-3 A storyboard illustrating static table view creation

Static cells are great for creating settings pages (or pages whose content doesn't come from a Core Data model or a web service or any such data source) as in Apple's Settings app.

> Static cells can be created only for table views that are from a `UITableViewController`. **You cannot create static cells for table views that are added as a subview of a** `UIViewController` **view.**

Prototype Cells

Prototype cells are similar to custom table view cells, but instead of creating prototype cells on separate nib files and loading them in the data source method `cellForRowAtIndexPath:`, you create them in IB on your storyboard and just set the data on your data source methods.

> You identify all prototype cells using a custom identifier, which ensures proper functioning of the table view cell queuing methods. Xcode warns you if a prototype cell in your storyboard doesn't have a cell identifier.

Custom Transitions

Storyboards make it easier to perform custom transitions from one view controller to another.

When segues are performed, the compiler generates necessary code to present or push the destination controller based on the transition style you set on your storyboard. You've found that two types of transition styles, Push and Modal, are supported natively by iOS. There's also a third type, Custom. Choose Custom and provide your own subclass of `UIStoryboardSegue` that handles your custom transition effects.

Create a subclass of `UIStoryBoardSegue` and override the `perform` method. In the `perform` method, access the pointer to the source view controller's main view's layer and do your custom transition animation (using Core Animation). When the animation is complete, push or present your destination view controller (you can get a pointer to this from the segue object). It's as simple as that.

To illustrate, I show you how to create a transition where the master view gets pushed down and the details view appears at the bottom of the screen.

Create a new application using the Master-Details template and open the `MainStoryboard`. Click on the only segue and change the type to Custom. Add a `UIStoryboardSegue` subclass, override the `perform` method, and paste the following code:

Custom Transition Using a Storyboard Segue (CustomSegue.m)

```
- (void) perform {

    UIViewController *src = (UIViewController *)self.sourceViewController;
    UIViewController *dest = (UIViewController
    *)self.destinationViewController;
```

```
    CGRect f = src.view.frame;
    CGRect originalSourceRect = src.view.frame;
    f.origin.y = f.size.height;

    [UIView animateWithDuration:0.3 animations:^{
      src.view.frame = f;

    } completion:^(BOOL finished){
      src.view.alpha = 0;
      dest.view.frame = f;
      dest.view.alpha = 0.0f;
      [[src.view superview] addSubview:dest.view];
      [UIView animateWithDuration:0.3 animations:^{

        dest.view.frame = originalSourceRect;
        dest.view.alpha = 1.0f;
      } completion:^(BOOL finished) {

        [dest.view removeFromSuperview];
        src.view.alpha = 1.0f;
        [src.navigationController pushViewController:dest animated:NO];
      }];
    }];
  }
```

That's it. You can do all kinds of crazy stuff with the layer pointers of the source and destination view controllers. Justin Mecham has open-sourced a great example of a doorway transition (ported from Ken Matsui) on GitHub (https://github.com/jsmecham/DoorwaySegue). You can also create your own transition effects by manipulating the layer pointers of the source and destination view controllers. With segues, creating a custom transition effect is far easier.

iOS 7 introduces much more sophisticated APIs for performing custom transitions, collection view layout transitions, and interactive transitions like the pervasive pop gesture of the navigation controller. A separate chapter is dedicated to this topic. You learn about custom transitions in Chapter 20.

Advantages

When you use storyboards, co-developers (and clients/project managers) can more easily understand the app's flow. Instead of going through multiple nib files and cross-referencing the instantiation code for understanding the flow, co-developers can open the storyboard file and see the complete flow. This alone is a compelling reason to use them.

Disadvantages – Merge Conflict Hell

Storyboards present another, rather annoying, problem for teams. The default application template in Xcode has one storyboard for the entire app. This means that when two developers work on the UI at the same time, merge conflicts become inevitable, and because storyboards are internally auto-generated XML, these merge conflicts are too complicated to fix. Even with the new XML file format that was designed to reduce merge conflicts, storyboards often get into a conflicted state.

The easiest way to solve this problem is to avoid it from happening in the first place. My recommendation is that you break your storyboards into multiple files, one file for every use case. In most cases, one developer will be working on a use case, and the chances of ending up with a merge conflict are lower.

Examples of use case-based storyboards for a Twitter client are `Login.Storyboard`, `Tweets.Storyboard`, `Settings.Storyboard`, and `Profile.Storyboard`. Instead of breaking up the storyboard into multiple nib files (going back to square one), use multiple storyboards to help solve the merge conflict issue while preserving the elegance of using storyboards. As I mentioned in the previous paragraph, a developer who is working on the Login use case will probably not be working on Tweets at the same time.

Summary

In this chapter you learned about storyboards and methods to implement custom transitions using storyboards. You also learned about unwinding segues introduced in iOS 6 and the advantages and disadvantages of using storyboards in your app. In addition, you learned how to work around the disadvantages of storyboards.

Further Reading

Apple Documentation

The following documents are available in the iOS Developer Library at `developer.apple.com` or through the Xcode Documentation and API Reference.

What's New in iOS 6

TableView Programming Guide

TableViewSuite

UIViewController Programming Guide

WWDC Sessions

The following session videos are available at `developer.apple.com`:

WWDC 2011, "Session 309: Introducing Interface Builder Storyboarding"

WWDC 2012, "Session 407: Adopting Storyboards in Your App"

Other Resources

enormego / EGOTableViewPullRefresh

`https://github.com/enormego/EGOTableViewPullRefresh`

jsmecham / DoorwaySegue

`https://github.com/jsmecham/DoorwaySegue`

## Chapter 5

Get a Handle on Collection Views

Before iOS 6, developers used `UITableView` for almost any type of display that shows a collection of items. Although Apple has been using a UI that looks similar to the collection views in photo apps for quite a long time (since the original iPhone), that UI wasn't available for use by third-party developers. We depended on third-party frameworks such as three20, or we hacked our own to display a collection list. For the most part, these solutions worked well, but implementing animation when items are added or deleted or implementing a custom layout like Cover Flow was difficult. Fortunately, iOS 6 introduced a brand new controller for the iOS just for displaying a collection of items. Collection view controllers are yet another building block UI that's similar to table view controllers. In this chapter, I introduce you to collection views in general, and you find out how to display a list of items using a collection view.

Collection Views

iOS 6 introduced the new controller named `UICollectionViewController`. Collection views provide a much more elegant way to display items in a grid than was previously possible using `UIKit`. Collection views were available on the Mac OS X software development kit (`NSCollectionView`); however, the iOS 6 collection view (`UICollectionView`) is very different from the Mac equivalent. In fact, the iOS 6 `UICollectionViewController`/`UICollectionView` is more similar to `UITableViewController`/`UITableView`, and since you already know how to use a `UITableViewController`, you will be at home with `UICollectionViewController`.

In this chapter, I explain the different classes that you need to know about, and I walk you though an app that displays a directory of images in a collection view. As I walk you through this example, I compare `UICollectionViewController` with the `UITableViewController` that you know well by now (because learning by association helps you remember better).

Classes and Protocols

In the following sections, I describe the most important classes and protocols that you need to know when you implement a collection view.

UICollectionViewController

The first and most important class is the `UICollectionViewController`. This class functions similarly to `UITableViewController`. It manages the collection view, stores the data required, and handles the data source and delegate protocol.

UICollectionViewCell

`UICollectionViewCell` is similar to the good old `UITableViewCell`. You normally don't have to create a `UICollectionViewCell`. You'll be calling the `dequeueCellWithReuseIdentifier:indexPath:` method to get one from the collection view. Collection views behave like table views inside a storyboard. You create `UICollectionViewCell` types (such as prototype table view cells) in Interface Builder (IB) inside the `UICollectionView` object instead. Every `UICollectionViewCell` is expected to have a `CellIdentifier`, failing which you'll get a compiler warning. The `UICollectionViewController` uses this `CellIdentifier` to enqueue and dequeue cells. The `UICollectionViewCell` is also responsible for maintaining and updating itself for selected and highlighted states. I show you how to do this in the "Adding a Selected State to the Cell" later in this chapter.

The following code snippet shows how to get a `UICollectionViewCell`:

```
MKPhotoCell *cell = (MKPhotoCell*)
[collectionView dequeueReusableCellWithReuseIdentifier:
@"MKPhotoCell" forIndexPath:indexPath];
```

The snippet is straight from the sample code for this chapter that I walk you through later.

> **If you're not using storyboards, you can call the** `registerClass:forCellWithReuseIdentifier:` **in** `collectionView` **to register a nib file.**

UICollectionViewDataSource

You might have guessed that `UICollectionViewDataSource` should be similar to `UITableViewDataSource`, and it is. The data source protocol has methods that should be implemented by the `UICollectionViewController`'s subclass. In the first example, you implement some of the methods in the data source to display photos in the collection view.

UICollectionViewDelegate

The delegate protocol named `UICollectionViewDelegate` has methods that should be implemented if you want to handle selection or highlighting events in your collection views. In addition, the `UICollectionViewCell` can also show context-sensitive menus such as the Cut/Copy/Paste menu. The action handlers for these methods are also passed to the delegate.

UICollectionViewDelegateFlowLayout

In addition to the preceding classes, you need to learn and understand the `UICollectionViewDelegateFlowLayout` protocol, which enables you to do advanced layout customization. I show how do to this in the "Advanced Customization with Collection View Custom Layout" section later in this chapter.

Example

For this example, you start by creating a single view application in Xcode. The first step is to choose iPad as the target device and click Next; then choose a location to save your project.

Editing the Storyboard

Open the `MainStoryboard.storyboard` file and delete the only `ViewController` inside it. Drag a `UICollectionViewController` from the object library. Ensure that this controller is set as the initial view controller for the storyboard.

Now, open the only view controller's header file in your project. Change the base class from `UIViewController` to `UICollectionViewController` and implement the protocols `UICollectionViewDataSource` and `UICollectionViewDelegate`. Go back to the storyboard and change the class type of the collection view controller to `MKViewController` (or whatever is necessary, depending on your class prefix).

Build and run your app. If you follow the steps properly, you should see a black screen in your iOS simulator.

> You also could use the empty application template and add a storyboard with a collection view. But that approach requires changes to App Delegate and your `Info.plist` because the Xcode's empty application template doesn't use storyboards.

Adding Your First Collection View Cell

The application's output isn't impressive, is it? You can spice it up, however. Go ahead and add a class that is a subclass of `UICollectionViewCell`. In the sample code, I call it `MKPhotoCell`. Open your storyboard and select the only collection view cell inside your collection view controller. Change the class to `MKPhotoCell`. This is shown in Figure 5-1.

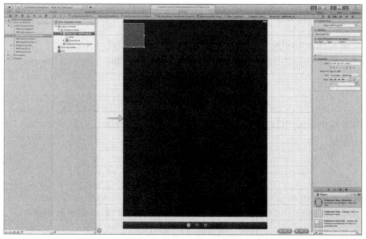

Figure 5-1 Xcode showing a storyboard with a collection view and a prototype collection view cell

Open the Attributes Inspector (the fourth tab) in the Utilities and set the Identifier to `MKPhotoCell`. This step is very important. You will use this identifier to dequeue cells later in code. Add a blank `UIView` as the subview of your collection view cell and change the background color to red (so you can see it).

Implementing Your Data Source

Now implement the data source as shown here:

```
- (NSInteger)numberOfSectionsInCollectionView:(UICollectionView *)
  collectionView {

  return 1;
}

-  (NSInteger)collectionView:(UICollectionView *)
collectionView numberOfItemsInSection:(NSInteger)section {

  return 100;
}

// The cell that is returned must be retrieved from a call to
// -dequeueReusableCellWithReuseIdentifier:forIndexPath:
-  (UICollectionViewCell *)collectionView:(UICollectionView *)
   collectionView cellForItemAtIndexPath:(NSIndexPath *)indexPath {

  MKPhotoCell *cell = (MKPhotoCell*)
    [collectionView dequeueReusableCellWithReuseIdentifier:@"MKPhotoCell"
    forIndexPath:indexPath];
  return cell;
}
```

Next, build and run the app. You'll see a grid of 100 cells, each painted in red. Impressive. All it took was a few lines of code. What is even more impressive is that the grid rotates and aligns itself when you turn your iPad to landscape.

Using the Sample Photos

Now you can replace the red-colored subview with something more interesting. For this example, you're going to display photos from a directory. Copy some photos (about 50) to your project. If you'd like, you can use the photos from the sample code.

Remove the red-colored subview you added in the previous section and add a `UIImageView` and a `UILabel` to the `UICollectionViewCell`. Add outlets in your `UICollectionViewCell` subclass and connect them appropriately in Interface Builder.

Preparing Your Data Source

Prepare your data source by iterating through the files in your directory. Add this to your `viewDidLoad` method in the collection view controller subclass. You can find it in the `MKViewController` class in the sample code:

```
self.photosList = [[NSFileManager defaultManager]
  contentsOfDirectoryAtPath:[self photosDirectory] error:nil];
```

The `photosDirectory` method is defined as follows:

```
-(NSString*) photosDirectory {
  return [[[NSBundle mainBundle] resourcePath]
 stringByAppendingPathComponent:@"Photos"];
}
```

Now, update your data source methods to return data based on this information. Your previous code returned 1 section and 100 items in that section. Change the value 100 to the number of photos you added to your project in the previous section. This is the size of the `photosList` array.

UICollectionViewDataSource Methods (MKViewController.m)

```
-  (UICollectionViewCell *)collectionView:(UICollectionView *)
    collectionView cellForItemAtIndexPath:(NSIndexPath *)indexPath {

  MKPhotoCell *cell = (MKPhotoCell*) [collectionView
  dequeueReusableCellWithReuseIdentifier:@"MKPhotoCell"
  forIndexPath:indexPath];

  NSString *photoName = [self.photosList objectAtIndex:indexPath.row];
  NSString *photoFilePath = [[self photosDirectory]
    stringByAppendingPathComponent:photoName];
  cell.nameLabel.text =[photoName stringByDeletingPathExtension];
  UIImage *image = [UIImage imageWithContentsOfFile:photoFilePath];
  UIGraphicsBeginImageContext(CGSizeMake(128.0f, 128.0f));
  [image drawInRect:CGRectMake(0, 0, 128.0f, 128.0f)];
  cell.photoView.image = UIGraphicsGetImageFromCurrentImageContext();
  UIGraphicsEndImageContext();

  return cell;
}
```

Build and run the app now. When you do, you'll see the photos neatly organized in rows and columns. Note, however, that the app creates a UIImage from a file in the `collectionView:cellForItemAtIndexPath:` method. Loading a UIImage from a file in `cellForItemAtIndexPath:` method is going to hurt performance.

Improving Performance

You can improve the performance by using a background Grand Central Dispatch (GCD) queue to create the images and optionally cache them for performance. Both of these methods are implemented in the sample code. For an in-depth understanding of GCD, read Chapter 23 in this book.

Supporting Landscape and Portrait Photos

The previous example was good, but not great. Both portrait and landscape images were cropped to 128 × 128 and looked a bit pixelated. The next task is to create two cells, one for landscape images and another for portrait images. Because both portrait and landscape images differ only by the image orientation, you don't need an additional UICollectionViewCell subclass.

Create another `UICollectionViewCell` in your storyboard and change the class to `MKPhotoCell`. Change the image view's size so that the orientation is portrait and the older cells' orientation is landscape. You can use 180 × 120 for the landscape cell and 120 × 180 for the portrait cell. Change the `CellIdentifier` to something like `MKPhotoCellLandscape` and `MKPhotoCellPortrait`. You'll be dequeuing one of these based on the image size.

> You can also use the `UICollectionViewDelegateFlowLayout` **method** `collectionView:` `layout:sizeForItemAtIndexPath:` **and return the size of your photo. But dynamically sizing the photos at runtime during layout is time-consuming, so I recommend having different cells for different orientation. The** `UICollectionViewDelegateFlowLayout` **protocol's delegate method is helpful if the item sizes are completely random.**

When you're done, your storyboard should look like Figure 5-2.

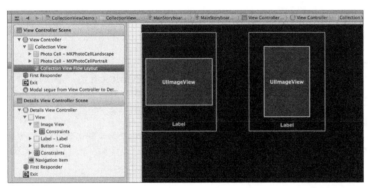

Figure 5-2 Storyboard showing two cells, one for landscape images and another for portrait images

Determining the Orientation

You now have to determine which cell to dequeue based on the image's orientation. You can get the image orientation by inspecting the size property of the `UIImage` after you create it. If the image is wider than its height, it's a landscape photo; if not, it's in portrait orientation. Of course, if you're going to compute this in your `collectionView:cellForItemAtIndexPath:` method, your scrolling speed will be affected. The sample code precomputes the size and hence the orientation and stores the orientation in a separate array when the view loads, all on a background GCD queue. The following code listing shows the modified `collectionView:cellForItemAtIndexPath:`.

collectionView:cellForItemAtIndexPath: Method in MKViewController.m

```
- (UICollectionViewCell *)collectionView:(UICollectionView *)
    collectionView cellForItemAtIndexPath:(NSIndexPath *)indexPath {

    static NSString *CellIdentifierLandscape = @"MKPhotoCellLandscape";
    static NSString *CellIdentifierPortrait = @"MKPhotoCellPortrait";
```

```
    int orientation = [[self.photoOrientation objectAtIndex:
      indexPath.row] integerValue];

    MKPhotoCell *cell = (MKPhotoCell*)
      [collectionView dequeueReusableCellWithReuseIdentifier:
                               orientation == PhotoOrientationLandscape ?
                               CellIdentifierLandscape:CellIdentifierPortrait
                               forIndexPath:indexPath];

    NSString *photoName = [self.photosList objectAtIndex:indexPath.row];
    NSString *photoFilePath = [[self photosDirectory]
      stringByAppendingPathComponent:photoName];
    cell.nameLabel.text =[photoName stringByDeletingPathExtension];

    __block UIImage* thumbImage = [self.photosCache objectForKey:photoName];
    cell.photoView.image = thumbImage;

    if(!thumbImage) {
      dispatch_async(dispatch_get_global_queue
      (DISPATCH_QUEUE_PRIORITY_HIGH, 0), ^{

        UIImage *image = [UIImage imageWithContentsOfFile:photoFilePath];
        if(orientation == PhotoOrientationPortrait) {
          UIGraphicsBeginImageContext(CGSizeMake(180.0f, 120.0f));
          [image drawInRect:CGRectMake(0, 0, 120.0f, 180.0f)];
          thumbImage = UIGraphicsGetImageFromCurrentImageContext();
          UIGraphicsEndImageContext();
        } else {

          UIGraphicsBeginImageContext(CGSizeMake(120.0f, 180.0f));
          [image drawInRect:CGRectMake(0, 0, 180.0f, 120.0f)];
          thumbImage = UIGraphicsGetImageFromCurrentImageContext();
          UIGraphicsEndImageContext();
        }

        dispatch_async(dispatch_get_main_queue(), ^{

          [self.photosCache setObject:thumbImage forKey:photoName];
          cell.photoView.image = thumbImage;
        });
      });
    }

    return cell;
}
```

This code listing completes the data source. But you're not done yet. Go ahead and implement the delegate method. The delegate method tells you which collection view was **tapped**, and you display the photo in a separate form sheet.

Adding a Selected State to the Cell

First, add a `selectedBackgroundView` to your cell:

```
-(void) awakeFromNib {

   self.selectedBackgroundView = [[UIView alloc] initWithFrame:self.frame];
   self.selectedBackgroundView.backgroundColor = [UIColor
colorWithWhite:0.3
   alpha:0.5];

   [super awakeFromNib];
}
```

Now, implement the following three delegate methods and return YES for the delegate that asks whether the collection view should select or highlight the said item.

UICollectionViewDelegate Methods in MKViewController.m

```
-  (BOOL)collectionView:(UICollectionView *)collectionView
   shouldHighlightItemAtIndexPath:(NSIndexPath *)indexPath {

   return YES;
}

-  (BOOL)collectionView:(UICollectionView *)collectionView
      shouldSelectItemAtIndexPath:(NSIndexPath *)indexPath {

   return YES;
}
```

Finally, build and run the app. Tap on a photo. At this point, you should see a selection as shown in Figure 5-3.

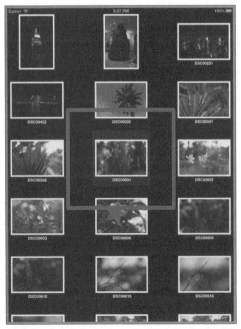

Figure 5-3 Screenshot of iOS Simulator showing cell selection

Handling Taps on Your Collection View Cell

Creating that example was easy, right? Now, you can show a detailed view when you tap on this item. To do so, create a details controller in your storyboard. Also, add necessary outlets in your controller subclass (a `UIImageView` and a `UILabel`) and a property to pass the photo to be displayed from the collection view.

To create a segue, Ctrl-click the collection view controller and drag it to your details controller.

Select the segue and set the identifier of your segue to something like `MainSegue`. Also change the style to "modal" and presentation to "Form Sheet." You can leave the transition style to the default value or change it to whatever you prefer.

Implement the last delegate method, `collectionView:didSelectItemAtIndexPath:`, as shown here:

collectionView:didSelectItemAtIndexPath: in MKViewController.m

```
- (void) collectionView: (UICollectionView *)collectionView
    didSelectItemAtIndexPath: (NSIndexPath *)indexPath {

    [self performSegueWithIdentifier:@"MainSegue" sender:indexPath];
}
```

The next step is to implement the `prepareForSegue` method:

prepareForSegue Method in MKViewController.m

```
- (void) prepareForSegue: (UIStoryboardSegue *)segue sender: (id)sender {

    NSIndexPath *selectedIndexPath = sender;
    NSString *photoName = [self.photosList
    objectAtIndex:selectedIndexPath.row];

    MKDetailsViewController *controller = segue.destinationViewController;
    controller.photoPath = [[self photosDirectory]
      stringByAppendingPathComponent:photoName];
}
```

Build and run the app and tap on a photo. You should see the Details view showing the photo you tapped.

That step completes handling the delegate. Wait! That's not the end of the road! You can also add multiple sections to the collection view and set decoration views and backgrounds to them. I show you how to do this now.

Adding a "Header" and a "Footer"

A collection view doesn't call these features headers and footers. When you open the `UICollectionViewDataSource` header file, you probably won't find any methods that can be used to add these elements. The collection view's default layout class, `UICollectionViewFlowLayout` supports adding headers and footers. The headers and footers are called as supplementary views, and there are two kinds: `UICollectionElementKindSectionHeader` and `UICollectionElementKindSectionFooter`.

If you are writing a custom layout, you can support your own supplementary views. You add a supplementary view in your storyboard by selecting the `CollectionView` and enabling the "Section Header" or "Section Footer" and dragging a `UICollectionReusableView` from the object browser to your collection view. The next important step is to set an identifier to your supplementary view. You can do this by selecting the newly dragged `UICollectionReusableView` and setting the identifier on the Utilities panel.

Next, you add methods (in fact, just one method) to your `UICollectionViewController` subclass, as follows:

Collection View Datasource Method to Provide a Footer

```
- (UICollectionReusableView *)collectionView:(UICollectionView
    *)collectionView
            viewForSupplementaryElementOfKind:(NSString *)kind
                              atIndexPath:(NSIndexPath *)indexPath {


    static NSString *SupplementaryViewIdentifier =
    @"SupplementaryViewIdentifier";

    return [collectionView dequeueReusableSupplementaryViewOfKind:
     UICollectionElementKindSectionFooter
     withReuseIdentifier:SupplementaryViewIdentifier
     forIndexPath:indexPath];
}
```

This code fragment is slightly different from a table view where you create a section header in the data source method and return it. The collection view controller takes it to the next level by adding dequeue support to supplementary views as well.

Advanced Customization with Collection View Custom Layout

The most important differentiator of a collection view and a table view is that a collection view doesn't know how to lay out its items. It delegates the layout mechanics to a `UICollectionViewLayout` subclass. The default layout is a *flow* layout provided by the `UICollectionViewFlowLayout` class. This class allows you to tweak various settings through `UICollectionViewDelegateFlowLayout` protocol. For instance, you can change the `scrollDirection` property from vertical (the default) to horizontal and instantly get a horizontally scrolling collection view.

The most important part of `UICollectionViewLayout` is that by subclassing it, you can create your own custom layouts. For example, a "CoverFlowLayout". `UICollectionViewFlowLayout` happens to be one such subclass that is provided by default from Apple. For this reason, collection view flow layout calls the "headers" and "footers" supplementary views. It's a header if you're using the built-in flow layout. It could be something else if you're using a custom layout such as a Cover Flow layout.

Collection views are powerful because the layout is handled independent of the class, and this alone means that you'll probably see a huge array of custom UI components, including springboard imitations, custom home screens, visually rich layouts such as Cover Flow, masonry layouts, and such. In the following sections, I show you how to use a subclass of `UICollectionViewLayout` and create a masonry layout using collection views.

Masonry Layout

A masonry layout is one in which the items in a collection view resemble a stone mortar wall. Although the stones may be of irregular sizes, the wall remains a square or rectangle. In a masonry layout, the item size (usually the height) might vary, but the "layout engine" calculates the position and lays out the items so that it resembles a mortar wall. Masonry layout was pioneered by Pinterest and is now used by various sites. It's more commonly used to display a large quantity of items tiled on two-dimensional space.

A masonry layout sample is shown on the right in **Figure 5-4.** `UICollectionViewDelegateFlowLayout` protocol has a `collectionView:layout:sizeForItemAtIndexPath:` method. You normally implement this method if your item's size is a variable. But when you do that, the default collection view layout will look something like the layout in Figure 5.4. The default layout calculates the maximum height of all the items in a given row and starts laying out the next row items so that every item is at a distance of *at least* maximum height + padding from the item in the previous row.

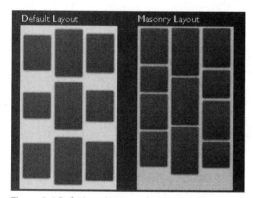

Figure 5-4 Default and Masonry layout in a collection view

In both images in Figure 5.4, the first row items are of height (approximately) 110, 150, and 135 points, respectively. The second row items are of height 85, 145, and 90 points, respectively. The default layout positions the tiles in the following order. The maximum heights of items in the two rows are calculated. In this case, the first row's maximum height is 150, and the second row's is 145. The second row items will be laid out between heights of 160 (150 points + a 10-point padding) and 305 (160 + 145 points for the tallest item) assuming a padding of 10 points. The default collection view flow layout vertically centers the items. Because the height of the first item is 85 points, it is positioned at the vertical center of the height of the second row, 145. This creates an additional padding of 60 points, 30 on top and 30 to the bottom of the item. The height of the second item is exactly 145; there is not any additional padding. The height of the third item is 90 points, and that creates a padding of 55 points, 27.5 each on the top and bottom of the item. This padding adds discordance to the layout.

To lay out items as shown on the left of Figure 5-4, you should manually calculate the positions of all items in your collection view. You do this by implementing your own layout subclassing `UICollectionViewLayout`. You can also implement custom layouts by subclassing `UICollectionViewFlowLayout`, but in this case, it's easier to subclass `UICollectionViewLayout`. To do a Cover Flow-like custom layout, you can use `UICollectionViewFlowLayout` as your starting point .

I show you how to do this using an example. To get started, create a simple collection view-based application like the previous example and display 100 same-sized cells. You learned how to do this earlier in this chapter, in the first example.

When you do this, your output should look similar to Figure 5-5.

Figure 5-5 Collection View displaying same-sized cell

Now change the size (height) of the items at random. To do that, implement the
`UICollectionViewDelegateFlowLayout` method, as follows:

Changing the Item Size Using the Flow Layout Delegate

```
- (CGSize)collectionView:(UICollectionView *)collectionView
  layout:(UICollectionViewLayout*)collectionViewLayout
  sizeForItemAtIndexPath:(NSIndexPath *)indexPath {

  CGFloat randomHeight = 100 + (arc4random() % 140);
  return CGSizeMake(100, randomHeight); // 100 to 240 pixels tall
}
```

When you implement this method, your output should look like Figure 5-6.

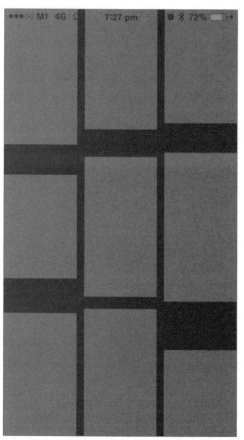

Figure 5-6 Collection View displaying cells of random height

Now create a subclass of `UICollectionViewLayout` and add it to your project. I call it `MKMasonryViewLayout`. For this example, you are going to write just enough code to change the layout to look like Figure 5-7.

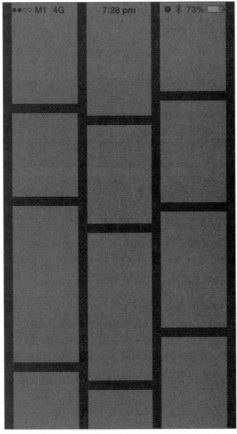

Figure 5-7 Collection View displaying cells of random height but in Masonry layout

The layout subclass has two main responsibilities. The first is to calculate the frames of every item in the collection view. The second responsibility is to calculate the overall content size of the collection view. After you calculate the frame, calculating the content size should be fairly easy. These two responsibilities are handled in the following method. All custom layouts are performed by modifying the layout attribute. In this case, you modify the frame of the layout attribute.

Methods to Be Overridden by the Layout Subclass

```
- (void) prepareLayout;
- (NSArray *)layoutAttributesForElementsInRect:(CGRect)rect;
- (CGSize) collectionViewContentSize;
```

The approach is to calculate the frames of all the items and cache the frame sizes somewhere. The first method, `prepareLayout`, is called by the collection view before it starts the layout process. In this method, you calculate the frames. In this case, you introduce a couple of variables: `numberOfColumns` and `interItemSpacing`.

In this example, the item's width is dependent on these two variables. The item's height is a variable that should be provided by the data source. You can write a protocol for this. I call it `MKMasonryViewLayoutDelegate` in the sample code. In this delegate, write a protocol method called `collectionView:layout:height ForItemAtIndexPath:`

If you were to follow Apple's naming convention for layout delegates, the name should be `MKMasonryViewDelegateLayout`. **For purposes of this example, I find Apple's naming convention here unconventional.**

You now have the necessary foundation for calculating the frames. The logic here is pretty straightforward. The x and `width` values of the item are calculated based on the variables `numberOfColumns` and `interItemSpacing`.

Calculating the Item Width

```
CGFloat fullWidth = self.collectionView.frame.size.width;
CGFloat availableSpaceExcludingPadding = fullWidth - (self.interItemSpacing
 * (self.numberOfColumns + 1));
        CGFloat itemWidth = availableSpaceExcludingPadding /
        self.numberOfColumns;
```

Calculating the Item's x Origin

```
CGFloat x = self.interItemSpacing + (self.interItemSpacing + itemWidth) *
 currentColumn;
```

You get the height by calling the protocol method you wrote. You get the item's y origin by progressively adding the height of all the previous items in that column. I use the `latYValueForColumn` dictionary to maintain the height of all the columns in the layout.

Calculating the Item's y Origin

```
CGFloat y = [self.lastYValueForColumn[@(currentColumn)] doubleValue];
CGFloat height =
  [((id<MKMasonryViewLayoutDelegate>)self.collectionView.delegate)
                    collectionView:self.collectionView
                    layout:self
                    heightForItemAtIndexPath:indexPath];
```

(continued)

```
itemAttributes.frame = CGRectMake(x, y, itemWidth, height);
y+= height;
y += self.interItemSpacing;
self.lastYValueForColumn[@(currentColumn)] = @(y);
```

You start off by getting the y origin from the dictionary lastYValueForColumn. This initially is 0. You then add the height and interItemSpacing to this y and save this value back to lastYValueForColumn. During the next iteration, this value is read back again, and you position the next item at this y origin.

After you finish the calculation, you store them in a dictionary called layoutInfo. You will use this layoutInfo to calculate the items in the next method, layoutAttributesForElementsInRect:. As the collection view is scrolled, this method is called with the rectangle that is visible. Using the CGRectIntersectsRect method, you filter out the items that are within the rectangle and return an array of items that are within the specified rectangle. This approach is illustrated in the following code:

Calculating the Items Within a Given Rectangle

```
- (NSArray *)layoutAttributesForElementsInRect:(CGRect)rect {

  NSMutableArray *allAttributes =
[NSMutableArray arrayWithCapacity:self.layoutInfo.count];

  [self.layoutInfo enumerateKeysAndObjectsUsingBlock:
^(NSIndexPath *indexPath,

UICollectionViewLayoutAttributes *attributes,

                                          BOOL *stop) {

    if (CGRectIntersectsRect(rect, attributes.frame)) {
      [allAttributes addObject:attributes];
    }
  }];
  return allAttributes;
}
```

The next and the final task is to calculate the collection view's content size. You already store the y origin in the dictionary lastYValueForColumn. The value of the next y origin is the height of that column if there is no next item. This means self.lastYValueForColumn[@0] gives the height of the first column and self.lastYValueForColumn[@1] gives the height of the second column and so on. The biggest of all these values becomes the content size of the collection view. This calculation is shown in the following code snippet:

Calculating the contentSize of the Collection View

```
-(CGSize) collectionViewContentSize {

  NSUInteger currentColumn = 0;
  CGFloat maxHeight = 0;
  do {
    CGFloat height = [self.lastYValueForColumn[@(currentColumn)]
    doubleValue];
```

```
    if(height > maxHeight)
      maxHeight = height;
    currentColumn ++;
  } while (currentColumn < self.numberOfColumns);

  return CGSizeMake(self.collectionView.frame.size.width, maxHeight);
}
```

The complete code is available from the book's website in the file `MKMasonryViewLayout.m`.

That wasn't hard, right? In this example, you modified the `frame` of the layout attribute. You can also change the `center`, `size`, `transform3D`, `bounds`, `transform`, `alpha`, `zIndex`, and `hidden` properties. You use the `transform` and `transform3D` to create layouts like Cover Flow. I show you how to do this in the next section.

Cover Flow Layout

Apple first introduced Cover Flow in iTunes. Although Cover Flow is completely removed from iOS and iTunes on Mac, it's still used in Finder (Mac OS X 10.9).

Creating a Cover Flow layout is easier than creating the previous layout. This time, you start off by subclassing `UICollectionViewFlowLayout` instead. Think of a Cover Flow layout as a horizontal flow layout where the item's `transform3D` is modified. In fact, in the custom layout method, you are going to do exactly that. You are going to modify the `transform3D` property (along with `zIndex` and `alpha`) and rotate the images around the y-axis.

For this layout, you are not going to precalculate or cache the frames as in the previous example. Doing this in the `layoutAttributesForElementsInRect:` method is faster and easier.

For all the visible items in the collection view, you are going to update the `transform3D`. The rotation angle is proportional to the distance of the item from the center. That is, if the item is at the center, the rotation angle is 0. The angle ranges from 0 to M_PI_4 ($\pi/4$) depending on the distance from the center. The bold part of the following code snippet illustrates how to use the `transform3D` property.

Cover Flow Layout (MKCoverFlowLayout.m)

```
- (NSArray*)layoutAttributesForElementsInRect:(CGRect)rect
{
  NSArray* array = [super layoutAttributesForElementsInRect:rect];
  CGRect visibleRect;
  visibleRect.origin = self.collectionView.contentOffset;
  visibleRect.size = self.collectionView.bounds.size;
  float collectionViewHalfFrame = self.collectionView.frame.size.
  width/2.0f;

  for (UICollectionViewLayoutAttributes* attributes in array) {
```

(continued)

```
    if (CGRectIntersectsRect(attributes.frame, rect)) {
      CGFloat distance = CGRectGetMidX(visibleRect) - attributes.center.x;
      CGFloat normalizedDistance = distance / collectionViewHalfFrame;
      if (ABS(distance) < collectionViewHalfFrame) {
        CGFloat zoom = 1 + ZOOM_FACTOR*(1 - ABS(normalizedDistance));
        CATransform3D rotationAndPerspectiveTransform =
        CATransform3DIdentity;
        rotationAndPerspectiveTransform.m34 = 1.0 / -500;
        rotationAndPerspectiveTransform =
        CATransform3DRotate(rotationAndPerspectiveTransform,
      (normalizedDistance) * M_PI_4, 0.0f, 1.0f, 0.0f);
        CATransform3D zoomTransform = CATransform3DMakeScale(zoom, zoom,
        1.0);
        attributes.transform3D = CATransform3DConcat(zoomTransform,
        rotationAndPerspectiveTransform);
        attributes.zIndex = ABS(normalizedDistance) * 10.0f;
        CGFloat alpha = (1 - ABS(normalizedDistance)) + 0.1;
        if(alpha > 1.0f) alpha = 1.0f;
        attributes.alpha = alpha;
      } else {

        attributes.alpha = 0.0f;
      }
    }
  }
}
  return array;
}
```

The complete code is available from the book's website in the file MKCoverFlowLayout.m.

Summary

In this chapter, you learned about a powerful controller class that will revolutionize the way you write custom controls and layouts. Unlike a table view, a collection view lets you control the layout by delegating it. This alone should be a compelling reason to use collection views instead of table views in your apps.

Further Reading

Apple Documentation

The following documents are available in the iOS Developer Library at developer.apple.com or through the Xcode Documentation and API Reference.

UICollectionViewController Class Reference

UICollectionViewLayout Class Reference

Cocoa Auto Layout Guide

WWDC Sessions

The following session videos are available at `developer.apple.com`.

WWDC 2012, "Session 205: Introducing Collection Views"

WWDC 2012, "Session 219: Advanced Collection Views and Building Custom Layouts"

WWDC 2012, "Session 202: Introduction to Auto Layout for iOS and OS X"

WWDC 2012, "Session 228: Best Practices for Mastering Auto Layout"

WWDC 2012, "Session 232: Auto Layout by Example"

Chapter 6

Stay in Bounds with Auto Layout

Cocoa Auto Layout was introduced a couple of years ago with Mac OS X SDK. Apple has a long-standing history of testing new software development kit features on Mac SDKs and bringing them to iOS the following year. Cocoa Auto Layout debuted in iOS 6.

Auto Layout is a *constraint-based layout* engine that *automatically* adjusts and calculates the frames based on constraints you set. The layout model used prior to iOS 6 is called the "springs and struts" model. Although that model worked efficiently for the most part, you still had to write code to lay out your subviews when trying to do custom layout during a rotation. Also, because Auto Layout is constraint-based and not frame-based, you can size your UI elements based on content automatically. This means that translating your UI to a new language just got easier. For example, German is a *long* language. The English word *pen* is *Kugelschreiber* in German. Auto Layout automatically makes your labels bigger to accommodate the longer text. You learn about the Auto Layout feature that makes this possible in the "Intrinsic Size and Localization" section later in this chapter. Using Auto Layout also obviates the need to create multiple nib files, one for every language.

Whenever Apple introduces a new technology, you can use it in one of the following ways. You can either convert the whole application to use the new technology or migrate it partially or write new code using the new technology. With iOS 5, you followed one of these ways for Automatic Reference Counting (ARC) conversion and storyboards. In iOS 6 and iOS 7, this applies to Auto Layout. You can migrate an existing application completely to Auto Layout, use it only for new UI elements, or do a partial conversion.

As late as 2012, some developers were still supporting iOS 5. Because Auto Layout was not supported on iOS 5, most developers didn't make the effort to learn Auto Layout. Moreover, the benefits gained by using Auto Layout in iOS 6 were minimal. With iOS 7, the metrics of almost all built-in controls have changed. The view is now taller by 20 points because all views are full screen by default. The `UISegmentedControl` is much thinner, and so is `UISlider`. The width of a `UISwitch` is smaller in iOS 7 and so on. Supporting iOS 6 and iOS 7 *without* Auto Layout is something I do not recommend. Even if migrating to Auto Layout means doing a complete rewrite, you should do so because you will eventually end up removing lots of boilerplate layout code. Moreover, Xcode 5 has made it dramatically easier to work with Auto Layout. This chapter is all about Auto Layout. The purpose of this chapter is to introduce you to the constraint-based layout engine and Auto Layout in Xcode 5.

Auto Layout in Xcode 4

I have a good news for you: If you ever tried Auto Layout in Xcode 4, you can safely unlearn it. In fact, the previous edition of this book did not even cover Auto Layout in Xcode 4, and I recommended that you use Auto Layout with Visual Format Language. Visual Format Language was not without its flaws. It was completely unfriendly with code refactoring. The layout code completely broke if you renamed or refactored an IBOutlet. It's that time again. With Xcode 5, you can safely unlearn Visual Format Layout as well. To master Auto Layout in Xcode 5, start as if you are learning about Auto Layout for the first time.

Getting Started with Auto Layout

Thinking in terms of Auto Layout might be difficult if you are a "pro" at using the struts and springs method. But for a newbie to iOS, Auto Layout might be easy. My first recommendation is to "unlearn" the springs and struts method.

Why do I say that? The human brain is an associative machine. When you actually lay out your UI elements, your mental model (or your designer's mental model) is indeed "thinking" in terms of visual constraints. This happens not only when you are designing user interfaces; you do it subconsciously when you're shifting furniture in your living room. You do it when you park your car in a parking lot. You do it when you hang a photo frame on your wall. In fact, when you were doing UI layout using the old springs and struts method, you were converting the relative, constraint-based layout in your mind to what springs and struts can offer, and you were writing code to do layout that cannot be done automatically. With Auto Layout, you can express how you think—by applying visual constraints. Auto Layout also lets you express constraints between views, whereas the older springs and struts method allowed you to specify layout relationships only between a view and its super view. Letting you express constraints between sibling views makes it much easier to create more meaningful layouts.

Here's what your designer may normally say:

- "Place the button 10 pixels from the bottom."
- "Labels should have equal spacing between them."
- "The image should be centered horizontally at all times."
- "The distance between the button and the image should always be 100 pixels."
- "The button should always be at least 44 pixels tall."
- and so on . . .

Before Auto Layout, you had to manually convert these design criteria to auto-resize mask constraints, and for criteria that could not be easily done with auto-resize masks, you wrote layout code in the `layoutSubviews` method. With Auto Layout, these constraints can be expressed naturally.

For example, a constraint such as "Labels should have equal spacing between them" cannot be implemented with the springs and struts method. You have to write layout code by calculating the frame sizes of the superview and calculating the frames of all the labels. The word *them* in this constraint makes the constraint relative to another label within the same view (a sibling view instead of the super view).

What's New in Auto Layout in Xcode 5

Everything about Auto Layout in Xcode 5 is new. The complete workflow is different and much better. The first change you will notice when you use Auto Layout in Xcode 5 is that Xcode 5 does not automatically add constraints for you unless you explicitly ask it to add them. When you enabled Auto Layout in Xcode 4, the IDE automatically generated Auto Layout constraints that were translated from your auto resizing masks. In Xcode 5, this happens at runtime. In your Interface Builder, you do not see any automatically added constraints. When you add your first constraint manually, Xcode 5 turns off the `translatesAutoresizingMaskIntoConstraints` for that view.

In Xcode 4, for every constraint you added, the IDE generated additional constraints to prevent layout bugs. This happened even before you finished adding all your constraints and was arguably the biggest problem working with Auto Layout in Xcode 4. Xcode 5, in contrast, puts you in control of layout. No additional constraints are added automatically. Instead, when the constraints you add don't sufficiently define the layout, Xcode 5's Storyboard/Interface Builder compiler generates compilation warnings and optionally provides tips to fix them. A whole lot better than Xcode 4! To top this all, in Xcode 5, when you have not added any constraints, the IDE automatically adds fixed position and size constraints automatically at runtime.

Now it's time to try Auto Layout in Xcode 5. To get started, create a new Single View application and Xcode. Change the view controller's simulated metrics size from Retina 4-inch Full Screen to Retina 3.5-inch Full Screen. Then go ahead and add a date picker to this view. Align it so that the date picker is at the bottom edge of the view. Don't add any constraints at this point. Your storyboard should look like the one in Figure 6-1.

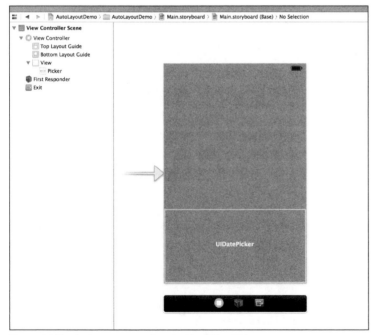

Figure 6-1 Xcode showing a storyboard with a UIDatePicker

Now, run this application on a 4-inch device/simulator. Your output should look something like that in Figure 6-2.

Notice the 88-pixel space left between the bottom edge of the screen and the date picker. This is clearly not what you want.

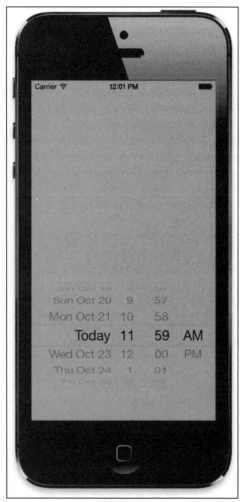

Figure 6-2 Output of the Date Picker running on 3.5-inch simulator

Now, if your view controller's simulated metrics size is set to 4-inch Full Screen and you run it on a 3.5-inch device, you see that the bottom 88 pixels of your date picker are out of the bounds of the view, which again is probably not what you want.

Auto Layout in Xcode 5

You can fix this layout issue by using Auto Layout. In Xcode 5, you see four new menu buttons when you open an Interface Builder document or a storyboard, as illustrated in Figure 6-3.

The first button enables you to add alignment constraints. The second enables you to add standard constraints such as size and position relative to other views. The third button provides you options to let Xcode 5 automatically generate constraints or update frames of your subviews to their correct locations based on your constraints. The last one lets you set an option that determines who all inherits your constraints. By default, both the Siblings and Ancestors and the Descendants options are checked. I recommend you leave the first option,

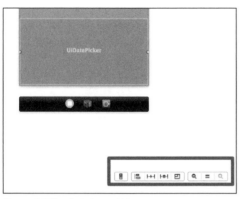

Figure 6-3 The Auto Layout Menu

Siblings and Ancestors, *unchecked.* Leaving this option checked makes it hard to align your subviews within a view. Your subview's origin gets fixed, and the ancestor view moves around when you try to resize your subview, which, in my opinion, is quite annoying. You will use the other three menu buttons when creating your layouts.

Now you can add some constraints to make the date picker stick to the bottom of your view. To do this, you use the second menu button. Click on it and add the three constraints by clicking on the I-beams, as shown in Figure 6-4.

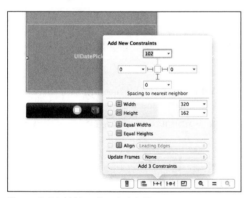

Figure 6-4 Add New Constraints Popover

At this point, superficially, this looks almost similar to your springs and struts method, but it is not. The main difference is that this panel adds your constraint relative to the nearest neighbor and not relative to the super view. After adding these constraints, you see that the date picker is aligned to the bottom of the screen regardless of the screen resolution.

Intrinsic Sizes

Most control elements in iOS have something called an *intrinsic size.* To understand how this size affects your Auto Layout constraint, replace the date picker with a text view and add the same set of constraints as you did for the date picker view. When you do, you see an error and a warning, as shown in Figure 6-5.

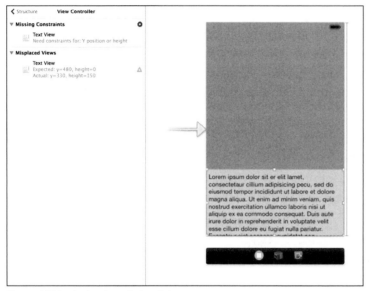

Figure 6-5 Xcode showing misplaced views

The error says that the view is missing a constraint and expects you to have a constraint for either the *y* position of your text view or its height. Note that you didn't see this issue when you used a date picker. The reason is that some controls like the date picker have an intrinsic size. In fact, when you select a date picker and open the size inspector, you are not able to edit its height. Auto Layout takes this as one of the constraints when you use a date picker. When you are working with text views, you instead have to provide a height (or *y* position) constraint explicitly.

Now click the second menu button and add a height constraint. The result is shown in Figure 6-6.

Similar to the date picker, sliders, progress views, activity indicator views, segmented controls, and switches also have intrinsic sizes.

Intrinsic Size and Localization

The intrinsic size of a `UILabel` (and `UIButton`) is interesting. Its intrinsic size reflects its content. The longer the title, the wider the label's intrinsic size. This, coupled with Auto Layout, makes it easy to create buttons and labels that automatically adjust their sizes when their content changes in a different language.

Design Time and Runtime Layout

Xcode 5 doesn't automatically add constraints (unless you explicitly ask for them). Instead, what happens is, if the constraints that you added don't unambiguously describe the layout, Xcode 5 shows a list of warnings or errors and also highlights the ambiguity in your layout. In most cases, anything that would cause a user interface to be laid out wrong is treated as a warning, and a runtime crash (primarily due to ambiguous constraints) is treated as an error. Xcode 5 also shows two types of highlights. A yellow, solid-line bounding box shows the design time layout of the offending control, and a red, dotted-line bounding box shows the runtime layout of the offending control.

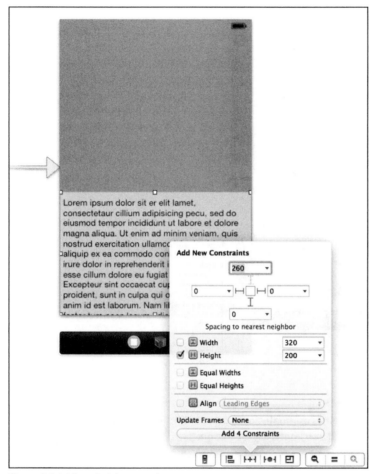

Figure 6-6 Adding a height constraint

At this point, go ahead and add a label to your view. Position it at rectangle (20, 200, 280, 21). Set the label text to "Welcome to Xcode 5." Now add a constraint by clicking the second menu button; then set the spacing to the left neighbor to 20. Your document should look like the one shown in Figure 6-7. You added just one constraint, and this clearly is not enough to unambiguously describe the layout. Xcode 5 shows a warning and an error. The warning says that the label is misplaced. It's misplaced because the constraints don't adequately define the layout. The error says that there is a missing constraint for y.

Now if you look at the highlights around your control, you see a yellow, solid-line bounding box and a red, dotted-line bounding box. The yellow-colored bounding box highlights the offending control and shows its size and position at design time. The red-colored bounding box shows the size and position of the control at runtime. In this case, the label has an intrinsic size, which is the size that is just right to show the entire text without clipping. That is exactly what the red bounding box highlights. In some other cases, when there are multiple ambiguous constraints, the red dotted-line bounding box shows you the final size and position of the control at runtime.

Go ahead and add a couple more constraints to fix the spacing of the label to the right and top nearest neighbor. That should fix the Auto Layout issue.

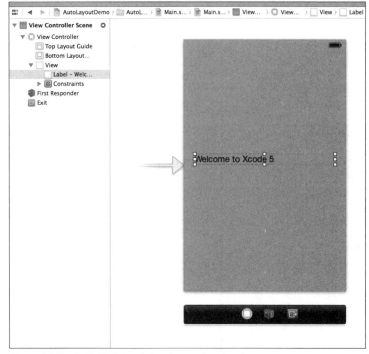

Figure 6-7 Xcode highlighting design time and runtime layout

Instead of adding these two missing constraints manually, click on the third menu button and choose Add Missing Constraints to have them added automatically by Xcode.

Updating Frames Automatically

The next step is to remove all the constraints added to the label. Then click the first menu button and add the Horizontal Center in Container and Vertical Center in Container constraints. Now your document should look like Figure 6-8. Again, you see a red, dotted-line bounding box showing you the runtime layout of this label. You also see a yellow line with a number. In this example, it is 29.5. This number tells you that the label's y position will be higher by 29.5 points at runtime. This type of highlight is shown when one of your views is misplaced. To fix the layout in this case, you should open the size inspector and reposition the view to the frame suggested by Auto Layout's red, dotted-line bounding box. The same can be done automatically by clicking the third menu button and choosing Update Frames.

Top and Bottom Layout Guides

In iOS 7, every view controller has two properties called `topLayoutGuide` and `bottomLayoutGuide`. Xcode 5 shows them as objects in your Interface Builder document. You can align your views with respect to these two guides as well. Previously, you used to align your views to the top and bottom edges of the container view. You now align your views to the `topLayoutGuide` and the `bottomLayoutGuide`, respectively. The advantages are that these guides include the height of the status bar or the bottom tab bar if one is present,

Figure 6-8 Xcode showing design and runtime layout for a label

and they automatically align your controls based on your constraints, taking into account the height of the status bar, navigation bar, and tab bar, respectively. The advantages of these guides are even more important when your app has to run on both iOS 6 and iOS 7 because in iOS 7, despite having a status bar, navigation bar, and tab bar, because the bars are translucent, your view controller still extends below them, unlike in iOS 6. Because your view controller extends below the translucent bar its size varies from iOS 6 to iOS 7.

Layout Previews in Assistant Editor

Xcode 5's assistant editor can also show you a preview of what your user interface is going to look like. You can simulate an iOS 6 device or an iOS 7 device and portrait or landscape orientation. Just open the assistant editor, click the top-level breadcrumb menu's first crumb, and select Preview. You then see the preview of your user interface in this assistant editor pane. The preview pane is updated in real time whenever you make changes to the document.

Debugging Auto Layout at Design Time

Debugging Auto Layout is made much easier in Xcode 5. Previously, you had to run the app on all different kinds of devices and wait for it to crash. Xcode 5 lets you visually see and handle most of these crashes at design time. This is a huge improvement that alone makes your workflow better.

Auto Layout has two main types of issues. The first type occurs when your set of constraints is just not enough to determine the layout in all possible orientation/screen sizes. The second type of issue occurs when you have too many constraints that conflict with each other in at least one orientation.

The first type can be broadly classified into two categories, ambiguous frames and misplaced views. An ambiguous frame happens when you have not added constraints to unambiguously determine the position and size of a view at runtime. A misplaced view, is a side effect of ambiguous frames. Xcode warns you of a misplaced view when the size and position of the view at design time might not match its size or position at runtime. Adding necessary constraints usually fixes this problem.

Conflicting constraints happen when you have added constraints that conflict with each other. When this happens, the Auto Layout engine tries to break constraints at runtime (and logs a message that lists the broken constraint) and attempts the layout. If it fails, you get a runtime crash. Conflicting constraints are the only kind of layout error not caught at design time. In most cases, constraints conflict only in one orientation. This can happen when constraints that you added seem to work exactly as you want in portrait orientation but conflict with each other in landscape orientation. For example, a height constraint of 480 points on a table view might look good on a 3.5-inch device, but when you run this on a 4-inch device, you see an 88-point area to the bottom of the screen below the table view. The Auto Layout engine starts throwing errors when you rotate your device (either 3.5-inch or 4-inch) to landscape. In landscape mode, the screen height is limited to 320 points. Setting the table view's height constraint to 480 points is now a conflicting constraint.

Most developers who used Auto Layout in Xcode 4 tend to add more constraints than necessary, just to be safe. In Xcode 5, you don't have to. In fact, these extra constraints are the number one reason for Auto Layout runtime crashes in Xcode 5.

In the "Debugging Layout Errors" section, later in this chapter, you learn how to handle Auto Layout runtime crashes and debug your layout.

Working with Scroll Views in Auto Layout

A scroll view scrolls the content by changing the origin of the bounds. When you use Auto Layout and pin the left, top, bottom, and right of a scroll view to the edge, you are essentially pinning the edge of the content view. That means even if the scroll view's `contentSize` is bigger than the frame size, because of the constraint, altering bounds has no effect. To make this work with Auto Layout, add your scroll view into another view and pin the scroll view to the edges of the container's super view.

Alternatively, you can also set the `translatesAutoResizingMasksToConstraints` property to `NO`.

Working with Frames and Auto Layout

When you use Auto Layout, you are no longer able to update frames of a subview. Even if you try to call the `setFrame` method, nothing happens. The Auto Layout engine has the ultimate authority over the size and position of the subview. In case you still need to update a frame at runtime (assuming that you are creating a custom UI that needs this), you should instead be looking at updating the corresponding `NSLayoutConstraint`'s constant parameter instead of updating the frame directly.

Visual Format Language

In this section, you learn about Visual Format Language and how to lay out a UI by code using that language, the Auto Layout way, instead of calculating frames and repositioning them. So far you've been working with Interface Builder to lay out your subviews. Visual Format Language is helpful if you're using code to create and lay out the UI elements.

Adding a layout constraint to a view is easy. You start by creating an `NSLayoutConstraint` instance and adding it as a constraint to the view. You can create an `NSLayoutConstraint` instance using the Visual Format Language or using the class method, as follows:

Adding a Constraint to a View

```
[self.view addConstraint:
[NSLayoutConstraint constraintWithItem:self.myLabel
                             attribute:NSLayoutAttributeRight
                             relatedBy:NSLayoutRelationEqual
                                toItem:self.myButton
                             attribute:NSLayoutAttributeLeft
                            multiplier:10.0
                              constant:100.0]];
```

Adding a Constraint Using Visual Format Language

```
NSDictionary *viewsDictionary = NSDictionaryOfVariableBindings(self.
myLabel,self.myButton);
NSArray *constraints = [NSLayoutConstraint
                        constraintsWithVisualFormat:@"[myLabel]-100-
                        [myButton]"
                        options:0 metrics:nil
                        views:viewsDictionary];
[self.view addConstrints:constraints];
```

The second method is more expressive and lets you specify a constraint using an *ASCII Art* style. In the preceding example, you created a constraint that ensures that the space between the label and the button is always 100 pixels. The following is the visual format language syntax for specifying this constraint:

```
[myLabel]-100-[myButton]
```

Well, that was easy. Now, take a look at a more complicated example. The Visual Format Language is powerful and expressive, almost like regular expressions, yet readable. You can connect the labels and buttons to the superview using the following syntax:

```
|-[myLabel]-100-[myButton]-|
```

You can add a nested constraint to the button as follows:

```
|-[myLabel]-100-[myButton (>=30)]-|
```

This constraint ensures that `myButton` is at least 30 pixels at all orientations.

You can even add multiple constraints to the same button:

```
|-[myLabel]-100-[myButton (>=30, <=50)]-|
```

This constraint ensures that `myButton` is at least 30 pixels but not more than 50 pixels at all orientations.

Now, when you add conflicting constraints like the one here

```
|-[myLabel]-100-[myButton (>=30, ==50)]-|
```

Auto Layout gracefully tries to satisfy the requirement by ignoring the conflicting constraint. But when your UI layout cannot be performed without ambiguities, Auto Layout crashes. I show you two methods to handle debugging errors in the next couple of sections.

Disadvantages of Visual Format Language

Visual Format Layout is great if you don't like Interface Builder and storyboards. But otherwise, because the language depends on the Objective-C runtime to create your constraint (since your outlet names are written inside strings), it poses a serious difficulty and unexplained crashes should you refactor your IBOutlets.

In the past, I recommended that you use Visual Format Layout. But that was because Xcode 4 was horribly buggy and the workflow was harder to use. With Xcode 5, I highly recommend that you don't use Visual Format Language. It's unintuitive and counterproductive if you refactor your outlets later.

Debugging Layout Errors

When Auto Layout throws exceptions, you see something like the following on your console. Create an iPhone application using the single view template and add a table view as a subview to your main view. Add five constraints, as shown in Figure 6-9.

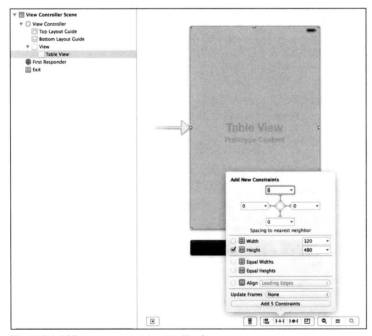

Figure 6-9 Adding new constraints using Xcode

The height constraint here is redundant. It doesn't manifest itself as a problem when the device is in portrait mode. However, when you turn the device into landscape mode, you see a warning on the console that looks similar to this:

```
Unable to simultaneously satisfy constraints.
Probably at least one of the constraints in the following list is one you
don't want. Try this: (1) look at each constraint and try to figure out
which you don't expect; (2) find the code that added the unwanted constraint
or constraints and fix it. (Note: If you're seeing
NSAutoresizingMaskLayoutConstraints that you don't understand, refer to
the documentation for the UIView property
translatesAutoresizingMaskIntoConstraints)
(
    "<NSLayoutConstraint:0x8a8ee10 V:[UITableView:0xb1edc00(480)]>",
    "<NSLayoutConstraint:0x8a93730 V:|-(0)-
[UITableView:0xb1edc00]    (Names: '|':UIView:0x8a8f840 )>",
    "<NSLayoutConstraint:0x8a93790 V:[UITableView:0xb1edc00]-
(0)-[_UILayoutGuide:0x8a93270]>",
    "<_UILayoutSupportConstraint:0x8a7c3e0
V:[_UILayoutGuide:0x8a93270(0)]>",
    "<_UILayoutSupportConstraint:0x8a93870
_UILayoutGuide:0x8a93270.bottom == UIView:0x8a8f840.bottom>",
    "<NSAutoresizingMaskLayoutConstraint:0x8e94cb0 h=--& v=--&
H:[UIView:0x8a8f840(320)]>"
)

Will attempt to recover by breaking constraint
<NSLayoutConstraint:0x8a8ee10 V:[UITableView:0xb1edc00(480)]>

Break on objc_exception_throw to catch this in the debugger.
The methods in the UIConstraintBasedLayoutDebugging category
on UIView listed in <UIKit/UIView.h> may also be helpful.
```

The console log lists the constraints Auto Layout is trying to satisfy. In the preceding case, there's a constraint that the height of the table view should be at least 480 pixels tall. When the device was in portrait mode, the layout engine was able to lay out the UI with all the constraints, but when you rotated it to landscape mode, it logged the layout error message because in landscape mode, the height of the iPhone is less than the 480 points and the constraints provided could not be satisfied. The log also says that it will try to recover by breaking a constraint.

Summary

In this chapter, you learned about a powerful way to express your layout in terms of constraints. In iOS 6, you dealt with multiple screen sizes. In iOS 7, you have to deal with multiple screen sizes and multiple operating systems.

For example, your view inside a navigation controller would be 416 points tall on a 3.5-inch device running iOS 6 and 504 points tall on a 4-inch device. Now, in iOS 7, the height of your status bar and navigation bars does not affect your contained view size. Your view gets extended below those translucent bars. That means the same view is now 480 points on a 3.5-inch device and 568 points on a 4-inch device. That almost makes it as hard as catering to four different screen sizes. Auto Layout is the only way to go, and it's the future of layout. After you practice a month or so, you will start appreciating the power of Auto Layout and will almost never again need to write layout code manually.

Further Reading

Apple Documentation

The following documents are available in the iOS Developer Library at `developer.apple.com` or through the Xcode Documentation and API Reference.

Cocoa Auto Layout Programming Guide

(Note that in this guide most of the code samples are Mac applications, but that shouldn't be a problem. Every layout code, whether written or designed in Xcode 5, will also work as such in iOS.)

Technical Note TN2154: UIScrollView And Autolayout

`https://developer.apple.com/library/ios/technotes/tn2154/_index.html`

WWDC Sessions

The following session videos are available at `developer.apple.com`.

WWDC 2012, "Session 202: Introduction to Auto Layout for iOS and OS X"

WWDC 2012, "Session 205: Introducing Collection Views"

WWDC 2012, "Session 219: Advanced Collection Views and Building Custom Layouts"

WWDC 2012, "Session 228: Best Practices for Mastering Auto Layout"

WWDC 2012, "Session 232: Auto Layout by Example"

WWDC 2013, "Session 405: Interface Builder Core Concepts"

WWDC 2013, "Session 406: Taking Control of Auto Layout in Xcode 5"

## Chapter 7

Better Custom Drawing

Your users expect a beautiful, engaging, and intuitive interface. It's up to you to deliver. No matter how powerful your features, if your interface seems "clunky," you're going to have a hard time making the sale. This is about more than just pretty colors and flashy animations. A truly beautiful and elegant user interface is a key part of a user-centric application. Keeping your focus on delighting your user is the key to building exceptional applications.

One of the tools you need to create an exceptional user interface is custom drawing. In this chapter, you learn the mechanics of drawing in iOS, with focus on flexibility and performance. This chapter doesn't cover iOS UI design. For information on how to design iOS interfaces, start with Apple's *iOS Human Interface Guidelines* and *iOS Application Programming Guide*, available in the iOS developer documentation.

In this chapter, you find out about the several drawing systems in iOS, with a focus on UIKit and Core Graphics. By the end of this chapter, you will have a strong grasp of the UIKit drawing cycle, drawing coordinate systems, graphic contexts, paths, and transforms. You'll know how to optimize your drawing speed through correct view configuration, caching, pixel alignment, and use of layers. You'll be able to avoid bloating your application bundle with avoidable prerendered graphics.

With the right tools, you can achieve your goal of a beautiful, engaging, and intuitive interface, while maintaining high performance, low memory usage, and small application size.

iOS's Many Drawing Systems

iOS has several major drawing systems: UIKit, Core Graphics (Quartz), Core Animation, Core Image, and OpenGL ES. Each is useful for a different kind of problem.

- **UIKit**—This is the highest-level interface, and the only interface in Objective-C. It provides easy access to layout, compositing, drawing, fonts, images, animation, and more. You can recognize UIKit elements by the prefix `UI`, such as `UIView` and `UIBezierPath`. UIKit also extends `NSString` to simplify drawing text with methods such as `drawInRect:withAttributes:`.

- **Core Graphics (also called Quartz 2D)**—The primary drawing system underlying UIKit, Core Graphics is what you use most frequently to draw custom views. Core Graphics is highly integrated with `UIView` and other parts of UIKit. Core Graphics data structures and functions can be identified by the prefix `CG`.

- **Core Animation**—This system provides powerful two- and three-dimensional animation services. It is also highly integrated into `UIView`. Chapter 8 covers Core Animation in detail.

- **Core Image**—A Mac technology first available in iOS 5, Core Image provides very fast image filtering such as cropping, sharpening, warping, and just about any other transformation you can imagine.

■ **OpenGL ES**—Most useful for writing high-performance games—particularly 3D games—Open GL ES is a subset of the OpenGL drawing language. For other applications on iOS, Core Animation is generally a better choice. OpenGL ES is portable between most platforms. A discussion of OpenGL ES is beyond the scope of this book, but many good books are available on the subject.

UIKit and the View Drawing Cycle

When you change the frame or visibility of a view, draw a line, or change the color of an object, the change is not immediately displayed on the screen. This sometimes confuses developers who incorrectly write code like this:

```
progressView.hidden = NO; // This line does nothing
[self doSomethingTimeConsuming];
progressView.hidden = YES;
```

It's important to understand that the first line (`progressView.hidden = NO`) does absolutely nothing useful. This code does not cause the Progress view to be displayed while the time-consuming operation is in progress. No matter how long this method runs, you will never see the view displayed. Figure 7-1 shows what actually happens in the drawing loop.

All drawing occurs on the main thread, so as long as your code is running on the main thread, nothing can be drawn. That is one of the reasons you should never execute a long-running operation on the main thread. It not only prevents drawing updates but also prevents event handling (such as responding to touches). As long as your code is running on the main thread, your application is effectively "hung" to the user. This isn't noticeable as long as you make sure that your main thread routines return quickly.

You may now be thinking, "Well, I'll just run my drawing commands on a background thread." You generally can't do that because drawing to the current UIKit context isn't thread-safe. Any attempt to modify a view on a background thread leads to undefined behavior, including drawing corruption and crashes. (See the section "Caching and Background Drawing," later in the chapter, for more information on how you can draw in the background.)

This behavior is not a problem to be overcome. The consolidation of drawing events is one part of iOS's capability to render complex drawings on limited hardware. As you see throughout this chapter, much of UIKit is dedicated to avoiding unnecessary drawing, and this consolidation is one of the first steps.

So how do you start and stop an activity indicator for a long-running operation? You use dispatch or operation queues to put your expensive work in the background, while making all your UIKit calls on the main thread, as shown in the following code.

ViewController.m (TimeConsuming)

```
- (IBAction)doSomething:(id)sender {
    [sender setEnabled:NO];
    [self.activity startAnimating];

    dispatch_queue_t bgQueue = dispatch_get_global_queue(
                    DISPATCH_QUEUE_PRIORITY_DEFAULT, 0);
```

```
dispatch_async(bgQueue, ^{
  [self somethingTimeConsuming];

  dispatch_async(dispatch_get_main_queue(), ^{
    [self.activity stopAnimating];
    [sender setEnabled:YES];
  });
});
}
```

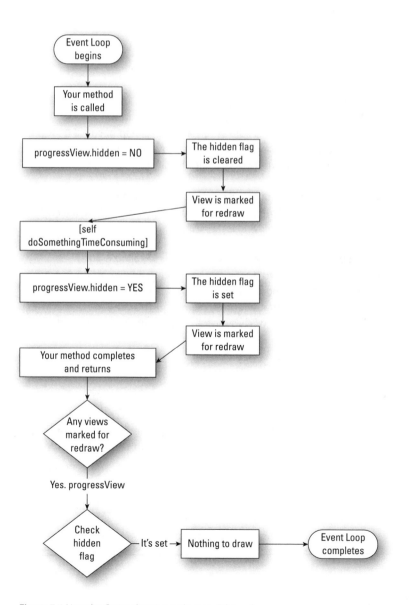

Figure 7-1 How the Cocoa drawing cycle consolidates changes

When the `IBAction` is called, you start animating the activity indicator. You then put a call to `somethingTimeConsuming` on the default background dispatch queue. When that finishes, you put a call to `stopAnimating` on the main dispatch queue. Dispatch and operation queues are covered in Chapter 9.

To summarize:

- iOS consolidates all drawing requests during the run loop and draws them all at once.

- You must not block the main thread to do complex processing.

- You must not draw into the main view graphics context except on the main thread. You need to check each UIKit method to ensure it does not have a main thread requirement. Some UIKit methods can be used on background threads as long as you're not drawing into the main view context.

View Drawing Versus View Layout

`UIView` separates the layout ("rearranging") of subviews from drawing (or "display"). Doing so is important for maximizing performance because layout is generally cheaper than drawing. Layout is cheap because `UIView` caches drawing operations onto GPU-optimized bitmaps. These bitmaps can be moved around, shown, hidden, rotated, and otherwise transformed and composited very inexpensively using the GPU.

When you call `setNeedsDisplay` on a view, it is marked "dirty" and will be redrawn during the next drawing cycle. Don't call it unless the content of the view has really changed. Most UIKit views automatically manage redrawing when their data is changed, so you generally don't need to call `setNeedsDisplay` except on custom views.

When a view's subviews need to be rearranged because of an orientation change or scrolling, UIKit calls `setNeedsLayout`. This, in turn, calls `layoutSubviews` on the affected views. By overriding `layoutSubviews`, you can make your application much smoother during rotation and scrolling events. You can rearrange your subviews' frames without necessarily having to redraw them, and you can hide or show views based on orientation. You can also call `setNeedsLayout` if your data changes in ways that need only layout updates rather than drawing.

Custom View Drawing

Views can provide their content by including subviews, including layers, or implementing `drawRect:`. Typically, if you implement `drawRect:`, you don't mix this with layers or subviews, although doing so is legal and sometimes useful. Most custom drawing is done with UIKit or Core Graphics, although OpenGL ES has become easier to integrate when needed.

Two-dimensional drawing generally breaks down into several operations:

- Lines

- Paths (filled or outlined shapes)

- Text

- Images

- Gradients

Two-dimensional drawing does not include manipulation of individual pixels because that is destination-dependent. You can achieve this result using a bitmap context, but not directly with UIKit or Core Graphics functions.

Both UIKit and Core Graphics use a "painter" drawing model. This means that each command is drawn in sequence, overlaying previous drawings during that event loop. Order is very important in this model, and you must draw back-to-front. Each time `drawRect:` is called, it's your responsibility to draw the entire area requested. The drawing "canvas" is not preserved between calls to `drawRect:`.

Drawing with UIKit

In the "old days" before iPad, most custom drawing had to be done with Core Graphics because there was no way to draw arbitrary shapes with UIKit. In iPhoneOS 3.2, Apple added `UIBezierPath` and made it much easier to draw entirely in Objective-C. UIKit still lacks support for lines, gradients, shading, and some advanced features such as anti-aliasing controls and precise color management. Even so, UIKit is now a very convenient way to manage the most common custom drawing needs.

The simplest way to draw rectangles is to use `UIRectFrame` or `UIRectFill`, as shown in the following code:

```
- (void)drawRect:(CGRect)rect {
  [[UIColor redColor] setFill];
  UIRectFill(CGRectMake(10, 10, 100, 100));
}
```

Notice how you first set the pen color using `-[UIColor setFill]`. Drawing is done into a graphics context provided by the system before calling `drawRect:`. That context includes a lot of information including stroke color, fill color, text color, font, transform, and more. At any given time, just one stroke pen and one fill pen are available, and you use their colors to draw everything. The "Managing Graphics Contexts" section, later in this chapter, covers how to save and restore contexts, but for now just note that drawing commands are order-dependent, and that includes commands that change the pens.

> The graphics context provided to `drawRect:` is specifically a view graphics context. There are other types of graphics contexts, including PDF and bitmap contexts. All of them use the same drawing techniques, but a view graphics context is optimized for drawing onto the screen.

Paths

UIKit includes much more powerful drawing commands than its rectangle functions. It can draw arbitrary curves and lines using `UIBezierPath`. A Bézier curve is a mathematical way of expressing a line or curve using a small number of control points. Most of the time, you don't need to worry about the math because `UIBezierPath` has simple methods to handle the most common paths: lines, arcs, rectangles (optionally rounded), and ovals. With these methods, you can quickly draw most shapes needed for UI elements. The following code is an example of a simple shape scaled to fill the view, as shown in Figure 7-2. You draw this several ways in the upcoming examples.

Figure 7-2 Output of FlowerView

FlowerView.m (Paths)

```
- (void)drawRect:(CGRect)rect {
  CGSize size = self.bounds.size;
  CGFloat margin = 10;
  CGFloat radius = rintf(MIN(size.height - margin,
                             size.width - margin) / 4);

  CGFloat xOffset, yOffset;
  CGFloat offset = rintf((size.height - size.width) / 2);
  if (offset > 0) {
    xOffset = rint(margin / 2);
    yOffset = offset;
  }
  else {
    xOffset = -offset;
    yOffset = rint(margin / 2);
  }

  [[UIColor redColor] setFill];
  UIBezierPath *path = [UIBezierPath bezierPath];
  [path addArcWithCenter:CGPointMake(radius * 2 + xOffset,
                                     radius + yOffset)
                  radius:radius
              startAngle:-M_PI
```

```
                endAngle:0
              clockwise:YES];
  [path addArcWithCenter:CGPointMake(radius * 3 + xOffset,
                                     radius * 2 + yOffset)
                  radius:radius
              startAngle:-M_PI_2
                endAngle:M_PI_2
              clockwise:YES];
  [path addArcWithCenter:CGPointMake(radius * 2 + xOffset,
                                     radius * 3 + yOffset)
                  radius:radius
              startAngle:0
                endAngle:M_PI
              clockwise:YES];
  [path addArcWithCenter:CGPointMake(radius + xOffset,
                                     radius * 2 + yOffset)
                  radius:radius
              startAngle:M_PI_2
                endAngle:-M_PI_2
              clockwise:YES];
  [path closePath];
  [path fill];
}
```

`FlowerView` creates a path made up of a series of arcs and fills it with red. Creating a path doesn't cause anything to be drawn. A `UIBezierPath` is just a sequence of curves, like an `NSString` is a sequence of characters. Only when you call `fill` is the curve drawn into the current context.

Note the use of the M_PI (π) and M_PI_2 (π/2) constants. Arcs are described in radians, so π and fractions of π are important. `math.h` defines many such constants that you should use rather than recomputing them. Arcs measure their angles clockwise, with 0 radians pointing to the right, π/2 radians pointing down, π (or - π) radians pointing left, and - π/2 radians pointing up. You can use 3 π/2 for up if you prefer, but I find `–M_PI_2` easier to visualize than `3*M_PI_2`. If radians give you a headache, you can make a function out of it:

```
CGFloat RadiansFromDegrees(CGFloat d) {
  return d * M_PI / 180;
}
```

Generally, I recommend just getting used to radians rather than doing so much math, but if you need unusual angles, working in degrees can be easier.

When calculating `radius` and `offset`, you use `rintf` (round to closest integer) to ensure that you're point-aligned (and therefore pixel-aligned). That helps improve drawing performance and avoids blurry edges. Most of the time, that's what you want, but in cases in which an arc meets a line, it can lead to off-by-one drawing errors. Usually, the best approach is to move the line so that all the values are integers, as discussed in the following section.

Understanding Coordinates

Subtle interactions among coordinates, points, and pixels can lead to poor drawing performance and blurry lines and text. Consider the following code:

```
CGContextSetLineWidth(context, 3.);

// Draw 3pt horizontal line from {10,100} to {200,100}
CGContextMoveToPoint(context, 10., 100.);
CGContextAddLineToPoint(context, 200., 100.);
CGContextStrokePath(context);

// Draw 3pt horizontal line from {10,105.5} to {200,105.5}
CGContextMoveToPoint(context, 10., 105.5);
CGContextAddLineToPoint(context, 200., 105.5);
CGContextStrokePath(context);
```

Figure 7-3 shows the output of this program on a non-Retina display, scaled to make the differences more obvious.

Figure 7-3 Comparison of line from {10,100} and line from {10,105.5}

The line from {10, 100} to {200, 100} is much more blurry than the line from {10, 105.5} to {200, 105.5}. The reason is due to the way iOS interprets coordinates.

When you construct a CGPath, you work in so-called *geometric coordinates*. These are the same kinds of coordinates that mathematicians use, representing the zero-dimensional point at the intersection of two grid lines. It's impossible to draw a geometric point or a geometric line, because they're infinitely small and thin. When iOS draws, it has to translate these geometric objects into *pixel coordinates*. These two-dimensional boxes can be set to a specific color. A pixel is the smallest unit of display area that the device can control.

Figure 7-4 shows the geometric line from {10, 100} to {200, 100}.

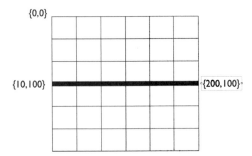

Figure 7-4 Geometric line from {10, 100} to {200, 100}

When you call CGContextStrokePath, iOS centers the line along the path. Ideally, the line would be three pixels wide, from $y = 98.5$ to $y = 101.5$, as shown in Figure 7-5.

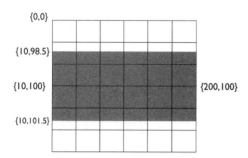

Figure 7-5 Ideal three-pixel line

This line is impossible to draw, however. Each pixel must be a single color, and the pixels at the top and bottom of the line include two colors. Half is the stroke color, and half is the background color. iOS solves this problem by averaging the two. This same technique is used in anti-aliasing. The result is shown in Figure 7-6.

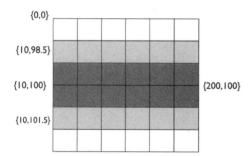

Figure 7-6 Anti-aliased three-pixel line

On the screen, this line looks slightly blurry. The solution to this problem is to move horizontal and vertical lines to the half-point so that when iOS centers the line, the edges fall along pixel boundaries, or to make your line an even width.

You can also encounter this problem with nonintegral line widths, or if your coordinates aren't integers or half-integers. Any situation that forces iOS to draw fractional pixels will cause blurriness.

Fill is not the same as stroke. A stroke line is centered on the path, but fill colors all the pixels up to the path. If you fill the rectangle from {10,100} to {200,103}, each pixel is filled correctly, as shown in Figure 7-7.

The discussion so far has equated points with pixels. On a Retina display, they are not equivalent. The iPhone 4 has four pixels per point and a scale factor of two. That subtly changes things, but generally for the better. Because all the coordinates used in Core Graphics and UIKit are expressed in points, all integral line widths are effectively an even number of pixels. For example, if you request a 1-point stroke width, this is the same as a 2-pixel stroke width. To draw that line, iOS needs to fill one pixel on each side of the path. That's an integral number of pixels, so no anti-aliasing occurs. You can still encounter blurriness if you use coordinates that are neither integers nor half-integers.

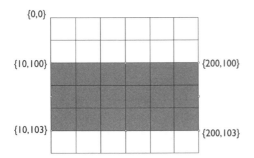

Figure 7-7 Filling the rectangle from {10,100} to {200,103}

Offsetting by a half-point is unnecessary on a Retina display, but it doesn't hurt. As long as you intend to support non-Retina displays, you need to apply a half-point offset for drawing horizontal and vertical lines.

All this discussion applies only to horizontal and vertical lines. Sloping or curved lines should be anti-aliased so that they're not jagged, so you generally have no reason to offset them.

Resizing and contentMode

Returning to `FlowerView` found in the earlier section, "Paths," if you rotate the device as shown in Figure 7-8, you can see that the view is distorted briefly before suddenly redrawing, even though you have code that adjusts for the size of the view.

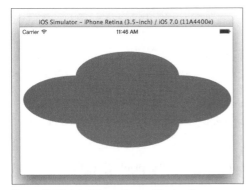

Figure 7-8 Rotated FlowerView, mid-animation

iOS optimizes drawing by taking a snapshot of the view and adjusting it for the new frame. The `drawRect:` method may not be called. The property `contentMode` determines how the view is adjusted. The default, `UIViewContentModeScaleToFill`, scales the image to fill the new view size, changing the aspect ratio if needed. That's why the shape is distorted.

In iOS 7, the view generally is redrawn after the rotation even if you don't request it. In older versions of iOS, the view is not redrawn after rotation by default.

There are a lot of ways to automatically adjust the view. You can move it around without resizing it, or you can scale it in various ways that preserve or modify the aspect ratio. The key is to make sure that any mode you use exactly matches the results of your `drawRect:` in the new orientation. Otherwise, your view will "jump" the next time you redraw. This usually works as long as your `drawRect:` doesn't consider its `bounds` during drawing. In `FlowerView`, you use the `bounds` to determine the size of your shape, so it's hard to get automatic adjustments to work correctly.

Use the automatic modes if you can because they can improve performance. When you can't, ask the system to call `drawRect:` when the frame changes by using `UIViewContentModeRedraw`, as shown in the following code:

```
- (void)awakeFromNib {
    self.contentMode = UIViewContentModeRedraw;
}
```

Transforms

iOS platforms have access to a nice GPU that can do matrix operations quickly. If you can convert your drawing calculations into matrix operations, you can leverage the GPU and get excellent performance. Transforms are just such a matrix operation.

iOS has two kinds of transforms: affine and 3D. `UIView` handles only affine transforms, so the discussion focuses on it right now. An affine transform is a way of expressing rotation, scaling, shear, and translation (shifting) as a matrix. These transforms are shown in Figure 7-9.

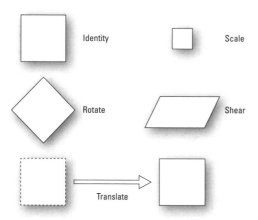

Figure 7-9 Affine transforms

A single transform combines any number of these operations into a 3 × 3 matrix. iOS has functions to support rotation, scaling, and translation. If you want shear, you have to write the matrix yourself. (You can also use `CGAffineTransformMakeShear` from Jeff LaMarche; see "Further Reading" at the end of the chapter.)

Transforms can dramatically simplify and speed up your code. Often it's much easier and faster to draw in a simple coordinate space around the origin and then to scale, rotate, and translate your drawing to the place where you want it. For instance, `FlowerView` includes a lot of code like this:

```
CGPointMake(radius * 2 + xOffset, radius + yOffset)
```

That's a lot of typing, a lot of math, and a lot of things to keep straight in your head. What if, instead, you just draw it in a 4 × 4 box as shown in Figure 7-10?

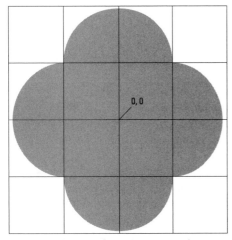

Figure 7-10 Drawing FlowerView in a 4 × 4 box

Now all the interesting points fall on nice, easy coordinates like {0,1} and {1,0}. The following code shows how to draw using this transform. Compare the highlighted sections with the `FlowerView` code earlier in this chapter.

FlowerTransformView.m (Transforms)

```
static inline CGAffineTransform
CGAffineTransformMakeScaleTranslate(CGFloat sx, CGFloat sy,
                                    CGFloat dx, CGFloat dy) {
    return CGAffineTransformMake(sx, 0.f, 0.f, sy, dx, dy);
}

- (void)drawRect:(CGRect)rect {
    CGSize size = self.bounds.size;
    CGFloat margin = 10;

    [[UIColor redColor] set];
    UIBezierPath *path = [UIBezierPath bezierPath];
```

```
    [path addArcWithCenter:CGPointMake(0, -1)
                    radius:1
                startAngle:-M_PI
                  endAngle:0
                 clockwise:YES];
    [path addArcWithCenter:CGPointMake(1, 0)
                    radius:1
                startAngle:-M_PI_2
                  endAngle:M_PI_2
                 clockwise:YES];
    [path addArcWithCenter:CGPointMake(0, 1)
                    radius:1
                startAngle:0
                  endAngle:M_PI
                 clockwise:YES];
    [path addArcWithCenter:CGPointMake(-1, 0)
                    radius:1
                startAngle:M_PI_2
                  endAngle:-M_PI_2
                 clockwise:YES];
    [path closePath];

    CGFloat scale = floorf((MIN(size.height, size.width)
                           - margin) / 4);

    CGAffineTransform transform;
    transform = CGAffineTransformMakeScaleTranslate(scale,
                                                    scale,
                                                    size.width/2,
                                                    size.height/2);
    [path applyTransform:transform];
    [path fill];
}
```

When you're done constructing your path, you compute a transform to move it into your view's coordinate space. You scale it by the size you want divided by the size it currently is (4), and you translate it to the center of the view. The utility function CGAffineTransformMakeScaleTranslate isn't just for speed (although it is faster). It's easier to get the transform correct this way. If you try to build up the transform one step at a time, each step affects later steps. Scaling and then translating is not the same as translating and then scaling. If you build the matrix all at once, you don't have to worry about that.

This technique can be used to draw complicated shapes at unusual angles. For instance, to draw an arrow pointing to the upper right, you can generally draw it more easily pointing to the right and then rotate it.

You have a choice between transforming the path using applyTransform: and transforming the whole view by setting the transform property. Which is best depends on the situation, but I usually prefer to transform the path rather than the view when practical. Modifying the view's transform makes the results of frame and bounds more difficult to interpret, so I avoid this approach when I can. As you see in the following section, you can also transform the current context, which sometimes is the best approach.

Drawing with Core Graphics

Core Graphics, sometimes called Quartz 2D or just Quartz, is the main drawing system in iOS. It provides destination-independent drawing, so you can use the same commands to draw to the screen, layer, bitmap, PDF, or printer. Anything starting with CG is part of Core Graphics. Figure 7-11 and the following code provide an example of a simple scrolling graph.

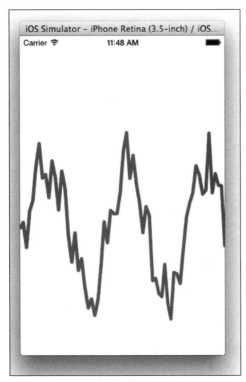

Figure 7-11 Simple scrolling graph

GraphView.m (Graph)

```
@interface GraphView ()
@property (nonatomic, readwrite, strong) NSMutableArray *values;
@property (nonatomic, readwrite, strong) dispatch_source_t timer;
 @end

@implementation GraphView

const CGFloat kXScale = 5.0;
const CGFloat kYScale = 100.0;

static inline CGAffineTransform
CGAffineTransformMakeScaleTranslate(CGFloat sx, CGFloat sy,
    CGFloat dx, CGFloat dy) {
  return CGAffineTransformMake(sx, 0.f, 0.f, sy, dx, dy);
```

```objc
}

- (void)awakeFromNib {
  [self setContentMode:UIViewContentModeRight];
  self.values = [NSMutableArray array];

  __weak id weakSelf = self;
  double delayInSeconds = 0.25;
  self.timer =
      dispatch_source_create(DISPATCH_SOURCE_TYPE_TIMER, 0, 0,
          dispatch_get_main_queue());
  dispatch_source_set_timer(
      self.timer, dispatch_walltime(NULL, 0),
      (unsigned)(delayInSeconds * NSEC_PER_SEC), 0);
  dispatch_source_set_event_handler(self.timer, ^{
    [weakSelf updateValues];
  });
  dispatch_resume(self.timer);
}

- (void)updateValues {
  double nextValue = sin(CFAbsoluteTimeGetCurrent())
      + ((double)rand()/(double)RAND_MAX);
  [self.values addObject:
      [NSNumber numberWithDouble:nextValue]];
  CGSize size = self.bounds.size;
  CGFloat maxDimension = MAX(size.height, size.width);
  NSUInteger maxValues =
      (NSUInteger)floorl(maxDimension / kXScale);

  if ([self.values count] > maxValues) {
    [self.values removeObjectsInRange:
        NSMakeRange(0, [self.values count] - maxValues)];
  }

  [self setNeedsDisplay];
}

- (void)dealloc {
  dispatch_source_cancel(_timer);
}

- (void)drawRect:(CGRect)rect {
  if ([self.values count] == 0) {
    return;
  }

  CGContextRef ctx = UIGraphicsGetCurrentContext();
  CGContextSetStrokeColorWithColor(ctx,
                                   [[UIColor redColor] CGColor]);
  CGContextSetLineJoin(ctx, kCGLineJoinRound);
  CGContextSetLineWidth(ctx, 5);
```

(continued)

```
    CGMutablePathRef path = CGPathCreateMutable();

    CGFloat yOffset = self.bounds.size.height / 2;
    CGAffineTransform transform =
        CGAffineTransformMakeScaleTranslate(kXScale, kYScale,
                                             0, yOffset);

    CGFloat y = [[self.values objectAtIndex:0] floatValue];
    CGPathMoveToPoint(path, &transform, 0, y);

    for (NSUInteger x = 1; x < [self.values count]; ++x) {
      y = [[self.values objectAtIndex:x] floatValue];
      CGPathAddLineToPoint(path, &transform, x, y);
    }

    CGContextAddPath(ctx, path);
    CGPathRelease(path);
    CGContextStrokePath(ctx);
  }
  @end
```

Every quarter second, this code adds a new number to the end of the data and removes an old number from the beginning. Then it marks the view as dirty with `setNeedsDisplay`. The drawing code sets various advanced line drawing options not available with `UIBezierPath` and creates a `CGPath` with all the lines. It then transforms the path to fit into the view, adds the path to the context, and strokes it.

> Core Graphics uses the Core Foundation memory management rules. Core Foundation objects require manual retain and release, even under ARC. Note the use of `CGPathRelease`. For full details, see Chapter 10.

You may be tempted to cache the `CGPath` here so that you don't have to compute it every time. That's a good instinct, but in this case, it wouldn't help. iOS already avoids calling `drawRect:` except when the view is dirty, which happens only when the data changes. When the data changes, you need to calculate a new path. Caching the old path in this case would just complicate the code and waste memory.

Mixing UIKit and Core Graphics

Within `drawRect:`, UIKit and Core Graphics can generally intermix without issue, but outside of `drawRect:`, you may find that things drawn with Core Graphics appear upside down. UIKit uses an upper-left origin (ULO) coordinate system, whereas Core Graphics uses a lower-left origin (LLO) system by default. As long as you use the context returned by `UIGraphicsGetCurrentContext` inside of `drawRect:`, everything is fine because this context is already flipped. But if you create your own context using functions like `CGBitmapContextCreate`, it'll be LLO. You can either do your math backward or flip the context:

```
    CGContextTranslateCTM(ctx, 0.0f, height);
    CGContextScaleCTM(ctx, 1.0f, -1.0f);
```

This code moves (translates) the height of the context and then flips it using a negative scale. When going from UIKit to Core Graphics, the transform is reversed:

```
CGContextScaleCTM(ctx, 1.0f, -1.0f);
CGContextTranslateCTM(ctx, 0.0f, -height);
```

First flip it and then translate it.

Managing Graphics Contexts

Before calling `drawRect:`, the drawing system creates a graphics context (`CGContext`). A context includes a lot of information such as a pen color, text color, current font, transform, and more. Sometimes you may want to modify the context and then put it back the way you found it. For instance, you may have a function to draw a specific shape with a specific color. The system has only one stroke pen, so when you change the color, this would change things for your caller. To avoid side effects, you can push and pop the context using `CGContextSaveGState` and `CGContextRestoreGState`.

Do not confuse this with the similar-sounding `UIGraphicsPushContext` and `UIGraphicsPopContext`. They do not do the same thing. `CGContextSaveGState` remembers the current state of a context. `UIGraphicsPushContext` changes the current context. Here's an example of `CGContextSaveGState`:

```
[[UIColor redColor] setFill];
CGContextSaveGState(UIGraphicsGetCurrentContext());
[[UIColor blackColor] setFill];
CGContextRestoreGState(UIGraphicsGetCurrentContext());
UIRectFill(CGRectMake(10, 10, 100, 100)); // Red
```

This code sets the fill pen color to red and saves off the context. It then changes the pen color to black and restores the context. When you draw, the pen is red again.

The following code illustrates a common error:

```
[[UIColor redColor] setStroke];
// Next line is nonsense
UIGraphicsPushContext(UIGraphicsGetCurrentContext());
[[UIColor blackColor] setStroke];
UIGraphicsPopContext();
UIRectFill(CGRectMake(10, 10, 100, 100)); // Black
```

In this case, you set the pen color to red and then switch context to the current context, which does nothing useful. You then change the pen color to black and pop the context back to the original (which effectively does nothing). You now will draw a black rectangle, which is almost certainly not what was meant.

The purpose of `UIGraphicsPushContext` is not to save the current *state* of the context (pen color, line width, and so on), but to switch contexts entirely. Say you are in the middle of drawing something into the current view context, and now want to draw something completely different into a bitmap context. If you want to use UIKit to do any of your drawing, you'd want to save off the current UIKit context, including all the drawing that had been done, and switch to a completely new drawing context. That's what `UIGraphicsPushContext` does. When you finish creating your bitmap, you pop the stack and get your old context back. That's what `UIGraphicsPopContext` does. This only matters in cases in which you want to draw into the new bitmap context with UIKit. As long as you use Core Graphics functions, you don't need to push or pop contexts because Core Graphics functions take the context as a parameter.

This operation is pretty useful and common. It's so common that Apple has made a shortcut for it called `UIGraphicsBeginImageContext`. This shortcut takes care of pushing the old context, allocating memory for a new context, creating the new context, flipping the coordinate system, and making it the current context. Most of the time, that's just what you want.

Here's an example of how to create an image and return it using `UIGraphicsBeginImageContext`. The result is shown in Figure 7-12.

Figure 7-12 Text drawn with reverseImageForText:

MYView.m (Drawing)

```
- (UIImage *)reverseImageForText:(NSString *)text {
    const size_t kImageWidth = 200;
    const size_t kImageHeight = 200;
    CGImageRef textImage = NULL;
    UIFont *font = [UIFont preferredFontForTextStyle:UIFontTextStyleHeadline];
    UIColor *color = [UIColor redColor];

    UIGraphicsBeginImageContext(CGSizeMake(kImageWidth,
                                           kImageHeight));
```

```
[text drawInRect:CGRectMake(0, 0, kImageWidth, kImageHeight)
    withAttributes:@{
                     NSFontAttributeName: font,
                     NSForegroundColorAttributeName: color
                     }];

textImage =
    UIGraphicsGetImageFromCurrentImageContext().CGImage;

UIGraphicsEndImageContext();

return [UIImage imageWithCGImage:textImage
                          scale:1.0
              orientation:UIImageOrientationUpMirrored];

}
```

Optimizing UIView Drawing

`UIView` and its subclasses are highly optimized, so when possible, use them rather than custom drawing. For instance, `UIImageView` is faster and uses less memory than anything you're likely to put together in an afternoon with Core Graphics. The following sections cover a few issues to keep in mind when using `UIView` to keep it drawing as well as it can.

Avoid Drawing

The fastest drawing is the drawing you never do. iOS goes to great lengths to avoid calling `drawRect:`. It caches an image of your view and moves, rotates, and scales it without any intervention from you. Using an appropriate `contentMode` lets the system adjust your view during rotation or resizing without calling `drawRect:`. The most common cause for `drawRect:` running is your calling `setNeedsDisplay`. Avoid calling `setNeedsDisplay` unnecessarily. Remember, though, `setNeedsDisplay` just schedules the view to be redrawn. Calling `setNeedsDisplay` many times in a single event loop is no more expensive, practically, than calling it once, so don't coalesce your calls. iOS is already doing that for you.

Caching and Background Drawing

If you need to do a lot of calculations during your drawing, cache the results when you can. At the lowest level, you can cache the raw data you need rather than asking for it from your delegate every time. Beyond that, you can cache static elements such as `CGFont` or `CGGradient` objects so that you generate them only once. Fonts and gradients are useful to cache this way because they're often reused.

Although caching small objects can be very useful, be careful of caching images. When an image is drawn on the screen or into a bitmap context, the system has to create a fully decoded representation of it. At 4 bytes per pixel, the memory can add up very quickly. `UIImage` can sometimes purge this bitmap representation when memory is tight, but it can't do that for images you've custom drawn into a bitmap context. In many cases, especially on Retina iPads, it is better to redraw than to cache the result.

Much of this caching or precalculation can be done in the background. You may have heard that you must always draw on the main thread, but this advice isn't completely true. Several UIKit functions must be called only on the main thread, such as `UIGraphicsBeginImageContext`, but you are free to create a `CGBitmapContext` object on any thread using `CGBitmapCreateContext` and draw into it. Since iOS 4, you can use UIKit drawing methods like `drawAtPoint:` on background threads as long as you draw into your own `CGContext` and not the main view graphics context (the one returned by `UIGraphicsGetCurrentContext`). You should access a given `CGContext` only on a single thread, however.

Custom Drawing Versus Prerendering

There are two major approaches to managing complex drawing. You can draw everything programmatically with `CGPath` and `CGGradient`, or you can prerender everything in a graphics program such as Adobe Photoshop and display it as an image. If you have an art department and plan to have extremely complex visual elements, Photoshop is often the only way to go.

Prerendering has a lot of disadvantages, however. First, it introduces resolution dependence. You may need to manage 1-scale and 2-scale versions of your images and possibly different images for iPad and iPhone. Having to do so complicates workflow and bloats your product. Prerendering can make minor changes difficult and lock you into precise element sizes and colors if every change requires a round trip to the artist. Many artists are still unfamiliar with how to draw stretchable images and how to best provide images to be composited for iOS.

Apple originally encouraged developers to prerender because early iPhones couldn't compute gradients fast enough. Since the iPhone 3GS, speed has been less of an issue, and each generation makes custom drawing more attractive.

Today, I recommend custom drawing when you can do it in a reasonable amount of code. This is usually the case for small elements like buttons. When you do use prerendered artwork, I suggest that you keep the art files fairly "flat" and composite in code. For instance, you may use an image for a button's background but handle the rounding and shadows in code. That way, as you want to make minor tweaks, you don't have to rerender the background.

A middle ground in this is automatic Core Graphics code generation with tools such as PaintCode and Opacity. These tools are not panaceas. Typically, the code generated is not ideal, and you may have to modify it, complicating the workflow if you want to regenerate the code. That said, I recommend investigating these tools if you are doing a lot of UI design. See "Further Reading" at the end of this chapter for links to sites with information on these tools.

Pixel Alignment and Blurry Text

One of the most common causes of subtle drawing problems is pixel misalignment. If you ask Core Graphics to draw at a point that is not aligned with a pixel, it performs anti-aliasing as discussed in "Understanding Coordinates" earlier in this chapter. This means it draws part of the information on one pixel and part on another, giving the illusion that the line is between the two. This illusion makes things smoother but also makes them fuzzy. Anti-aliasing also takes processing time, so it slows down drawing. When possible, make sure that your drawing is pixel-aligned to avoid this situation.

Prior to the Retina display, pixel-aligned meant integer coordinates. Since iOS 4, coordinates are in points, not pixels. There are two pixels to the point on the current Retina display, so half-points (1.5, 2.5) are also pixel-aligned. In the future, there might be four or more pixels to the point, and it could be different from device to device. Even so, unless you need pixel accuracy, it is easiest to just make sure you use integer coordinates for your frames.

Generally, it's the frame origin that matters for pixel alignment. This causes an unfortunate problem for the `center` property. If you set the center to an integral coordinate, your origin may be misaligned. This alignment issue is particularly noticeable with text, especially with `UILabel`. Figure 7-13 demonstrates this problem. It is subtle and somewhat difficult to see in print, so you can also demonstrate it with the program `BlurryText` available with the online files for this chapter.

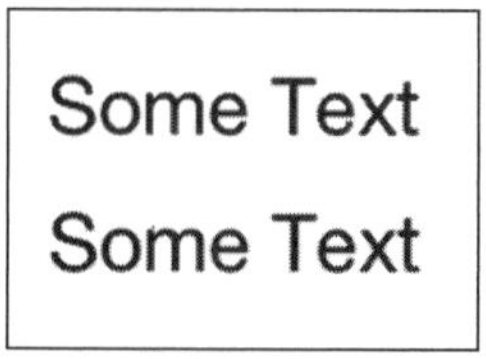

Some Text

Some Text

Figure 7-13 Text that is pixel-aligned (top) and unaligned (bottom)

You can choose from two solutions. First, odd font sizes (13 rather than 12, for example) typically align correctly. If you make a habit of using odd font sizes, you can often avoid the problem. To be certain you avoid the problem, you need to make sure that the frame is integral either by using `setFrame:` instead of `setCenter:` or by using a `UIView` category like `setAlignedCenter::`

```
- (void)setAlignedCenter:(CGPoint)center {
    self.center = center;
    self.frame = CGRectIntegral(self.frame);
}
```

Because `setAlignedCenter:` effectively sets the frame twice, it's not the fastest solution, but it is very easy and fast enough for most problems. `CGRectIntegral()` returns the smallest integral rectangle that encloses the given rectangle.

As non-Retina displays phase out, blurry text will be less of an issue as long as you set `center` to integer coordinates. For now, though, it is still a concern.

Alpha, Opaque, Hidden

Views have three properties that appear related but that are actually orthogonal: `alpha`, `opaque`, and `hidden`.

The `alpha` property determines how much information a view contributes to the pixels within its frame. So an `alpha` of 1 means that all of the view's information is used to color the pixel. An `alpha` of 0 means that none of the view's information is used to color the pixel. Remember, nothing is really transparent on an iPhone screen. If you set the entire screen to transparent pixels, the user isn't going to see the circuit board or the ground. In the end, it's just a matter of what color to draw the pixel. So, as you raise and lower the `alpha`, you're changing how much this view contributes to the pixel versus views "below" it.

Marking a view opaque or not doesn't actually make its content more or less transparent. Opaque is a promise that the drawing system can use for optimization. When you mark a view as opaque, you're promising the drawing system that you will draw every pixel in your rectangle with fully opaque colors. Doing so allows the drawing system to ignore views below yours, and that can improve performance, particularly when applying transforms. You should mark your views opaque whenever possible, especially views that scroll, such as UITableViewCell. However, if any partially transparent pixels are in your view, or if you don't draw every pixel in your rectangle, setting opaque can have unpredictable results. Setting a nontransparent backgroundColor ensures that all pixels are drawn.

Closely related to opaque is clearsContextBeforeDrawing. This is YES by default, and sets the context to transparent black before calling drawRect:. This setting avoids any garbage data in the view. It's a pretty fast operation, but if you're going to draw every pixel anyway, you can get a small benefit by setting it to NO.

Finally, hidden indicates that the view should not be drawn at all and is generally equivalent to an alpha of 0. The hidden property cannot be animated, so it's common to hide views by animating alpha to 0.

Hidden and transparent views don't receive touch events. The meaning of transparent is not well defined in the documentation, but through experimentation, I've found that it's an alpha less than 0.1. Do not rely on this particular value, but the point is that "nearly transparent" is generally treated as transparent. You cannot create a "transparent overlay" to catch touch events by setting the alpha very low.

You can make a view transparent and still receive touch events by setting its alpha to 1, opaque to NO, and backgroundColor to nil or [UIColor clearColor]. A view with a transparent background is still considered visible for the purposes of hit detection.

Summary

iOS has a rich collection of drawing tools. This chapter focused on Core Graphics and its Objective-C descendant, UIKit. By now, you should have a good understanding of how systems interact and how to optimize your iOS drawing.

Chapter 8 discusses Core Animation, which puts your interface in motion. Also covered is CALayer, a powerful addition to UIView and CGLayer and an important tool for your drawing toolbox even if you're not animating. Chapter 19 takes you even further, bringing physics-like animation behaviors to your views.

iOS 5 added Core Image to iOS for tweaking pictures. iOS also has ever-growing support for OpenGL ES for drawing advanced 3D graphics and textures. OpenGL ES is a book-length subject of its own, so it isn't tackled here, but you can get a good introduction in Apple's *OpenGL ES Programming Guide for iOS* (see the "Further Reading" section).

Further Reading

Apple Documentation

The following documents are available in the iOS Developer Library at developer.apple.com or through the Xcode Documentation and API Reference.

Drawing and Printing Guide for iOS

iOS Human Interface Guidelines

iOS App Programming Guide

OpenGL ES Programming Guide for iOS

Quartz 2D Programming Guide

Technical Q&A QA1708: Improving Image Drawing Performance on iOS

View Programming Guide for iOS

Other Resources

LaMarche, Jeff. "iPhone Development." Jeff has several articles that provide a lot of insight into using `CGAffineTransform`.

`iphonedevelopment.blogspot.com/search/label/CGAffineTransform`

PaintCode. Vector editor that exports Core Graphics code. Particularly well suited to common UI elements.

`www.paintcodeapp.com`

Opacity. Another vector editor that exports Core Graphics code.

`likethought.com/opacity`

Chapter 8

Layers Like an Onion: Core Animation

The iPhone has made animation central to the mobile experience. Views slide in and out, applications zoom into place, pages fly into the bookmark list. Apple has made animation not just a beautiful part of the experience, but a better way to let the user know what's happening and what to expect. When views slide into place from right to left, it's natural to press the left-pointing button to go back to where you were. When you create a bookmark and it flies to the toolbar, it's obvious where you should look to get back to that bookmark. These subtle cues are a critical part of making your user interface intuitive as well as engaging. To facilitate all this animation, iOS devices include a powerful GPU and frameworks that let you harness that GPU easily.

In this chapter, you discover the two main animation systems of iOS: view animations and the Core Animation framework. You find out how to draw with Core Animation layers and how to move layers around in two and three dimensions. Common decorations such as rounded corners, colored borders, and shadows are trivial with `CALayer`, and you discover how to apply them quickly and easily. You learn how to create custom automatic animations, including animating your own properties. Finally, because Core Animation is all about performance, you find out how to manage layers in multithreaded applications.

This chapter focuses on animations for view-based programming. These frameworks are ideal for most iOS applications except games. Game development is outside the scope of this book, and it's usually best served by built-in frameworks such as OpenGL ES or third-party frameworks such as Cocos2D. For more information on OpenGL ES, see the OpenGL ES for iOS portal at `developer.apple.com`. For more information on Cocos2D, see `cocos2d-iphone.org`.

View Animations

`UIView` provides rich animation functionality that's easy to use and well optimized. Most common animations can be handled with `+animateWithDuration:animations:` and related methods. You can use `UIView` to animate `frame`, `bounds`, `center`, `transform`, `alpha`, `backgroundColor`, and `contentStretch`. Most of the time, you'll animate `frame`, `center`, `transform`, and `alpha`.

You're likely familiar with basic view animations, so I just touch on the high points in this section and then move on to more advanced layer-based drawing and animation.

Let's start with a simple animation of a ball that falls when you tap the view. `CircleView` just draws a circle in its frame. The following code creates the animation shown in Figure 8-1.

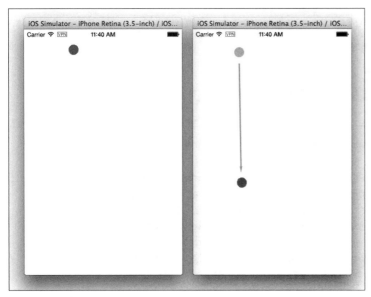

Figure 8-1 CircleView animation

ViewAnimationViewController.m (ViewAnimation)

```
- (void)viewDidLoad {
  [super viewDidLoad];
  self.circleView = [[CircleView alloc] initWithFrame:
                      CGRectMake(0, 0, 20, 20)];
  self.circleView.center = CGPointMake(100, 20);
  [[self view] addSubview:self.circleView];

  UITapGestureRecognizer *g;
  g = [[UITapGestureRecognizer alloc]
       initWithTarget:self
       action:@selector(dropAnimate)];
  [[self view] addGestureRecognizer:g];
}

...

- (void)dropAnimate {
  [UIView animateWithDuration:3 animations:^{
    self.circleView.center = CGPointMake(100, 300);
  }];
}
```

This is the simplest kind of view-based animation, and it can handle most common problems, particularly animating size, location, and opacity. It's also common to animate `transform` to scale, rotate, or translate the view over time. Less commonly, you can animate `backgroundColor` and `contentStretch`. Animating the background color is particularly useful in HUD-style interfaces to move between mostly transparent and mostly opaque backgrounds. This technique can be more effective than just animating the overall `alpha`.

Chaining animations is also straightforward, as shown in the following code:

```
- (void)dropAnimate {
  [UIView
   animateWithDuration:3 animations:^{
     self.circleView.center = CGPointMake(100, 300);
   }
   completion:^(BOOL finished){
     [UIView animateWithDuration:1 animations:^{
       self.circleView.center = CGPointMake(250, 300);
     }];
   }];
}
```

Now the ball will drop and then move to the right. But there's a subtle problem with this code. If you tap the screen while the animation is in progress, the ball will jump to the lower left and then animate to the right. That's probably not what you want. The issue is that every time you tap the view, this code runs. If an animation is in progress, it's canceled and the `completion` block runs with `finished==NO`. You look at how to handle that issue next.

Managing User Interaction

The problem mentioned in the preceding section is caused by a user experience mistake: allowing the user to send new commands while you're animating the last command. Sometimes that's what you want, but in this case, it isn't. Any time you create an animation in response to user input, you need to consider this issue.

When you animate a view, by default it automatically stops responding to user interaction. So, while the ball is dropping, tapping it doesn't generate any events. In this example, however, tapping the main view causes the animation. There are two solutions. First, you can change your user interface so that tapping the ball causes the animation:

```
[self.circleView addGestureRecognizer:g];
```

The other solution is to ignore taps while the ball is animating. The following code shows how to disable the `UIGestureRecognizer` in the gesture recognizer callback and then enable it when the animation completes:

```
- (void)dropAnimate:(UIGestureRecognizer *)recognizer {
  [UIView
   animateWithDuration:3 animations:^{
     recognizer.enabled = NO;
     self.circleView.center = CGPointMake(100, 300);
   }
   completion:^(BOOL finished){
     [UIView
      animateWithDuration:1 animations:^{
        self.circleView.center = CGPointMake(250, 300);
      }
```

```
        completion:^(BOOL finished) {
            recognizer.enabled = YES;
        }];
    }];
}
```

This technique is nice because it minimizes side effects to the rest of the view, but you might want to prevent all user interaction for the view while the animation runs. In that case, you replace `recognizer.enabled` with `self.view.userInteractionEnabled`.

Drawing with Layers

View animations are powerful, so rely on them whenever you can, especially for basic layout animation. They also provide a small number of stock transitions that you can read about in the "Animations" section of the *View Programming Guide for iOS* available at `developer.apple.com`. If you have basic needs, these are great tools.

But you're here to go beyond the basic needs, and view animations have a lot of limitations. Their basic unit of animation is `UIView`, which is also the object that handles touch events, so you may not want to split it up just to animate a piece. `UIView` also doesn't support three-dimensional layout, except for basic z-ordering, so it can't create anything like Cover Flow. To move your UI to the next level, you need to use Core Animation.

Core Animation provides a variety of tools, several of which are useful even if you don't intend to animate anything. The most basic and important part of Core Animation is `CALayer`. This section explains how to draw using `CALayer` without animations. You explore animating later in the chapter.

In many ways, `CALayer` is very much like `UIView`. It has a location, size, transform, and content. You can override a draw method to draw custom content, usually with Core Graphics. There is a layer hierarchy exactly like the view hierarchy. You might ask, why even have separate objects?

The most important answer is that `UIView` is a somewhat heavier-weight object that manages drawing and event handling, particularly touch events. `CALayer` is all about drawing. In fact, `UIView` relies on a `CALayer` to manage its drawing, which allows the two to work very well together.

Every `UIView` has a `CALayer` to do its drawing. And every `CALayer` can have sublayers, just like every `UIView` can have subviews. Figure 8-2 shows the hierarchy.

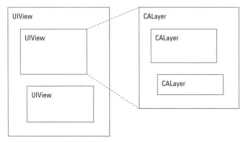

Figure 8-2 View and layer hierarchies

A layer draws whatever is in its `contents` property, which is a `CGImage` (see the note at the end of this section). It's your job to set this content somehow, and there are various ways of doing so. The simplest approach is to assign it directly, as shown here (and discussed more fully in "Setting Contents Directly" later in this chapter).

```
UIImage *image = ...;
CALayer *layer = ...;
layer.contents = (id)[image CGImage];
```

If you do not set the `contents` property directly, Core Animation goes through the following `CALayer` and delegate methods in the order presented in the following list to create it:

1. [CALayersetNeedsDisplay] —Your code needs to call this method. It marks the layer as dirty, requesting that `contents` be updated using the following steps in this list.

   Unless `setNeedsDisplay` is called, the `contents` property is never updated, even if it's `nil`.

2. [CALayer displayIfNeeded] —The drawing system automatically calls this method as needed. If the layer has been marked dirty by a call to `setNeedsDisplay`, the drawing system continues with the next steps.

3. [CALayer display] —This method is called by `displayIfNeeded` when appropriate. You shouldn't call it directly. The default implementation calls the delegate method `displayLayer:` if the delegate implements it. If not, `display` calls `drawInContext:`. You can override `display` in a subclass to set `contents` directly.

4. [delegate displayLayer:] —The default [CALayer display] calls this method if the delegate implements it. Its job is to set `contents`. If this method is implemented, even if it does nothing, no further custom drawing code is run.

5. [CALayer drawInContext:] —The default `display` method creates a view graphics context and passes it to `drawInContext:`. This is similar to [UIView drawRect:], but no UIKit context is set up for you automatically. To draw with UIKit, you need to call `UIGraphicsPushContext()` to make the passed context the current context. Otherwise, just use the passed context to draw with Core Graphics. The default `display` method takes the resulting context, creates a `CGImage` (see following note), and assigns it to `contents`. The default [CALayer drawInContext:] calls [delegate drawLayer:inContext:] if it's implemented. Otherwise, it does nothing. Note that you may call this directly. See the section "Drawing in Your Own Context," later in this chapter, for information on why you would call this directly.

6. [delegate drawLayer:inContext:] —If implemented, the default `drawInContext:` calls this method to update the context so that `display` can create a `CGImage`.

As you can see, there are several ways to set the contents of a layer. You can set it directly with `setContent:`, you can implement `display` or `displayLayer:`, or you can implement `drawInContext:` or `drawLayer:inContext:`. In the rest of this section, you find out about each approach.

The drawing system almost never automatically updates `contents` in the way that `UIView` is often automatically refreshed. For instance, `UIView` draws itself the first time it's put onscreen. `CALayer` does not. Marking a `UIView` as dirty with `setNeedsDisplay` automatically redraws all the subviews as well. Marking a `CALayer` as dirty with `setNeedsDisplay` doesn't impact sublayers. The point to remember is that the default behavior of a `UIView` is to draw when it thinks you need it. The default behavior of a `CALayer` is to never draw unless you explicitly ask for it. CALayer is a much lower-level object, and it's optimized to not waste time doing anything that isn't explicitly asked for.

> The contents **property is usually a** CGImage, **but this is not always the case. If you use custom drawing, Core Animation uses a private class,** CABackingStorage, **for** contents. **You can set** contents **to either a** CGImage **or the** contents **of another layer.**

Setting Contents Directly

Providing a content image (shown in the following code) is the easiest solution if you already have an image handy.

LayersViewController.m (Layers)

```
#import <QuartzCore/QuartzCore.h>
...
  UIImage *image = [UIImage imageNamed:@"pushing"];
  self.view.layer.contentsScale = [[UIScreen mainScreen] scale];
  self.view.layer.contentsGravity = kCAGravityCenter;
  self.view.layer.contents = (id)[image CGImage];
```

> **You must always import** QuartzCore.h **and link with** QuartzCore.framework **to use Core Animation. This step is easy to forget.**

The cast to id is needed because contents is defined as an id but actually expects a CGImageRef. To make this work with ARC, you need a cast. A common error is to pass a UIImage here instead of a CGImageRef. You don't get a compiler error or runtime warning. Your view is just blank.

By default, the contents are scaled to fill the view, even if that distorts the image. As with contentMode and contentStretch in UIView, you can configure CALayer to scale its image in different ways using contentsCenter and contentsGravity.

Implementing Display

The job of display or displayLayer: is to set contents to a correct CGImage. You can do this any way you'd like. The default implementation creates a CGContext, passes it to drawInContext:, turns the result into a CGImage, and assigns it to contents. The most common reason to override this implementation is if your layer has several states and you have an image for each. Buttons often work this way. You can create those images by loading them from your bundle and drawing them with Core Graphics or any other way you'd like.

Whether to subclass CALayer or use a delegate is really a matter of taste and convenience. UIView has a layer, and it must be that layer's delegate. In my experience, it's dangerous to make a UIView the delegate for any of the sublayers. Doing so can create infinite recursion when the UIView tries to copy its sublayers in certain operations such as transitions. So you can implement displayLayer: in UIView to manage its layer, or you can have some other object be the delegate for sublayers.

Having `UIView` implement `displayLayer:` seldom makes sense in my opinion. If your view content is basically several images, it's usually a better idea to use a `UIImageView` or a `UIButton` rather than a custom `UIView` with hand-loaded layer content. `UIImageView` is highly optimized for displaying images. `UIButton` is very good at switching images based on state and includes a lot of good user interface mechanics that are a pain to reproduce. Don't try to reinvent UIKit in Core Animation. UIKit likely does the job better than you can.

What can make more sense is to make your `UIViewController` the delegate for the layers, particularly if you aren't subclassing `UIView`. This way, you avoid extra objects and subclasses if your needs are pretty simple. Just don't let your `UIViewController` become overcomplicated.

Custom Drawing

As with `UIView`, you can provide completely custom drawing with `CALayer`. Typically, you draw with Core Graphics, but using `UIGraphicsPushContext`, you can also draw with UIKit.

See Chapter 7 for information on how to draw with Core Graphics and UIKit.

Using `drawInContext:` is just another way of setting `contents`. It's called by `display`, which is called only when the layer is explicitly marked dirty with `setNeedsDisplay`. The advantage of this approach over setting `contents` directly is that `display` automatically creates a `CGContext` appropriate for the layer. In particular, the coordinate system is flipped for you. (See Chapter 7 for a discussion of Core Graphics and flipped coordinate systems.) The following code shows how to implement the delegate method `drawLayer:inContext:` to draw the string "Pushing The Limits" in a layer using UIKit. Because Core Animation does not set a UIKit graphics context, you need to call `UIGraphicsPushContext` before calling UIKit methods, and `UIGraphicsPopContext` before returning.

DelegateView.m (Layers)

```
- (id)initWithFrame:(CGRect)frame {
    self = [super initWithFrame:frame];
    if (self) {
      [self.layer setNeedsDisplay];
      [self.layer setContentsScale:[[UIScreen mainScreen] scale]];
    }
    return self;
}

- (void)drawLayer:(CALayer *)layer inContext:(CGContextRef)ctx {
    UIGraphicsPushContext(ctx);
    [[UIColor whiteColor] set];
    UIRectFill(layer.bounds);

    UIFont *font = [UIFont preferredFontForTextStyle:UIFontTextStyleHeadline];
    UIColor *color = [UIColor blackColor];

    NSMutableParagraphStyle *style = [NSMutableParagraphStyle new];
    [style setAlignment:NSTextAlignmentCenter];
```

(continued)

```
    NSDictionary *attribs = @{NSFontAttributeName: font,
                              NSForegroundColorAttributeName: color,
                              NSParagraphStyleAttributeName: style};

    NSAttributedString *
    text = [[NSAttributedString alloc] initWithString:@"Pushing The Limits"
                                          attributes:attribs];

    [text drawInRect:CGRectInset([layer bounds], 10, 100)];
    UIGraphicsPopContext();
}
```

Note the call to `setNeedsDisplay` in `initWithFrame:`. As discussed earlier, layers don't automatically draw themselves when put onscreen. You need to mark them as dirty with `setNeedsDisplay`.

You may also notice the hand-drawing of the background rather than using the `backgroundColor` property. This is intentional. When you engage in custom drawing with `drawLayer:inContext:`, most automatic layer settings such as `backgroundColor` and `cornerRadius` are ignored. Your job in `drawLayer:inContext:` is to draw everything needed for the layer. There isn't helpful compositing going on for you as in `UIView`. If you want layer effects such as rounded corners together with custom drawing, put the custom drawing onto a sublayer and round the corners on the superlayer.

Drawing in Your Own Context

Unlike using `[UIView drawRect:]`, calling `[CALayer drawInContext:]` yourself is completely legal. You just need to generate a context and pass it in, which is nice for capturing the contents of a layer onto a bitmap or PDF so you can save it or print it. Calling `drawInContext:` this way is mostly useful if you want to composite this layer with something else, because, if all you want is a bitmap, you can just use `contents`.

`drawInContext:` draws only the current layer, not any of its sublayers. If you want to draw the layer and its sublayers, use `renderInContext:`, which also captures the current state of the layer if it's animating. `renderInContext:` uses the current state of the render tree that Core Animation maintains internally, so it doesn't call `drawInContext:`.

> If you are drawing the contents of a view hierarchy, you should also look at the `UISnapshotting` methods of `UIView`, particularly `drawViewHierarchyAtRect:`. This method is extremely fast if the goal is to take a snapshot of a view.

Moving Things Around

Now that you can draw in a layer, take a look into how to use those layers to create powerful animations.

Layers naturally animate. In fact, you need to do a small amount of work to prevent them from animating. Consider the following example.

LayerAnimationViewController.m (LayerAnimation)

```objc
- (void)viewDidLoad {
  [super viewDidLoad];
  CALayer *squareLayer = [CALayer layer];
  squareLayer.backgroundColor = [[UIColor redColor] CGColor];
  squareLayer.frame = CGRectMake(100, 100, 20, 20);
  [self.view.layer addSublayer:squareLayer];

  UIView *squareView = [UIView new];
  squareView.backgroundColor = [UIColor blueColor];
  squareView.frame = CGRectMake(200, 100, 20, 20);
  [self.view addSubview:squareView];

  [self.view addGestureRecognizer:
   [[UITapGestureRecognizer alloc]
    initWithTarget:self
    action:@selector(drop:)]];
}

- (void)drop:(UIGestureRecognizer *)recognizer {
  NSArray *layers = self.view.layer.sublayers;
  CALayer *layer = [layers objectAtIndex:0];
  [layer setPosition:CGPointMake(200, 250)];

  NSArray *views = self.view.subviews;
  UIView *view = [views objectAtIndex:0];
  [view setCenter:CGPointMake(100, 250)];
}
```

This example draws a small red sublayer and a small blue subview. When the view is tapped, both are moved. The view jumps immediately to the new location. The layer animates over a quarter-second. It's fast, but it's not instantaneous like the view.

CALayer implicitly animates all properties that support animation. You can prevent the animation by disabling actions:

```objc
[CATransaction setDisableActions:YES];
```

You learn about actions further in the "Auto-Animating with Actions" section, later in this chapter.

---

disableActions **is a poorly named method. Because it begins with a verb, you expect it to have a side effect (disabling actions) rather than returning the current value of the property. It should be** actionsDisabled **(or** actionsEnabled **to be parallel with** userInteractionEnabled**). Apple may remedy this name eventually, as it has with other misnamed properties. In the meantime, make sure to call** setDisableActions: **when you mean to change its value. You don't get a warning or error if you call** [CATransaction disableActions] **in a void context.**

Implicit Animations

You now know all the basics of animation. Just set layer properties, and your layers animate in the default way. But what if you don't like the defaults? For instance, you may want to change the duration of the animation. First, you need to understand transactions.

Most of the time when you change several layer properties, you want them all to animate together. You also don't want to waste the renderer's time calculating animations for one property change if the next property change affects it. For instance, `opacity` and `backgroundColor` are interrelated properties. Both affect the final displayed pixel color, so the renderer needs to know about both animations when working out the intermediate values.

Core Animation bundles property changes into atomic transactions (`CATransaction`). An implicit `CATransaction` is created for you the first time you modify a layer on a thread that includes a run loop. (If that last sentence piqued your interest, see the "Core Animation and Threads" section, later in this chapter.) During the run loop, all layer changes are collected, and when the run loop completes, all the changes are committed to the layer tree.

To modify the animation properties, you need to make changes to the current transaction. The following changes the duration of the current transaction to 2 seconds rather than the default quarter-second.

```
[CATransaction setAnimationDuration:2.0];
```

You can also set a completion block to run after the current transaction finishes animating using `[CATransaction setCompletionBlock:]`. You can use this to chain animations together, among other things.

Although the run loop creates a transaction for you automatically, you can also create your own explicit transactions using `[CATransaction begin]` and `[CATransaction commit]`. They allow you to assign different durations to different parts of the animation or to disable animations for only a part of the event loop.

> See the "Auto-Animating with Actions" section of this chapter for more information on how implicit animations are implemented and how you can extend them.

Explicit Animations

Implicit animations are powerful and convenient, but sometimes you want more control. That's where `CAAnimation` comes in. With `CAAnimation`, you can manage repeating animations, precisely control timing and pacing, and employ layer transitions. Implicit animations are implemented using `CAAnimation`, so everything you can do with an implicit animation can be done explicitly as well.

The most basic animation is a `CABasicAnimation`. It interpolates a property over a range using a timing function, as shown in the following code:

```
CABasicAnimation *anim = [CABasicAnimation
                            animationWithKeyPath:@"opacity"];
anim.fromValue = @1.0;
anim.toValue = @0.0;
anim.autoreverses = YES;
anim.repeatCount = INFINITY;
anim.duration = 2.0;
[layer addAnimation:anim forKey:@"anim"];
```

This code pulses the layer forever, animating the opacity from one to zero and back over 2 seconds. When you want to stop the animation, remove it:

```
[layer removeAnimationForKey:@"anim"];
```

An animation has key, fromValue, toValue, timingFunction, duration, and some other configuration options. The way it works is to make several copies of the layer, send setValue:forKey: messages to the copies, and then display. It captures the generated contents and displays them.

> If you have custom properties in your layer, you may notice that they're not set correctly during animation. The reason is that the layer is copied. You find out about this later in the "Animating Custom Properties" section of this chapter.

CABasicAnimations are basic, as the name implies. They're easy to set up and use, but they're not very flexible. If you want more control over the animation, you can move to CAKeyframeAnimation. The major difference is that instead of giving a fromValue and toValue, you now can give a path or sequence of points to animate through, along with individual timing for each segment. The *Animation Types and Timing Programming Guide* at developer.apple.com provides excellent examples. They're not technically difficult to set up. Most of the work is on the creative side to find just the right path and timing.

Model and Presentation

A common problem in animations is the dreaded "jump back." The mistake looks like this:

```
CABasicAnimation *fade;
fade = [CABasicAnimation animationWithKeyPath:@"opacity"];
fade.duration = 1;
fade.fromValue = @1.0;
fade.toValue = @0.0;
[circleLayer addAnimation:fade forKey:@"fade"];
```

This code fades the circle out over 1 second, just as expected, and then suddenly the circle reappears. To understand why this happens, you need to be aware of the difference between the model layer and the presentation layer.

The *model layer* is defined by the properties of the "real" CALayer object. Nothing in the preceding code modifies any property of circleLayer itself. Instead, CAAnimation makes copies of circleLayer and modifies them. These copies become the *presentation layer*. They represent roughly what is shown on the screen. There is technically another layer called the *render layer* that really represents what's on the screen, but it's internal to Core Animation, and you very seldom encounter it.

So what happens in the preceding code? CAAnimation modifies the presentation layer, which is drawn to the screen, and when it completes, all its changes are thrown away and the model layer is used to determine the new state. The model layer hasn't changed, so you snap back to where you started. The solution to this is to set the model layer, as shown here:

```
circleLayer.opacity = 0;
CABasicAnimation *fade;
fade = [CABasicAnimation animationWithKeyPath:@"opacity"];
. . .
[circleLayer addAnimation:fade forKey:@"fade"];
```

Sometimes this approach works fine, but sometimes the implicit animation in setOpacity: fights with the explicit animation from animationWithKeyPath:. The best solution to that problem is to turn off implicit animations if you're doing explicit animations:

```
[CATransaction begin];
[CATransaction setDisableActions:YES];
circleLayer.opacity = 0;
CABasicAnimation *fade;
fade = [CABasicAnimation animationWithKeyPath:@"opacity"];
. . .
[circleLayer addAnimation:fade forKey:@"fade"];
[CATransaction commit];
```

Sometimes you see people recommend setting removedOnCompletion to NO and fillMode to kCAFillModeBoth. This is not a good solution. It essentially makes the animation go on forever, which means the model layer is never updated. If you ask for the property's value, you continue to see the model value, not what you see on the screen. If you try to implicitly animate the property afterward, it doesn't work correctly because the CAAnimation is still running. If you ever remove the animation by replacing it with another with the same name, calling removeAnimationForKey: or removeAllAnimations, the old value snaps back. On top of all that, it wastes memory.

All this extra code becomes a bit of a pain, so you may like the following category on CALayer that wraps it all together and lets you set the duration and delay. Most of the time, I still prefer implicit animation, but this approach can make explicit animation a bit simpler.

CALayer+RNAnimation.m (LayerAnimation)

```
@implementation CALayer (RNAnimations)
- (void)setValue:(id)value
      forKeyPath:(NSString *)keyPath
        duration:(CFTimeInterval)duration
           delay:(CFTimeInterval)delay
```

```
{
    [CATransaction begin];
    [CATransaction setDisableActions:YES];
    [self setValue:value forKeyPath:keyPath];

    CABasicAnimation *anim;
    anim = [CABasicAnimation animationWithKeyPath:keyPath];
    anim.duration = duration;
    anim.beginTime = CACurrentMediaTime() + delay;
    anim.fillMode = kCAFillModeBoth;
    anim.fromValue = [[self presentationLayer] valueForKey:keyPath];
    anim.toValue = value;
    [self addAnimation:anim forKey:keyPath];

    [CATransaction commit];
}
@end
```

A Few Words on Timings

As in the universe at large, in Core Animation, time is relative. A second does not always have to be a second. Just like coordinates, time can be scaled.

CAAnimation conforms to the CAMediaTiming protocol, and you can set the speed property to scale its timing. Because of this, when considering timings between layers, you need to convert them just like you need to convert points that occur in different views or layers:

```
localPoint = [self convertPoint:remotePoint fromLayer:otherLayer];
localTime = [self convertTime:remotetime fromLayer:otherLayer];
```

This issue isn't very common, but it comes up when you're trying to coordinate animations. You might ask another layer for a particular animation and when that animation will end so that you can start your animation:

```
CAAnimation *otherAnim = [layer animationForKey:@"anim"];
CFTimeInterval finish = otherAnim.beginTime + otherAnim.duration;
myAnim.beginTime = [self convertTime:finish fromLayer:layer];
```

Setting beginTime like this is a nice way to chain animations, even if you hard-code the time rather than ask the other layer. To reference "now," just use CACurrentMediaTime().

These examples raise another issue, however. What value should your property have between now and when the animation begins? You would assume that it would be the fromValue, but that isn't how it works. It's the current model value because the animation hasn't begun. Typically, this is the toValue. Consider the following animation:

```
[CATransaction begin];
anim = [CABasicAnimation animationWithKeyPath:@"opacity"];
anim.fromValue = @1.0;
anim.toValue = @0.5;
anim.duration = 5.0;
anim.beginTime = CACurrentMediaTime() + 3.0;
```

```
[layer addAnimation:anim forKey:@"fade"];
layer.opacity = 0.5;
[CATransaction commit];
```

This animation does nothing for 3 seconds. During that time, the default property animation is used to fade `opacity` from 1.0 to 0.5. Then the animation begins, setting the opacity to its `fromValue` and interpolating to its `toValue`. So the layer begins with `opacity` of 1.0, fades to 0.5 over a quarter-second, and then 3 seconds later jumps back to 1.0 and fades again to 0.5 over 5 seconds. This result almost certainly isn't what you want.

You can resolve this problem by using `fillMode`. The default is `kCAFillModeRemoved`, which means that the animation has no influence on the values before or after its execution. This behavior can be changed to "clamp" values before or after the animation by setting the fill mode to `kCAFillModeBackwards`, `kCAFillModeForwards`, or `kCAFillModeBoth`. Figure 8-3 illustrates this effect.

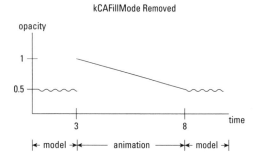

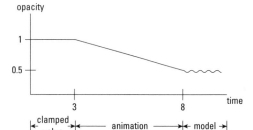

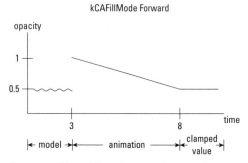

Figure 8-3 Effect of fill modes on media timing functions

In most cases, you want to set `fillMode` to `kCAFillModeBackwards` or `kCAFillModeBoth`.

Into the Third Dimension

Chapter 7 discussed how to use `CGAffineTransform` to make `UIView` drawing much more efficient. This technique limits you to two-dimensional transformations: translate, rotate, scale, and skew. With layers, however, you can apply three-dimensional transformations by adding perspective. This is often called 2.5D rather than 3D because it doesn't make layers into truly three-dimensional objects in the way that OpenGL ES does. But it does allow you to give the illusion of three-dimensional movement.

You rotate layers around an anchor point. By default, the anchor point is in the center of the layer, designated {0.5, 0.5}. It can be moved anywhere within the layer, making it convenient to rotate around an edge or corner. The anchor point is described in terms of a unit square rather than in points. So the lower-right corner is {1.0, 1.0}, no matter how large or small the layer is.

Here's a simple example of a three-dimensional box.

BoxViewController.h (Box)

```
@interface BoxViewController : UIViewController
@property (nonatomic, readwrite, strong) CALayer *topLayer;
@property (nonatomic, readwrite, strong) CALayer *bottomLayer;
@property (nonatomic, readwrite, strong) CALayer *leftLayer;
@property (nonatomic, readwrite, strong) CALayer *rightLayer;
@property (nonatomic, readwrite, strong) CALayer *frontLayer;
@property (nonatomic, readwrite, strong) CALayer *backLayer;
@end
```

BoxViewController.m (Box)

```
@implementation BoxViewController

const CGFloat kSize = 100.;
const CGFloat kPanScale = 1./100.;

- (CALayer *)layerWithColor:(UIColor *)color
                  transform:(CATransform3D)transform {
  CALayer *layer = [CALayer layer];
  layer.backgroundColor = [color CGColor];
  layer.bounds = CGRectMake(0, 0, kSize, kSize);
  layer.position = self.view.center;
  layer.transform = transform;
  [self.view.layer addSublayer:layer];
  return layer;
}

static CATransform3D MakePerspectiveTransform() {
  CATransform3D perspective = CATransform3DIdentity;
  perspective.m34 = -1./2000.;
  return perspective;
}
```

(continued)

```objc
- (void)viewDidLoad {
  [super viewDidLoad];

  CATransform3D transform;
  transform = CATransform3DMakeTranslation(0, -kSize/2, 0);
  transform = CATransform3DRotate(transform, M_PI_2, 1.0, 0, 0);
  self.topLayer = [self layerWithColor:[UIColor redColor]
                             transform:transform];

  transform = CATransform3DMakeTranslation(0, kSize/2, 0);
  transform = CATransform3DRotate(transform, M_PI_2, 1.0, 0, 0);
  self.bottomLayer = [self layerWithColor:[UIColor greenColor]
                                transform:transform];

  transform = CATransform3DMakeTranslation(kSize/2, 0, 0);
  transform = CATransform3DRotate(transform, M_PI_2, 0, 1, 0);
  self.rightLayer = [self layerWithColor:[UIColor blueColor]
                               transform:transform];

  transform = CATransform3DMakeTranslation(-kSize/2, 0, 0);
  transform = CATransform3DRotate(transform, M_PI_2, 0, 1, 0);
  self.leftLayer = [self layerWithColor:[UIColor cyanColor]
                              transform:transform];

  transform = CATransform3DMakeTranslation(0, 0, -kSize/2);
  transform = CATransform3DRotate(transform, M_PI_2, 0, 0, 0);
  self.backLayer = [self layerWithColor:[UIColor yellowColor]
                              transform:transform];

  transform = CATransform3DMakeTranslation(0, 0, kSize/2);
  transform = CATransform3DRotate(transform, M_PI_2, 0, 0, 0);
  self.frontLayer = [self layerWithColor:[UIColor magentaColor]
                               transform:transform];

  self.view.layer.sublayerTransform = MakePerspectiveTransform();

  UIGestureRecognizer *g = [[UIPanGestureRecognizer alloc]
                             initWithTarget:self
                             action:@selector(pan:)];
  [self.view addGestureRecognizer:g];
}

- (void)pan:(UIPanGestureRecognizer *)recognizer {
  CGPoint translation = [recognizer translationInView:self.view];
  CATransform3D transform = MakePerspectiveTransform();
  transform = CATransform3DRotate(transform,
                                  kPanScale * translation.x,
                                  0, 1, 0);
  transform = CATransform3DRotate(transform,
                                  -kPanScale * translation.y,
                                  1, 0, 0);
  self.view.layer.sublayerTransform = transform;
}
@end
```

`BoxViewController` shows how to build a simple box and rotate it based on panning. All the layers are created with `layerWithColor:transform:`. Notice that all the layers have the same `position`. They only appear to be in the shape of a box through transforms that translate and rotate them.

You apply a perspective `sublayerTransform` (a transform applied to all sublayers, but not the layer itself). I don't go into the math here, but the `m34` position of the 3D transform matrix should be set to `-1/EYE_DISTANCE`. For most cases, 2000 units works well, but you can adjust this to "zoom the camera."

You could also build this box by setting `position` and `zPosition` rather than translating, as shown in the following code. This technique may be more intuitive for some developers.

BoxTransformViewController.m (BoxTransform)

```objc
- (CALayer *)layerAtX:(CGFloat)x y:(CGFloat)y z:(CGFloat)z
                color:(UIColor *)color
            transform:(CATransform3D)transform {
  CALayer *layer = [CALayer layer];
  layer.backgroundColor = [color CGColor];
  layer.bounds = CGRectMake(0, 0, kSize, kSize);
  layer.position = CGPointMake(x, y);
  layer.zPosition = z;
  layer.transform = transform;
  [self.contentLayer addSublayer:layer];
  return layer;

}
- (void)viewDidLoad {
  [super viewDidLoad];
  CATransformLayer *contentLayer = [CATransformLayer layer];
  contentLayer.frame = self.view.layer.bounds;
  CGSize size = contentLayer.bounds.size;
  contentLayer.transform =
    CATransform3DMakeTranslation(size.width/2, size.height/2, 0);
  [self.view.layer addSublayer:contentLayer];

  self.contentLayer = contentLayer;

  self.topLayer = [self layerAtX:0 y:-kSize/2 z:0
                          color:[UIColor redColor]
                      transform:MakeSideRotation(1, 0, 0)];
  ...
}

- (void)pan:(UIPanGestureRecognizer *)recognizer {
  CGPoint translation = [recognizer translationInView:self.view];
  CATransform3D transform = CATransform3DIdentity;
  transform = CATransform3DRotate(transform,
                                  kPanScale * translation.x,
                                  0, 1, 0);
  transform = CATransform3DRotate(transform,
                                  -kPanScale * translation.y,
                                  1, 0, 0);
  self.view.layer.sublayerTransform = transform;
}
```

You now need to insert a CATransformLayer to work with. If you just use a CALayer, then zPosition is used only for calculating layer order. It's not used to determine location in space. The lack of a perspective transform makes the box look completely flat. CATransformLayer supports zPosition without requiring you to apply a perspective transform.

Decorating Your Layers

A major advantage of CALayer over UIView, even if you're working only in 2D, is the automatic border effects that CALayer provides. For instance, CALayer can automatically give you rounded corners, a colored border, and a drop shadow. All these features can be animated, which can provide some nice visual effects. For instance, you can adjust the position and shadow to give the illusion of clicking as the user presses and releases a layer. The following code creates the layer shown in Figure 8-4.

DecorationViewController.m (Decoration)

```
CALayer *layer = [CALayer layer];
layer.frame = CGRectMake(100, 100, 100, 100);
layer.cornerRadius = 10;
layer.backgroundColor = [[UIColor redColor] CGColor];
layer.borderColor = [[UIColor blueColor] CGColor];
layer.borderWidth = 5;
layer.shadowOpacity = 0.5;
layer.shadowOffset = CGSizeMake(3.0, 3.0);
[self.view.layer addSublayer:layer];
```

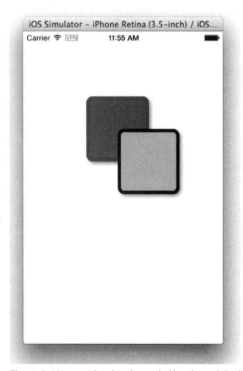

Figure 8-4 Layer with colored, rounded border and shadow

Auto-Animating with Actions

Most of the time, implicit animations do what you want, but sometimes you might like to configure them. You can turn off all implicit animations using CATransaction, but that applies only to the current transaction (generally the current run loop). To modify how an implicit animation behaves, and especially if you want it to always behave that way for this layer, you need to configure the layer's actions. Doing so enables you to configure your animations when you create the layer rather than apply an explicit animation every time you change a property.

Layer actions are fired in response to various changes on the layer, such as adding or removing the layer from the hierarchy or modifying a property. When you modify the position property, for instance, the default action is to animate it over a quarter-second. In the following examples, CircleLayer is a layer that draws a red circle in its center with the given radius.

ActionsViewController.m (Actions)

```
CircleLayer *circleLayer = [CircleLayer new];
circleLayer.radius = 20;
circleLayer.frame = self.view.bounds;
[self.view.layer addSublayer:circleLayer];
...
circleLayer.position = CGPointMake(100, 100);
```

You can modify this code so that changes in position always animate over 2 seconds:

```
CircleLayer *circleLayer = [CircleLayer new];
circleLayer.radius = 20;
circleLayer.frame = self.view.bounds;
[self.view.layer addSublayer:circleLayer];

CABasicAnimation *anim =
  [CABasicAnimation animationWithKeyPath:@"position"];
anim.duration = 2;
NSMutableDictionary *actions =
  [NSMutableDictionary dictionaryWithDictionary:
                                  [circleLayer actions]];
actions[@"position"] = anim;
circleLayer.actions = actions;
...
circleLayer.position = CGPointMake(100, 100);
```

Setting the action to [NSNull null] disables implicit animations for that property. A dictionary cannot hold nil, so you need to use the NSNull class.

Some special actions are needed when the layer is added to the layer tree (kCAOnOrderIn) and when it's removed (kCAOnOrderOut). For example, you can make a group animation of growing and fade-in like this:

```
CABasicAnimation *fadeAnim =
  [CABasicAnimation animationWithKeyPath:@"opacity"];
fadeAnim.fromValue = @0.4;
fadeAnim.toValue = @1.0;
```

```
CABasicAnimation *growAnim =
  [CABasicAnimation animationWithKeyPath:@"transform.scale"];
growAnim.fromValue = @0.8;
growAnim.toValue = @1.0;

CAAnimationGroup *groupAnim = [CAAnimationGroup animation];
groupAnim.animations = @[fadeAnim, growAnim];

  actions[kCAOnOrderIn] = groupAnim;
  circleLayer.actions = actions;
```

Actions are also important when dealing with transitions (kCATransition) when one layer is replaced with another. This technique is commonly used with a CATransition (a special type of CAAnimation). You can apply a CATransition as the action for the contents property to create special effects such as a slide show whenever the contents change. By default, the fade transition is used.

Animating Custom Properties

Core Animation implicitly animates several layer properties, but what about custom properties on CALayer subclasses? For instance, in the CircleLayer, you have a radius property. By default, radius is not animated, but contents is (using a fade CATransition). So changing the radius causes your current circle to cross-fade with your new circle. This probably isn't what you want. You want radius to animate just like position. Just a few steps are needed to make this animation work correctly, as shown in the following example.

CircleLayer.m (Actions)

```
@implementation CircleLayer
@dynamic radius;

- (id)init {
    self = [super init];
    if (self) {
      [self setNeedsDisplay];
    }

    return self;
}

- (void)drawInContext:(CGContextRef)ctx {
    CGContextSetFillColorWithColor(ctx,
                                    [[UIColor redColor] CGColor]);
    CGFloat radius = self.radius;
    CGRect rect;
    rect.size = CGSizeMake(radius, radius);
    rect.origin.x = (self.bounds.size.width - radius) / 2;
    rect.origin.y = (self.bounds.size.height - radius) / 2;
    CGContextAddEllipseInRect(ctx, rect);
    CGContextFillPath(ctx);
}
```

```objc
+ (BOOL)needsDisplayForKey:(NSString *)key {
  if ([key isEqualToString:@"radius"]) {
    return YES;
  }
  return [super needsDisplayForKey:key];
}

- (id < CAAction >)actionForKey:(NSString *)key {
  if ([self presentationLayer] != nil) {
    if ([key isEqualToString:@"radius"]) {
      CABasicAnimation *anim = [CABasicAnimation
                                 animationWithKeyPath:@"radius"];
      anim.fromValue = [[self presentationLayer]
                        valueForKey:@"radius"];
      return anim;
    }
  }

  return [super actionForKey:key];
}
@end
```

Let's start with a reminder of the basics. You call `setNeedsDisplay` in `init` so that the layer's `drawInContext:` is called the first time it's added to the layer tree. You override `needsDisplayForKey:` so that whenever `radius` is modified, you automatically redraw.

Now you come to your actions. You implement `actionForKey:` to return an animation with a `fromValue` of the currently displayed (`presentationLayer`) radius. This means that you'll animate smoothly if the animation is changed midflight.

It's critical to note that you implemented the `radius` **property using** `@dynamic` **here, not** `@synthesize`. `CALayer` **automatically generates accessors for its properties at runtime, and those accessors have important logic. It's vital that you not override these** `CALayer` **accessors by either implementing your own accessors or using** `@synthesize` **to do so.**

Note that using `@dynamic` **this way is not documented behavior.**

Core Animation and Threads

It's worth noting that Core Animation is tolerant of threading. You can generally modify `CALayer` properties on any thread, unlike `UIView` properties. `drawInContext:` may be called from any thread (although a given `CGContext` should be modified on only one thread at a time). Changes to `CALayer` properties are batched into transactions using `CATransaction`. This happens automatically if you have a run loop. If you don't have a run loop, you need to call `[CATransaction flush]` periodically. If at all possible, though, you should perform Core Animation actions on a thread with a run loop to improve performance.

Summary

Core Animation is one of the most important frameworks in iOS. It puts a fairly easy-to-use API in front of an incredibly powerful engine. It still has a few rough edges, however, and sometimes things need to be "just so" to make it work correctly (for example, implementing your properties with `@dynamic` rather than `@synthesize`). When it doesn't work correctly, debugging it can be challenging, so having a good understanding of how it works is crucial. I hope this chapter has made you confident enough with the architecture and the documentation to dive in and make some really beautiful apps.

Further Reading

Apple Documentation

The following documents are available in the iOS Developer Library at `developer.apple.com` or through the Xcode Documentation and API Reference.

> *Animation Types and Timing Programming Guide*
>
> *Core Animation Programming Guide*

Other Resources

> milen.me. (Milen Dzhumerov), "CA's 3D Model." An excellent overview of the math behind the perspective transform, including the magic -1/2000.
>
> `http://milen.me/technical/core-animation-3d-model/`

> *Cocoa with Love,* (Matt Gallagher), "Parametric Acceleration Curves in Core Animation." Explains how to implement timing curves that cannot be implemented with `CAMediaTimingFunction`, such as damped ringing and exponential decay.
>
> `cocoawithlove.com/2008/09/parametric-acceleration-curves-in-core.html`

## Chapter 9

Two Things at Once: Multitasking

There was a time, not too long ago, when ever-increasing processor speeds meant a free ride for developers. Each generation of hardware made all your code run faster with no extra work. All kinds of inefficiencies could be fixed by the magic of Moore's Law. That era is basically over. Every year, processors still get more powerful, but instead of getting faster, they are becoming more parallel. Even humble mobile devices like the iPhone have multicore CPUs. If you don't make use of multitasking in your app, you're giving up much of the processing power available to you.

In this chapter, you discover the best practices for multitasking and the major iOS frameworks for multitasking: run loops, threads, operations, and Grand Central Dispatch (GCD). If you're familiar with thread-based multitasking from other platforms, you find out how to reduce your reliance on explicit threads and make the best use of iOS's frameworks that avoid threads or handle threading automatically.

This chapter assumes that you have at least a passing familiarity with operation queues and Grand Central Dispatch, although you may never have used them in real code. If you've never heard of them, I suggest you skim the *Concurrency Programming Guide* before continuing. This chapter also assumes you have a basic understanding of blocks. If you're unfamiliar with blocks, you should read *Blocks Programming Topics*. Both of these documents are available at `developer.apple.com` or in the Xcode help system.

You can find the sample code for this chapter in the projects `SimpleGCD`, `SimpleOperation`, and `SimpleThread`.

Introduction to Multitasking and Run Loops

The most basic form of multitasking is the run loop. This is the form of multitasking that was originally developed for NeXTSTEP, and was the primary multitasking system until OS X 10.6 when Apple introduced Grand Central Dispatch. It is still a very important part of every Cocoa application.

Every Cocoa application is driven by a `do/while` loop that blocks waiting for an event, dispatches that event to interested listeners, and repeats until something tells it to stop. The object that handles this is called a *run loop* (NSRunLoop).

You almost never need to understand the internals of a run loop. There are mach ports and message ports and `CFRunLoopSourceRef` types and a variety of other arcana. These are incredibly rare in normal programs, even in very complex programs. What's important to understand is that the run loop is just a big `do/while` loop, running on one thread, pulling events off various queues and dispatching them to listeners one at a time on that same thread. This is the heart of a Cocoa application.

When your `applicationWillResignActive:` method, `IBAction`, or any other entry point to your program is called, an event fired somewhere that traced its way to a delegate call that you implemented. The system is waiting for you to return so it can continue. While your code is running on the main thread, scroll views can't scroll, buttons can't highlight, and timers can't fire. The entire UI is hanging, waiting for you to finish. Keep that in mind when you're writing your code.

I don't mean to say that everything is on the main run loop. Each thread has its own run loop. Animations generally run on background threads, as does much of `NSURLConnection` network handling. But the heart of the system runs on a single, shared run loop.

> Although each thread *has* a run loop, this doesn't mean that each thread *processes* its run loop. Run loops execute their `do/while` loop only in response to commands like `runUntilDate:`. The call to `UIApplicationMain` in `main.m` of almost every project runs the main run loop.

`NSTimer` relies on the run loop to dispatch messages. When you schedule an `NSTimer`, it asks the current run loop to dispatch a selector at a certain time. Each time the run loop iterates, it checks what time it is and fires any timers that have expired. Delay-action methods such as `performSelector:withObject:afterDelay:` are implemented by scheduling a timer.

Most of the time, all of this activity happens behind the scenes, and you don't need to worry about the run loop very much. `UIApplicationMain` sets up the main thread's run loop for you and keeps it running until the program terminates.

Although it's important to understand the basics of run loops, Apple has developed better approaches to handling common multitasking problems. Most of these are built on top of Grand Central Dispatch. The first step toward these better approaches is breaking your program into operations, as I discuss in the next section.

Developing Operation-Centric Multitasking

Many developers from other platforms are accustomed to thinking in terms of threads. iOS has good objects for dealing directly with threads, particularly `NSThread`, but I recommend avoiding them and moving toward operation-centric multitasking. Operations are a much more powerful abstraction than threads and can lead to substantially better and faster code if used correctly. Threads are expensive to create and sustain, so threaded designs often employ a small to moderate number of long-lived threads. These threads perform heavyweight operations such as "the network" or "the database" or "the calculations." Because they're focused on large areas of responsibility, they need access to lots of inputs and outputs, and that means locking. Locking is very expensive and a significant source of bugs.

Of course, threaded programs don't have to be written this way. It would be better if you spawned short-lived threads that focus on a single input and output, minimizing the need for locking. But threads are expensive to spawn, so you need to manage a thread pool. And small operations often have ordering dependencies, so ideally you create some kind of queue to keep things straight. Now, what if you could take the thread pool and the queue and let the operating system worry about them? That's exactly what `NSOperation` is for. It's the solution you come to by optimizing threaded programs to minimize locking. But operations can be better than that. Instead of worrying about the mechanics of semaphores and mutexes, you worry about the work you want to do.

An *operation* is an encapsulated unit of work, often expressed in the form of an Objective-C block. Operations support priority, dependencies, and cancellation, so they're ideal for situations in which you need to schedule some work that might not actually be required. For example, you may want to update images in a scrolling view, favoring images currently onscreen and canceling updates for images that have been scrolled offscreen. In the following example, you create a `UICollectionView` that contains random fractals. Calculating them is very expensive, so you generate them asynchronously. To improve display performance, you calculate the fractal at various resolutions, displaying a low-resolution fractal quickly and then improving the resolution if the user stops scrolling.

In this example, you use an `NSOperation` subclass to generate Julia sets (a kind of fractal) and display them in a `UICollectionView`. The work being done by the operation is somewhat complicated, so it's nice to put that work in its own file. For smaller operations, you could put the work into an inline block. The sample code is available in the project JuliaOp. Here is the basic structure of the operation's `main` method (I skipped the actual mathematics in the interest of space):

JuliaOperations.m (JuliaOp)

```
- (void)main {
  // ...
  // Configure bits[] to hold bitmap data
  // ...

  for (NSUInteger y = 0; y < height; ++y) {
    for (NSUInteger x = 0; x < width; ++x) {
      if (self.isCancelled) {
        return;
      }
      // ...
      // Calculate Julia values and update bits[]
      // May iterate up to 255 times per pixel.
      // ...
    }
  }

  // ...
  // Create bitmap and store in self.image
  // ...
}
```

Here are some key features of this operation:

▦ **All required data is given to the operation before it starts.** The operation doesn't need to interact with other objects during its run, so there's no need for locking.

▦ **When the operation is complete, it stores its result in a local ivar.** Again, this approach avoids the need for locking. You could use a delegate method here to update something outside the operation, but as you'll see shortly, completion blocks can be a simpler approach.

▦ **The operation periodically checks `isCancelled` so it can abort if requested.** This is an important point. Calling `cancel` on a running operation does not cause it to stop running. That would leave the system in an unknown state. It's up to you to check `isCancelled` and abort in whatever way is appropriate. It's possible for an operation to return YES for both `isExecuting` and `isCancelled`.

I describe cancellation in more detail shortly. First, I discuss how this operation is created and queued in `JuliaCell`. Each cell in the `UICollectionView` creates a separate operation for each resolution. The operation is just responsible for taking a set of input values and a resolution and returning an image. The operation doesn't know anything about how that image will be used. It doesn't even know that other resolutions are being calculated. In the `UICollectionViewController`, you configure each cell by passing the row as a random seed and a queue to work with.

CollectionViewController.m (JuliaOp)

```
- (UICollectionViewCell *)collectionView:(UICollectionView *)
collectionView
                cellForItemAtIndexPath:(NSIndexPath *)indexPath {
    JuliaCell *
    cell = [self.collectionView
            dequeueReusableCellWithReuseIdentifier:@"Julia"
            forIndexPath:indexPath];
    [cell configureWithSeed:indexPath.row queue:self.queue];
    return cell;
}
```

The queue is just a shared `NSOperationQueue` that you create in the `UICollectionViewController`. I explain `maxConcurrentOperationCount` and `countOfCores()` in the section "Setting Maximum Concurrent Operations."

```
self.queue = [[NSOperationQueue alloc] init];
self.queue.maxConcurrentOperationCount = countOfCores();
```

In `JuliaCell`, you configure the cell by creating a set of operations and adding them to the queue as follows:

JuliaCell.m (JuliaOp)

```
- (void)configureWithSeed:(NSUInteger)seed
                queue:(NSOperationQueue *)queue {
    // ...
    JuliaOperation *prevOp = nil;
    for (CGFloat scale = minScale; scale <= maxScale; scale *= 2) {
        JuliaOperation *op = [self operationForScale:scale seed:seed];
        if (prevOp) {
            [op addDependency:prevOp];
        }
        [self.operations addObject:op];
        [queue addOperation:op];
        prevOp = op;
    }
}
```

Notice that each operation is dependent on the previous operation. This hierarchy of dependencies ensures that the high-resolution image isn't scheduled before the low-resolution image. Every cell creates several operations, and all the operations from all the cells are put onto the same queue. `NSOperationQueue` automatically orders the operations to manage dependencies.

In `operationForScale:seed:`, you configure the actual operation for a given resolution, as shown here:

JuliaCell.m (JuliaOp)

```
- (JuliaOperation *)operationForScale:(CGFloat)scale
                                 seed:(NSUInteger)seed {
  JuliaOperation *op = [[JuliaOperation alloc] init];
  op.contentScaleFactor = scale;

  CGRect bounds = self.bounds;
  op.width = (unsigned)(CGRectGetWidth(bounds) * scale);
  op.height = (unsigned)(CGRectGetHeight(bounds) * scale);

  srandom(seed);

  op.c = (long double)random()/LONG_MAX + I*(long double)random()/LONG_MAX;
  op.blowup = random();
  op.rScale = random() % 20; // Biased, but simple is more important
  op.gScale = random() % 20;
  op.bScale = random() % 20;

  __weak JuliaOperation *weakOp = op;
  op.completionBlock = ^{
    if (! weakOp.isCancelled) {
      [[NSOperationQueue mainQueue] addOperationWithBlock:^{
        JuliaOperation *strongOp = weakOp;
        if (strongOp && [self.operations containsObject:strongOp]) {
          self.imageView.image = strongOp.image;
          self.label.text = strongOp.description;
          [self.operations removeObject:strongOp];
        }
      }];
    }
  };

  if (scale < 0.5) {
    op.queuePriority = NSOperationQueuePriorityVeryHigh;
  }
  else if (scale <= 1) {
    op.queuePriority = NSOperationQueuePriorityHigh;
  }
  else {
    op.queuePriority = NSOperationQueuePriorityNormal;
  }

  return op;
}
```

The operation has various data associated with it (c, blowup, and the various scales). These data are configured once, on the main thread, so there are no race conditions.

The completionBlock is a convenient way to handle the operation's results. The completion block is called whenever the block finishes, even if it's cancelled. That's why you check isCancelled before dealing with the results. Because the UI updates are processed on the main thread, it's possible that this cell will have scrolled off the screen and been reused prior to the UI update code running. That's why you check that the operation still exists and that it's still one of the operations that this cell cares about (`containsObject:`) before applying the UI changes. Exactly how much of this kind of double-checking code is required depends on your specific problem.

Finally, you set the priority based on the scale. This encourages low-resolution images to be generated prior to high-resolution images.

Because this is a `UICollectionView`, the cells may be reused at any time. When that happens, you want to cancel all the current operations, which is easily done with `cancel`:

```
- (void)prepareForReuse {
    [self.operations makeObjectsPerformSelector:@selector(cancel)];
    [self.operations removeAllObjects];
    self.imageView.image = nil;
    self.label.text = @"";
}
```

At the same time, you remove all the operations from the `operations` ivar, ensuring that any pending operations won't update the UI. Remember that all the methods in JuliaOp except for the `completionBlock` run on the main thread (and the `completionBlock` moves the UI calls to the main thread as well). This approach minimizes locking concerns, improving reliability and performance.

Setting Maximum Concurrent Operations

> iOS 7 fixes the `NSOperationQueue` problems discussed in this section. There is no longer a need to set `maxConcurrentOperationsCount` if you support only iOS 7.

Finally, there is the matter of `maxConcurrentOperationCount`. `NSOperationQueue` tries to manage the number of threads it generates, but in many cases, it doesn't do an ideal job. This is likely your fault (and my fault, but read on). The common cause of trouble is flooding the main queue with lots of small operations when you want to update the UI.

For instance, Apple suggests in the "Asynchronous Design Patterns" session from WWDC 2012 that you use fewer operations. If the user scrolls, you could cancel that one operation rather than the many I've generated (including their many small main-thread updates). The problem is that then you don't get any parallelism on a multicore machine, which is the whole point of using operation queues.

In my experience, creating small operations as I demonstrate in JuliaOp, even if you have to cancel them later, makes coding much simpler and the code much more robust and understandable. The problem is that `NSOperationQueue` will schedule too many of them at a time. If you explicitly tell `NSOperationQueue` not to schedule more CPU-bound operations than you have CPUs, simple code runs very fast.

My recommendation may change in the future, but for CPU-bound operations, I recommend setting `maxConcurrentOperationsCount` to the number of cores. You can determine this with the function `countOfCores`:

```
unsigned int countOfCores() {
  unsigned int ncpu;
  size_t len = sizeof(ncpu);
  sysctlbyname("hw.ncpu", &ncpu, &len, NULL, 0);

  return ncpu;
}
```

Multitasking with Grand Central Dispatch

Grand Central Dispatch (GCD) is at the heart of multitasking in iOS. It is used throughout the system layer for nearly everything. NSOperationQueue is implemented on top of GCD, and the basic queue concept is similar. Instead of adding an NSOperation to NSOperationQueue, you add a block to a dispatch queue.

Dispatch queues are more low level than operation queues, however. You have no way to cancel a block after it is added to a dispatch queue. Dispatch queues are strictly first-in-first-out (FIFO), so there's no way to apply priorities or reorder blocks within a queue. If you need those kinds of features, definitely use NSOperationQueue instead of trying to re-create them with GCD.

You can do many things with dispatch queues that you cannot do with operations. For example, GCD offers dispatch_after, allowing you to schedule the next operation rather than sleeping. The time is in nanoseconds, which can lead to some confusion because nearly every time interval in iOS is in seconds. Luckily, Xcode automatically provides a conversion snippet if you type dispatch_after and press Enter. Using nanoseconds is optimized for the hardware, not the programmer. Passing the time in seconds would require floating-point math, which is more expensive and less precise. GCD is a very low-level framework and does not waste many cycles on programmer convenience.

Introduction to Dispatch Queues

Most iOS developers have encountered dispatch_async at some point, mostly to run things on the main queue. This may be the extent of your experience with queues, however. In this section, you learn more about what queues are and how best to use them.

It is important to remember that a dispatch queue is just a queue. It is not a thread. Don't think of a queue as something that executes blocks. A queue is something that orders blocks. Calling dispatch_async does not cause a block to run. It just appends the block to a queue. Almost all GCD methods are like this; they just append blocks to the end of a queue. If you think about queues as though they were threads, you'll tend to make design mistakes.

Queue Targets and Priority

Queues are hierarchical in GCD. Only the global system queues are actually scheduled to run. You can access these queues with dispatch_get_global_queue and one of the following priority constants:

- DISPATCH_QUEUE_PRIORITY_HIGH

- DISPATCH_QUEUE_PRIORITY_DEFAULT

- DISPATCH_QUEUE_PRIORITY_LOW

- DISPATCH_QUEUE_PRIORITY_BACKGROUND

All these queues are concurrent. GCD schedules as many blocks as there are threads available from the HIGH queue. When the HIGH queue is empty, it moves on to the DEFAULT queue, and so on. The system creates and destroys threads as needed, based on the number of cores available and the system load.

When you create your own queue, it is attached to one of these global queues (its *target*). By default, it is attached to the DEFAULT queue. When a block reaches the front of your queue, the block is effectively moved to the end of its target queue. When it reaches the front of the global queue, it's executed. You can change the target queue with dispatch_set_target_queue.

After a block is added to a queue, it runs in the order it was added. There is no way to cancel it, and there is no way to change its order relative to other blocks on the queue. But what if you want a high-priority block to "skip to the head of the line"? As shown in the following code, you create two queues, a high-priority queue and a low-priority queue, and make the high-priority queue the target of the low-priority queue:

```
dispatch_queue_t
low = dispatch_queue_create("low", DISPATCH_QUEUE_SERIAL);

dispatch_queue_t
high = dispatch_queue_create("high", DISPATCH_QUEUE_SERIAL);

dispatch_set_target_queue(low, high);
```

Dispatching to the low-priority queue is normal:

```
dispatch_async(low, ^{ /* Low priority block */ });
```

To dispatch to the high-priority queue, suspend the low queue and resume it after the high-priority block finishes:

```
dispatch_suspend(low);
dispatch_async(high, ^{
  /* High priority block */
  dispatch_resume(low);
});
```

Suspending a queue prevents scheduling any blocks that were initially put on that queue, as well as any queues that target the suspended queue. It doesn't stop currently executing blocks, but even if the low-priority block is next in line for the CPU, it isn't scheduled until dispatch_resume is called.

You need to balance dispatch_suspend and dispatch_resume exactly like retain and release. Suspending the queue multiple times requires an equal number of resumes.

Creating Synchronization Points with Dispatch Barriers

GCD offers a rich system of serial and concurrent queues. With some thought, you can use them to create many things other than simple thread management. For instance, GCD queues can be used to solve many common locking problems at a fraction of the overhead.

A *dispatch barrier* creates a synchronization point within a concurrent queue. While it's running, no other block on the queue is allowed to run, even if it's concurrent and other cores are available. If that sounds like an exclusive (write) lock, it is. Nonbarrier blocks can be thought of as shared (read) locks. As long as all access to the resource is performed through the queue, barriers provide very cheap synchronization.

For comparison, you could manage multithreaded access with `@synchronize`, which takes an exclusive lock on its parameter, as shown in the following code:

```
- (id)objectAtIndex:(NSUInteger)index {
  @synchronized(self) {
    return [self.array objectAtIndex:index];
  }
}
- (void)insertObject:(id)obj atIndex:(NSUInteger)index {
  @synchronized(self) {
    [self.array insertObject:obj atIndex:index];
  }
}
```

`@synchronized` is easy to use, but very expensive even when there is little contention. There are many other approaches. Most are either complicated and fast or simple and slow. GCD barriers offer a nice tradeoff.

```
- (id)objectAtIndex:(NSUInteger)index {
  __block id obj;
  dispatch_sync(self.concurrentQueue, ^{
    obj = [self.array objectAtIndex:index];
  });
  return obj;
}
- (void)insertObject:(id)obj atIndex:(NSUInteger)index {
  dispatch_barrier_async(self.concurrentQueue, ^{
    [self.array insertObject:obj atIndex:index];
  });
}
```

All that is required is a `concurrentQueue` property, created by calling `dispatch_queue_create` with the `DISPATCH_QUEUE_CONCURRENT` option. In the reader (`objectAtIndex:`), you use `dispatch_sync` to wait for it to complete. The practice of creating and dispatching blocks in GCD has very little overhead, so this approach is much faster than using a mutex. The queue can process as many reads in tandem as it has cores available. In the writer, you use `dispatch_barrier_async` to ensure exclusive access to the queue while writing. When you make the call asynchronous, the writer returns quickly, but any future reads on the same thread are guaranteed to return the value the writer set. GCD queues are FIFO, so all requests on the queue before the write are completed first, then the write runs alone, and only then are requests that were placed on the queue after the write processed. This consistent ordering prevents writer starvation and ensures that immediately reading after a write always yields the correct result.

Dispatch Groups

A dispatch group is similar to dependencies in `NSOperation`. First, you create a group:

```
dispatch_group_t group = dispatch_group_create();
```

Notice that groups have no configuration of their own. They're not tied to any queue. They're just a group of blocks. You usually add blocks by calling `dispatch_group_async`, which behaves like `dispatch_async`.

```
dispatch_group_async(group, queue, block);
```

You then register a block to be scheduled when the group completes using `dispatch_group_notify`:

```
dispatch_group_notify(group, queue, block);
```

When all blocks in the group complete, `block` will be scheduled on `queue`. You can have multiple notifications for the same group, and they can be scheduled on different queues if you like. If you call `dispatch_group_notify` and there are no blocks on the queue, the notification fires immediately. You can avoid this situation by using `dispatch_suspend` on the queue while you configure the group and then `dispatch_resume` to start it.

Groups aren't actually tracking blocks as much as they're tracking *tasks*. You can increment and decrement the number of tasks directly using `dispatch_group_enter` and `dispatch_group_leave`. So, in effect, `dispatch_group_async` is really no more than this:

```
dispatch_async(queue, ^{
  dispatch_group_enter(group);
  dispatch_sync(queue, block);
  dispatch_group_leave(group);
});
```

Calling `dispatch_group_wait` blocks the current thread until the entire group completes, which is similar to a thread *join* in Java.

Summary

The future of iOS development is multitasking. Apps will need to do more operations in parallel to leverage multicore hardware and provide the best experience for users. Traditional threading techniques are still useful, but operation queues and Grand Central Dispatch are more effective and promise greater performance with less contention and less locking. Learning to manage your internal multitasking and behaving appropriately when multitasking with other applications are fundamental parts of today's iOS development.

Further Reading

Apple Documentation

The following documents are available in the iOS Developer Library at `developer.apple.com` or through the Xcode Documentation and API Reference.

Concurrency Programming Guide

Threading Programming Guide

WWDC Sessions

The following session videos are available at `developer.apple.com`.

WWDC 2011, "Session 320: Adopting Multitasking in Your App"

WWDC 2011, "Session 210: Mastering Grand Central Dispatch"

WWDC 2012, "Session 712: Asynchronous Design Patterns with Blocks, GCD, and XPC"

Other Resources

Ash, Mike. *NSBlog*. Mike Ash is one of the most prolific writers on low-level threading issues. Although some of this information is now dated, many of his blog posts are still required reading.

`http://mikeash.com/pyblog/`

- Friday Q&A 2009-07-10: "Type Specifiers in C, Part 3"
- Friday Q&A 2010-01-01: "NSRunLoop Internals"
- Friday Q&A 2010-07-02: "Background Timers"
- Friday Q&A 2010-12-03: "Accessors, Memory Management, and Thread Safety"

CocoaDev, "LockingAPIs." CocoaDev collects much of the accumulated wisdom of the Cocoa developer community. The "Locking APIs" page includes links and discussion of the available technologies and tradeoffs.

`http://cocoadev.com/LockingAPIs`

Part III

The Right Tool for the Job

## Chapter 10

Building a (Core) Foundation

As an iOS developer, you will spend most of your time using the UIKit and Foundation frameworks. UIKit provides user interface elements such as UIView and UIButton. Foundation provides basic data structures such as NSArray and NSDictionary. They can handle the vast majority of problems the average iOS application will encounter. But some things require lower-level frameworks. The names of these lower-level frameworks often start with the word "Core"—Core Text, Core Graphics, and Core Video, for example. What they all have in common are C-based APIs based on Core Foundation.

Core Foundation provides a C API that is similar to the Objective-C Foundation framework. It provides a consistent object model with reference counting and containers, just like Foundation, and simplifies passing common data types to low-level frameworks. As you see later in this chapter, Core Foundation is tightly coupled with Foundation, making it easy to pass data between C and Objective-C.

In this chapter, you find out about the Core Foundation data types and naming conventions. You learn about Core Foundation allocators and how they provide greater flexibility than +alloc provides in Objective-C. This chapter covers Core Foundation string and binary data types extensively. You discover Core Foundation collection types, which are more flexible than their Foundation counterparts and include some not found in Objective-C. Finally, you learn how to move data easily between Core Foundation and Objective-C using toll-free bridging. When you're finished, you will have the tools you need to use the powerful Core frameworks, as well as more flexible data structures to improve your own projects.

You can find all code samples in this chapter in main.m and MYStringConversion.c in the online files for this chapter.

Core Foundation Types

Core Foundation is made up primarily of *opaque types*, which are simply C structs. Opaque types are similar to classes in that they provide encapsulation and some inheritance and polymorphism. The similarity should not be overstated, however. Core Foundation is implemented in pure C, which has no language support for inheritance or polymorphism, so sometimes the class metaphor can become strained. But for general usage, Core Foundation can be thought of as an object model with CFType as its root "class."

Like Objective-C, Core Foundation deals with pointers to instances. In Core Foundation, these pointers are given the suffix Ref. For example, a pointer to a CFType is a CFTypeRef, and a pointer to a string is a CFStringRef. Mutable versions of opaque types include the word Mutable, so a CFMutableStringRef is the mutable form of a CFStringRef. Generally, you can treat mutable types as if they were a subclass of the nonmutable type, just as in Foundation.

> For simplicity, and to match the Apple documentation, this chapter uses the term `CFString` to refer to the thing a `CFStringRef` points to, even though Core Foundation does not define the symbol `CFString`.

Because Core Foundation is implemented in C, and C has no language support for inheritance or polymorphism, how does Core Foundation give the illusion of an object hierarchy? First, `CFTypeRef` is just a `void*`. This fact provides a crude kind of polymorphism because it allows arbitrary types to be passed to certain functions, particularly `CFCopyDescription`, `CFEqual`, `CFHash`, `CFRelease`, and `CFRetain`.

Except for `CFTypeRef`, opaque types are structs. A mutable and immutable pair is usually of the form

```
typedef const struct __CFString * CFStringRef;
typedef struct __CFString * CFMutableStringRef;
```

Using this form, the compiler can enforce `const` correctness to provide a kind of inheritance. It should be clear that this isn't real inheritance. There is no good way to provide arbitrary subclasses of `CFString` that the compiler will type-check. For example, consider the following code:

testTypeMismatch (CoreFoundation:main.m)

```
CFStringRef errName = CFSTR("error");
CFErrorRef error = CFErrorCreate(NULL, errName, 0, NULL);
CFPropertyListRef propertyList = error;
```

A `CFError` is not a `CFPropertyList`, so line 3 should generate a warning. It doesn't because `CFPropertyListRef` is defined as `CFTypeRef`, which is required because it has several "subclasses," including `CFString`, `CFDate`, and `CFNumber`. When something has several subclasses, it generally has to be treated as a `void*` (`CFTypeRef`) in Core Foundation. It isn't obvious which types are actually `void*` from looking at the code, but luckily it doesn't come up that often. Most types are defined as a specific `struct` or `const struct`.

Naming and Memory Management

As in Cocoa, naming conventions are critical in Core Foundation. The most important rule is the Create Rule: If a function has the word `Create` or `Copy` in its name, you are an owner of the resulting object and must eventually release your ownership using `CFRelease`. Like Cocoa, Core Foundation objects are reference counted and can have multiple "owners." When the last owner calls `CFRelease`, the object is destroyed.

No equivalent of `NSAutoreleasePool` is available in Core Foundation, so functions with `Copy` in the name are much more common than in Cocoa. Some functions, however, return a reference to an internal data structure or to a constant object. These functions generally include the word `Get` in their name (the Get Rule). The caller is not an owner and does not need to release them.

> `Get` in Core Foundation is not the same as `get` in Cocoa. Core Foundation functions including `Get` return an opaque type or a C type. Cocoa methods that begin with `get` update a pointer passed by reference.

There is no Automatic Reference Counting (ARC) in Core Foundation. Memory management in Core Foundation is similar to manual memory management in Cocoa:

- If you `Create` or `Copy` an object, you are an owner.

- If you do not `Create` or `Copy` an object, you are not an owner. If you want to prevent the object from being destroyed, you must become an owner by calling `CFRetain`.

- If you are an owner of an object, you must call `CFRelease` when you're done with it.

> `CFRelease` **is similar to** `release` **in Objective-C, but there are important differences. The most critical is that you cannot call** `CFRelease(NULL)`**. This difference is somewhat unfortunate, and many specialized versions of** `CFRelease` **exist that do allow you to pass** `NULL` (`CGGradientRelease`**, for instance).**

Some functions have both `Create` and `Copy` in their name. For example, `CFStringCreateCopy` creates a copy of another `CFString`. Why not just `CFStringCopy`? The reason is that `Create` tells you other things about the function than just the ownership rule. It indicates that the first parameter is a `CFAllocatorRef`, which lets you customize how the newly created object is allocated. In almost all cases, you pass `NULL` for this parameter, which specifies the default allocator: `kCFAllocatorDefault`. (I cover allocators in more depth in a moment.) When you know that the function is a creator, the name also tells you that it makes a copy of the passed string.

Conversely, a function with `NoCopy` in its name does not make a copy. For example, `CFStringCreateWithBytesNoCopy` takes a pointer to a buffer and creates a string without copying the bytes. So who is now responsible for releasing the buffer? That brings us back to allocators.

Allocators

A `CFAllocatorRef` is a strategy for allocating and freeing memory. In almost all cases, you want the default allocator, `kCFAllocatorDefault`, which is the same as passing `NULL`. It allocates and frees memory in "the normal way" according to Core Foundation. This way is subject to change, and you shouldn't rely on any particular behavior. It is rare to need a specialized allocator, but in a few cases, such a tool can be useful. Here are the standard allocators to give an idea of what they can do:

- `kCFAllocatorDefault`—This default allocator is equivalent to passing `NULL`.

- `kCFAllocatorSystemDefault`—This original default system allocator is available in case you have changed the default allocator using `CFAllocatorSetDefault`. Using it is very rarely necessary.

- `kCFAllocatorMalloc`—This allocator calls `malloc`, `realloc`, and `free`. It is particularly useful as a deallocator for `CFData` and `CFString` if you created the memory with `malloc`.

- `kCFAllocatorMallocZone`—This allocator creates and frees memory in the default `malloc` zone. It can be useful with garbage collection on the Mac but is almost never useful in iOS.

- `kCFAllocatorNull`—This allocator does nothing. Like `kCFAllocatorMalloc`, it can be useful with `CFData` or `CFString` as a deallocator if you don't want to free the memory.

■ `kCFAllocatorUseContext`—This allocator is used only by the `CFAllocatorCreate` function. When you create a `CFAllocator`, the system needs to allocate memory. Like all other `Create` methods, it requires an allocator. This special allocator tells `CFAllocatorCreate` to use the functions passed to it to allocate the `CFAllocator`.

See the later section "Backing Storage for Strings" for examples of how these allocators can be used in a practical problem.

Introspection

Core Foundation allows a variety of type introspections, primarily for debugging purposes. The most fundamental one is the `CFTypeID`, which uniquely identifies the opaque type of the object, similar to `Class` in Objective-C. You can determine the type of a Core Foundation instance by calling `CFGetTypeID`. The returned value is opaque and subject to change between versions of iOS. You can compare the `CFTypeID` of two instances, but most often you compare the result of `CFGetTypeID` to the value from a function like `CFArrayGetTypeID`. All opaque types have a related `GetTypeID` function.

As in Cocoa, Core Foundation instances have a description for debugging purposes, returned by `CFCopyDescription`. This function returns a `CFString` that you're responsible for releasing. `CFCopyTypeIDDescription` provides a similar string that describes a `CFTypeID`. Don't rely on the format or content of these descriptions because they're subject to change.

To write debugging output to the console, use `CFShow`. It displays the value of a `CFString` or the description of other types. To display the description of a `CFString`, use `CFShowStr`. For example, given the following definitions

```
CFStringRef string = CFSTR("Hello");
CFArrayRef array = CFArrayCreate(NULL, (const void**)&string, 1,
                                 &kCFTypeArrayCallBacks);
```

here are the results for each kind of `CFShow` call:

```
CFShow(array);
<CFArray 0x6d47850 [0x1445b38]>{type = immutable, count = 1,
   values = (
      0 : <CFString 0x410c [0x1445b38]>{contents = "Hello"}
   )}

CFShow(string);
Hello

CFShowStr(string);
Length 5
IsEightBit 1
HasLengthByte 0
HasNullByte 1
InlineContents 0
Allocator SystemDefault
Mutable 0
Contents 0x3ba7
```

Strings and Data

CFString is a Unicode-based storage container that provides rich and efficient functionality for manipulating, searching, and converting international strings. Closely related are the CFCharacterSet and CFAttributedString classes. CFCharacterSet represents a set of characters for efficiently searching, including, or excluding certain characters from a string. CFAttributedString combines a string with ranges of attributes. It is most commonly used to handle rich text but can be used for a variety of metadata storage.

CFString is closely related to NSString, and they are generally interchangeable, as you see in the "Toll-Free Bridging" section later in this chapter. This section focuses on the differences between CFString and NSString.

Constant Strings

In Cocoa, a literal NSString is indicated by an at-sign, as in @"string". In Core Foundation, a literal CFString is indicated by the macro CFSTR, as in CFSTR("string"). If you're using the Apple-provided gcc and the option -fconstant-cfstrings, this macro uses a special built-in compiler hook that creates constant CFString objects at compile time. Clang also has this built-in compiler hook. If you're using standard gcc, an explicit CFStringMakeConstantString function is used to create these objects at runtime.

Because CFSTR has neither Create nor Copy in its name, you don't need to call CFRelease on the result. You may, however, call CFRetain normally if you like. If you do, you should balance it with CFRelease as usual. This way, you can treat constant strings in the same way as programmatically created strings.

Creating Strings

A common way to generate a CFString is from a C string. Here is an example:

testCString (CoreFoundation:main.m)

```
const char *cstring = "Hello World!";
CFStringRef string = CFStringCreateWithCString(NULL, cstring,
                    kCFStringEncodingUTF8);
CFShow(string);
CFRelease(string);
```

Although many developers are most familiar with NULL-terminated C strings, you can store strings in other ways, and understanding them can be useful in improving code efficiency. In network protocols, it can be very efficient to encode strings as a length value followed by a sequence of characters. If parsers are likely to need only a part of the packet, using length bytes to skip over the parts you don't need is faster than reading everything looking for NULL. If this length encoding is 1 byte long, the buffer is a Pascal string and Core Foundation can use it directly, as shown in the following code:

testPascalString (CoreFoundation:main.m)

```
// A common type of network buffer
struct NetworkBuffer {
  UInt8 length;
  UInt8 data[];
};
```

(continued)

```
// Some data we pulled off of the network into the buffer
static struct NetworkBuffer buffer = {
  4, {'T', 'e', 'x', 't'}};

CFStringRef string =
  CFStringCreateWithPascalString(NULL,
                                 (ConstStr255Param)&buffer,
                                 kCFStringEncodingUTF8);
CFShow(string);
CFRelease(string);
```

If you have length some other way, or if the length is not 1 byte long, you can use
`CFStringCreateWithBytes` similarly:

```
CFStringRef string = CFStringCreateWithBytes(NULL,
                                             buffer.data,
                                             buffer.length,
                                             kCFStringEncodingUTF8,
                                             false);
```

The final `false` indicates this string does not have a *byte order mark* (BOM) at the beginning. The BOM indicates whether the string was generated on a big endian or little endian system. A BOM is not needed or recommended for UTF-8 encodings. This is one of many reasons to choose UTF-8 when possible.

> Core Foundation constants begin with a k, unlike their Cocoa counterparts. For example, the Core Foundation counterpart to NSUTF8StringEncoding is kCFStringEncodingUTF8.

Converting to C Strings

Although converting from C strings is simple, converting back to C strings can be deceptively difficult. There are two ways to get a C string out of a `CFString`: Request the pointer to the internal C string representation or copy the bytes out into your own buffer.

Obviously, the easiest and fastest way to get the C string is to request the internal C string pointer:

```
const char *
cstring = CFStringGetCStringPtr(string, kCFStringEncodingUTF8);
```

This approach appears to be the best of all worlds. It's extremely fast, and you don't have to allocate or free memory. Unfortunately, it may not work, depending on how the string is currently encoded inside the `CFString`. If an internal C string representation is not available, this routine returns `NULL` and you have to use `CFStringGetCString` and pass your own buffer, although it isn't obvious how large a buffer you need. Here's an example of how to solve this problem:

MYStringConversion.c (CoreFoundation)

```
char * MYCFStringCopyUTF8String(CFStringRef aString) {
  if (aString == NULL) {
```

```
      return NULL;
  }

  CFIndex length = CFStringGetLength(aString);
  CFIndex maxSize =
    CFStringGetMaximumSizeForEncoding(length,
                                  kCFStringEncodingUTF8);
  char *buffer = (char *)malloc(maxSize);
  if (CFStringGetCString(aString, buffer, maxSize,
                        kCFStringEncodingUTF8)) {
    return buffer;
  }
  free(buffer);
  return NULL;
}
```

testCopyUTF8String (CoreFoundation:main.m)

```
CFStringRef string = CFSTR("Hello");
char * cstring = MYCFStringCopyUTF8String(string);
printf("%s\n", cstring);
free(cstring);
```

MYCFStringCopyUTF8String is not the fastest way to convert a CFString to a C string because it allocates a new buffer for every conversion, but it's easy to use and quick enough for many problems. If you're converting a lot of strings and want to improve speed and minimize memory churn, you might use a function like this one that supports reusing a common buffer:

MYStringConversion.c (CoreFoundation)

```
#import <malloc/malloc.h> // For malloc_size()

const char * MYCFStringGetUTF8String(CFStringRef aString,
                                     char **buffer) {
  if (aString == NULL) {
    return NULL;
  }

  const char *cstr = CFStringGetCStringPtr(aString,
                                    kCFStringEncodingUTF8);
  if (cstr == NULL) {
    CFIndex length = CFStringGetLength(aString);
    CFIndex maxSize =
    CFStringGetMaximumSizeForEncoding(length,
                            kCFStringEncodingUTF8) + 1; // +1 for \0
    if (maxSize > malloc_size(buffer)) {
      *buffer = realloc(*buffer, maxSize);
    }
    if (CFStringGetCString(aString, *buffer, maxSize,
                        kCFStringEncodingUTF8)) {
      cstr = *buffer;
```

(continued)

```
        }
      }
    return cstr;
}
```

The caller of `MYCFStringGetUTF8String` is responsible for passing a reusable buffer. The buffer may point to `NULL` or to preallocated memory. Keep in mind that the returned C string points into either the `CFString` or into `buffer`, so invalidating either one can cause the returned C string to become invalid. In particular, passing the same buffer repeatedly to this function may invalidate old results. That's the trade-off for its speed. Here's how it would be used:

testGetUTF8String (CoreFoundation:main.m)

```
CFStringRef strings[3] = { CFSTR("One"), CFSTR("Two"), CFSTR("Three") };
char * buffer = NULL;
const char * cstring = NULL;
for (unsigned i = 0; i < 3; ++i) {
    cstring = MYCFStringGetUTF8String(strings[i], &buffer);
    printf("%s\n", cstring);
}
free(buffer);
```

If you need conversion to be as fast as possible, and you know the maximum string length, the following approach is even faster:

testFastUTF8Conversion (CoreFoundation:main.m)

```
CFStringRef string = ...;
const CFIndex kBufferSize = 1024;
char buffer[kBufferSize];
CFStringEncoding encoding = kCFStringEncodingUTF8;
const char *cstring;
cstring = CFStringGetCStringPtr(string, encoding);
if (cstring == NULL) {
  if (CFStringGetCString(string, buffer, kBufferSize, encoding)) {
  cstring = buffer;
  }
}
printf("%s\n", cstring);
```

Because this approach relies on a stack variable (`buffer`), wrapping it into a simple function call is difficult, but this way you avoid any extra memory allocations.

Other String Operations

To developers familiar with `NSString`, most of `CFString` should be fairly obvious. You can find ranges of characters, append, trim and replace characters, compare, search, and sort as in Cocoa. `CFStringCreateWithFormat` provides functionality identical to `stringWithFormat:`. I don't explore all the functions here. You can find them all in the documentation for `CFString` and `CFMutableString`.

Backing Storage for Strings

Generally, a CFString allocates the required memory to store its characters. This memory is called the *backing storage*. If you have an existing buffer, continuing to use it is sometimes more efficient or convenient than copying all the bytes into a new CFString. You might use it because you have a buffer of bytes you want to convert into a string or because you want to continue to have access to the raw bytes while also using convenient string functions.

In the first case, in which you already have a buffer, you generally use a function like CFStringCreateWithBytesNoCopy.

testBytesNoCopy (CoreFoundation:main.m)

```
const char *cstr = "Hello";
char *bytes = CFAllocatorAllocate(kCFAllocatorDefault,
                                  strlen(cstr) + 1, 0);
strcpy(bytes, cstr);
CFStringRef str =
  CFStringCreateWithCStringNoCopy(kCFAllocatorDefault, bytes,
                                  kCFStringEncodingUTF8,
                                  kCFAllocatorDefault);
CFShow(str);
CFRelease(str);
```

Because you passed the default allocator (kCFAllocatorDefault) as the destructor, the CFString owns the buffer and will free it when it's done using the default allocator. This example matches the earlier call to CFAllocatorAllocate. If you were to allocate the buffer with malloc, the code would look like this:

testBytesNoCopyMalloc (CoreFoundation:main.m)

```
const char *cstr = "Hello";
char *bytes = malloc(strlen(cstr) + 1);
strcpy(bytes, cstr);
CFStringRef str =
  CFStringCreateWithCStringNoCopy(NULL, bytes, kCFStringEncodingUTF8,
                                  kCFAllocatorMalloc);
CFShow(str);
CFRelease(str);
```

In both cases, the allocated buffer is freed when the string is destroyed. But what if you want to keep the buffer for other uses? Consider the following code:

testBytesNoCopyNull (CoreFoundation:main.m)

```
const char *cstr = "Hello";
char *bytes = malloc(strlen(cstr) + 1);
strcpy(bytes, cstr);
CFStringRef str =
  CFStringCreateWithCStringNoCopy(NULL, bytes, kCFStringEncodingUTF8,
                                  kCFAllocatorNull);
CFShow(str);
```

(continued)

```
CFRelease(str);
printf("%s\n", bytes);
free(bytes);
```

You pass kCFAllocatorNull as the destructor. You still release the string because you created it with a Create function. But now the buffer pointed to by bytes is still valid after the call to CFRelease. You are responsible for calling free on bytes when you're done with the buffer.

You have no guarantee that the buffer you pass will be the actual buffer used. Core Foundation may call the deallocator at any time and create its own internal buffer. Most critically, you must not modify the buffer after creating the string. If you have a buffer that you want to access as a CFString while allowing changes to it, you need to use CFStringCreateMutableWithExternalCharactersNoCopy. This function creates a mutable string that always uses the provided buffer as its backing store. If you change the buffer, you need to let the string know by calling CFStringSetExternalCharactersNoCopy. Using these functions bypasses many string optimizations, so use them with care.

CFData

CFData is the Core Foundation equivalent to NSData. It is much like CFString with similar creation functions, backing store management, and access functions. The primary difference is that CFData does not manage encodings like CFString. You can find the full list of functions in the CFData and CFMutableData references.

Collections

Core Foundation provides a rich set of object collection types. Most have Cocoa counterparts such as CFArray and NSArray. A few specialized Core Foundation collections such as CFTree have no Cocoa counterpart. Core Foundation collections provide greater flexibility in how they manage their contents. In this section, you learn about the Core Foundation collections that have Objective-C equivalents: CFArray, CFDictionary, CFSet, and CFBag. The other Core Foundation collections are seldom used, but I introduce them so that you're aware of what's available if you need it.

A Core Foundation collection can hold anything that can fit in the size of a pointer (32 bits for the ARM processor) and can perform any action when adding or removing items to or from the collection. The default behavior is similar to the Cocoa equivalents, and Core Foundation collections generally retain and release instances when adding and removing items. Core Foundation uses a structure of function pointers that defines how to treat items in the collection. Configuring these callbacks allows you to highly customize your collection. You can store non-objects such as integers, create weak collections that do not retain their objects, or modify how objects are compared for equality. The "Callbacks" section, later in this chapter, covers this topic. Each collection type has a default set of callbacks defined in the header. For example, the default callbacks for CFArray are kCFTypeArrayCallBacks. While introducing the major collections, I focus on these default behaviors.

CFArray

CFArray corresponds to NSArray and holds an ordered list of items. Creating a CFArray takes an allocator, a series of values, and a set of callbacks, as shown in the following code:

testCFArray (CoreFoundation:main.m)

```
CFStringRef strings[3] = { CFSTR("One"), CFSTR("Two"), CFSTR("Three") };
CFArrayRef array = CFArrayCreate(NULL, (void *)strings, 3,
                                 &kCFTypeArrayCallBacks);
CFShow(array);
CFRelease(array);
```

Creating a `CFMutableArray` takes an allocator, a size, and a set of callbacks. Unlike the `NSMutableArray` capacity, which is only an initial size, the size passed to `CFMutableArray` is a fixed maximum. To allocate an array that can grow, pass a size of 0:

CFMutableArrayRef array = CFArrayCreateMutable(NULL, 0, &kCFTypeArrayCallBacks);

CFDictionary

`CFDictionary` corresponds to `NSDictionary` and holds key-value pairs. Creating a `CFDictionary` takes an allocator, a series of keys, a series of values, a set of callbacks for the keys, and a set of callbacks for the values.

testCFDictionary (CoreFoundation:main.m)

```
#define kCount 3
CFStringRef keys[kCount]   = { CFSTR("One"), CFSTR("Two"), CFSTR("Three") };
CFStringRef values[kCount] = { CFSTR("Foo"), CFSTR("Bar"), CFSTR("Baz") };
CFDictionaryRef dict =
  CFDictionaryCreate(NULL,
                     (void *)keys,
                     (void *)values,
                     kCount,
                     &kCFTypeDictionaryKeyCallBacks,
                     &kCFTypeDictionaryValueCallBacks);
```

Creating a `CFMutableDictionary` is like creating a `CFMutableArray`, except it has separate callbacks for the keys and values. As with `CFMutableArray`, the size is fixed if given. For a dictionary that can grow, pass a size of 0.

CFSet, CFBag

`CFSet` corresponds to `NSSet` and is an unordered collection of unique objects. `CFBag` corresponds to `NSCountedSet` and allows duplicate objects. As with their Cocoa counterparts, uniqueness is defined by equality. The function that determines equality is one of the callbacks.

Like `CFDictionary`, `CFSet` and `CFBag` can hold `NULL` values by passing `NULL` as their callback structure pointer.

Other Collections

Core Foundation includes several collections that do not have a Cocoa counterpart:

- `CFTree` provides a convenient way to manage tree structures that might otherwise be stored less efficiently in a `CFDictionary`. A short example of `CFTree` is shown in the section "Toll-Free Bridging" later in this chapter.

- `CFBinaryHeap` provides a binary-searchable container, similar to a sorted queue.

- `CFBitVector` provides a convenient way to store bit values.

Full information on `CFTree` is available in Apple's *Collections Programming Topics for Core Foundation*. See the Apple documentation on `CFBinaryHeap` and `CFBitVector` for more information on their usage. They are not often used and aren't heavily documented.

Callbacks

Core Foundation uses a structure of function pointers that define how to treat items in the collection. The structure includes the following members:

- `retain`—Called when an item is added to the collection. The default behavior is similar to `CFRetain` (you learn what "similar" means in a moment). If it's `NULL`, no action is performed.

- `release`—Called when an item is removed from the collection and when the collection is destroyed. The default behavior is similar to `CFRelease`. If it is `NULL`, no action is performed.

- `copyDescription`—Called for each object in response to functions that want a human-readable description for the entire collection, such as `CFShow` or `CFCopyDescription`. The default value is `CFCopyDescription`. If this is `NULL`, the collection has some built-in logic to construct a simple description.

- `equal`—Called to compare a collection object with another object to determine whether they're equal. The default value is `CFEqual`. If this is `NULL`, the collection uses strict equality (`==`) of the values. If the items are pointers to objects (as is the usual case), this means that objects are only equal to themselves.

- `hash`—Applies only to hashing collections such as dictionaries and sets. This function is used to determine the hash value of an object. A hash is a fast way to compare objects. Given an object, a hash function returns an integer such that if two objects are equal, their hashes are equal. Using this function allows the collection to quickly determine unequal objects with a simple integer comparison, saving the expensive call to `CFEqual` for objects that are possibly equal. The default value is `CFHash`. If this value is `NULL`, the value (usually a pointer) is used as its own hash.

The default values for `retain` and `release` act like `CFRetain` and `CFRelease` but are actually pointers to the private functions `__CFTypeCollectionRetain` and `__CFTypeCollectionRelease`. The `retain` and `release` function pointers include the collection's allocator in case you want to create a new object rather than retain an existing one. This use is incompatible with `CFRetain` and `CFRelease`, which do not take an allocator. Usually, these private functions don't matter because, in most cases, you either leave `retain` and `release` as default or set them to `NULL`.

Each collection type has a default set of callbacks defined in the header. For example, the default callbacks for `CFArray` are `kCFTypeArrayCallBacks`. They can be used to easily modify default behavior. The following example creates a nonretaining array, which could also hold non-objects such as integers:

testNRArray (CoreFoundation:main.m)

```
CFArrayCallBacks nrCallbacks = kCFTypeArrayCallBacks;
nrCallbacks.retain = NULL;
nrCallbacks.release = NULL;
```

```
CFMutableArrayRef nrArray = CFArrayCreateMutable(NULL, 0, &nrCallbacks);
CFStringRef string =
  CFStringCreateWithCString(NULL, "Stuff", kCFStringEncodingUTF8);
CFArrayAppendValue(nrArray, string);
CFRelease(nrArray);
CFRelease(string);
```

Another example of callback configuration is to allow NULL values or keys. Dictionaries, sets, and bags can hold NULL values or keys if the retain and release callbacks are NULL (this also makes them nonretaining). These types have CF*Type*GetValueIfPresent functions to handle this case. For example, the function CFDictionaryGetValueIfPresent() allows you to determine whether the value was NULL versus missing, as shown in the following code:

testCFDictionaryNULL (CoreFoundation:main.m)

```
CFDictionaryKeyCallBacks cb = kCFTypeDictionaryKeyCallBacks;
cb.retain = NULL;
cb.release = NULL;
CFMutableDictionaryRef dict =
  CFDictionaryCreateMutable(NULL, 0, &cb, CFDictionaryCreateMutable(NULL,
0, &cb,
CFDictionarySetValue(dict, NULL, CFSTR("Foo"));

const void *value;
Boolean fooPresent =
  CFDictionaryGetValueIfPresent(dict, NULL, &value);
CFRelease(dict);
```

Other collections, such as CFArray, cannot hold NULL values. As in Foundation, you must use a special placeholder NULL constant called kCFNull. It is an opaque type (CFNull), so it can be retained and released.

Core Foundation collections are much more flexible than their Cocoa equivalents. As you see in the next section, however, you can bring this flexibility almost transparently to Cocoa through the power of toll-free bridging.

Toll-Free Bridging

One of the cleverest aspects of Core Foundation is its capability to transparently exchange data with Foundation. For example, any function or method that accepts an NSArray also accepts a CFArray with at most a bridge cast. A *bridge cast* is an instruction to the compiler about how to apply Automatic Reference Counting.

In many cases, you only need to use the __bridge modifier, as shown in the following code:

testTollFree (CoreFoundation:main.m)

```
NSArray *nsArray = [NSArray arrayWithObject:@"Foo"];
printf("%ld\n", CFArrayGetCount((__bridge CFArrayRef)nsArray));
```

This code essentially tells the compiler to do nothing special. It should simply cast `nsArray` as a `CFArrayRef` and pass it to `CFArrayGetCount`. There is no change to the reference count of `nsArray`.

In iOS 7 the situation has improved. Because Core Foundation functions have consistent names, in almost all cases the compiler can figure out how a given parameter should be memory managed, so the `__bridge` modifier isn't required:

```
printf("%ld\n", CFArrayGetCount((CFArrayRef)nsArray));
```

This approach works for most of Core Foundation. To use it in your own code, you need to first audit all your functions and verify that they correctly obey Core Foundation naming conventions. When they do, you can bracket their declarations like this:

MYStringConversion.h (CoreFoundation)

```
CF_IMPLICIT_BRIDGING_ENABLED

char * MYCFStringCopyUTF8String(CFStringRef aString);
const char * MYCFStringGetUTF8String(CFStringRef aString, char **buffer);

CF_IMPLICIT_BRIDGING_DISABLED
```

Functions declared inside this block do not require `__bridge` casts for their parameters.

Toll-free bridging works from Core Foundation to Foundation:

testTollFreeReverse (CoreFoundation:main.m)

```
CFMutableArrayRef cfArray =
  CFArrayCreateMutable(NULL, 0, &kCFTypeArrayCallBacks);
CFArrayAppendValue(cfArray, CFSTR("Foo"));
NSLog(@"%ld", [(__bridge id)cfArray count]);
CFRelease(cfArray);
```

Note that a `__bridge` cast is required here. You are not passing `cfArray` to an audited function. You are converting it to an `id` and sending it a message.

The `__bridge` cast (implicit or explicit) works as long as there is no Core Foundation memory management involved. In the preceding examples, you aren't assigning the results to variables or returning them. Consider this case, however:

```
- (NSString *)firstName {
  CFStringRef cfString = CFStringCreate...;
  return (???)cfString;
}
```

How can you cast `cfString` correctly? Before ARC, you would have cast this to an `NSString` and called `autorelease`. With ARC, you can't call `autorelease`, and ARC doesn't know that `cfString` has an extra retain on it from `CFStringCreate....` You again use a bridge cast, this time in the form of a function, as in this example:

```
    return CFBridgingRelease(cfString);
```

This function transfers ownership from Core Foundation to ARC. In the process, it reduces the retain count by one to balance the CFStringCreate.... You must use a bridge cast to achieve this result. Calling CFRelease before returning the object would destroy the object.

When transferring an object from ARC to Core Foundation, you use CFBridgingRetain, which increases the retain count by one, as shown in the following code:

```
CFStringRef cfStr = CFBridgingRetain([nsString copy]);
nsString = nil; // Ownership now belongs to cfStr
...
CFRelease(cfStr);
```

The bridging functions can also be written in a typecast style as follows:

```
NSString *nsString = CFBridgingRelease(cfString);
NSString *nsString = (__bridge_transfer id)cfString;

CFStringRef cfString = CFBridgingRetain(nsString);
CFStringRef cfString = (__bridge_retained CFTypeRef)nsString;
```

> CFTypeRef **is a generic pointer to a Core Foundation object, and** id **is a generic pointer to an Objective-C object. You could also use explicit types here such as** CFStringRef **and** NSString*.

The function form is shorter and, in my opinion, easier to understand. CFBridgingRelease and CFBridgingRetain should be used only when an object is being transferred between ARC and Core Foundation. They're not replacements for CFRetain or CFRelease or a way to "trick" the compiler into adding an extra retain or release on Objective-C objects. After an object is transferred from Core Foundation to ARC, the Core Foundation variable should not be used again and should be set to NULL. Conversely, when converting from ARC to Core Foundation, you need to immediately set the ARC variable to nil. You have transferred ownership from one variable to another, so the old variable needs to be treated as invalid.

Toll-free bridging not only is convenient for moving information between C and Objective-C, but also enables Cocoa developers to make use of certain Core Foundation functions that have no Objective-C equivalent. For example, CFURLCreateStringByAddingPercentEscapes allows much more powerful transformations than the equivalent NSURL stringByAddingPercentEscapesUsingEncoding:.

Even types that aren't explicitly toll-free bridged are still bridged to NSObject. This means that you can store Core Foundation objects (even ones with no Cocoa equivalent) in Cocoa collections, as shown in this example:

testTreeInArray (CoreFoundation:main.m)

```
CFTreeContext ctx = {0, (void*)CFSTR("Info"), CFRetain,
                     CFRelease, CFCopyDescription};
CFTreeRef tree = CFTreeCreate(NULL, &ctx);
```

(continued)

```
NSArray *array = @[(__bridge id)tree];
CFRelease(tree);
NSLog(@"Array=%@", array);
```

Toll-free bridging is implemented in a fairly straightforward way. Every Objective-C object structure begins with an ISA pointer to a `Class`:

```
typedef struct objc_class *Class;
typedef struct objc_object {
  Class isa;
} *id;
```

Core Foundation opaque types begin with a `CFRuntimeBase`, and the first element of that is also an ISA pointer:

```
typedef struct __CFRuntimeBase {
  uintptr_t _cfisa;
  uint8_t _cfinfo[4];
#if __LP64__
  uint32_t _rc;
#endif
} CFRuntimeBase;
```

`_cfisa` points to the toll-free bridged Cocoa class. These are subclasses of the equivalent Cocoa class; they forward Objective-C method calls to the equivalent Core Foundation function call. For instance, `CFString` is bridged to the private toll-free bridging class `NSCFString`.

If there is no explicit bridging class, `_cfisa` points to `__NSCFType`, which is a subclass of `NSObject`, and forwards calls like `retain` and `release`.

To handle Objective-C classes passed to Core Foundation functions, all public toll-free functions look something like this:

```
CFIndex CFStringGetLength(CFStringRef str) {
  CF_OBJC_FUNCDISPATCHV(__kCFStringTypeID, CFIndex, (NSString *)str,
  length);
  __CFAssertIsString(str);
  return __CFStrLength(str);
}
```

`CF_OBJC_FUNCDISPATCHV` checks the `_cfisa` pointer (see the following note). If it matches the Core Foundation bridging class for the given `CFTypeID`, it passes the call along to the real Core Foundation function. Otherwise, it translates the call into an Objective-C message (`length` in this case, which is converted into a selector by the macro) and does not return.

The majority of Core Foundation is open source. You can download it from `opensource.apple.com`. Look for the CF project in OS X. The preceding code comes from `CFString.h`.

The implementation of `CF_OBJC_FUNCDISPATCHV` **is not open source. In this section, you can assume that it is similar to the previous macro from OS X,** `CF_OBJC_FUNCDISPATCH0`, **which was open source.**

Summary

Core Foundation bridges the gap between C and Objective-C code, providing powerful data structures for C and near-transparent data passing to and from low-level code. As Apple releases more low-level Core frameworks that require these types, Core Foundation is an increasingly important part of an iOS developer's toolkit.

Core Foundation data structures are generally more flexible than their Cocoa equivalents. They provide better control over how memory is managed through allocators and often include functions for more specialized problems such as handling Pascal strings or configurable URL percent substitutions. Core Foundation collections can be configured to be nonretaining and can even store non-objects such as integers.

Although Objective-C is extremely powerful, you can still generally write code that is faster and more efficient in pure C, which is why the lowest-level APIs are all C APIs. For those parts of your programs that require the kind of performance you can get only from C, Core Foundation provides an excellent collection of abstract data types that you can easily exchange with the higher-level parts of your program. The vast majority of problems in iOS are best solved in Cocoa and Objective-C, but for those places that C is appropriate, Core Foundation is a powerful tool.

Further Reading

Apple Documentation

The following documents are available in the iOS Developer Library at `developer.apple.com` or through the Xcode Documentation and API Reference.

Collections Programming Topics for Core Foundation

Core Foundation Design Concepts

Data Formatting Guide for Core Foundation

Dates and Times Programming Guide for Core Foundation

Transitioning to ARC Release Notes: "Managing Toll-Free Bridging"

Memory Management Programming Guide for Core Foundation

Property List Programming Topics for Core Foundation

Strings Programming Guide for Core Foundation

Other Resources

Clang Documentation. "Automatic Reference Counting"

`clang.llvm.org/docs/AutomaticReferenceCounting.html`

ridiculous_fish, "Bridge." An entertaining introduction to toll-free bridging internals by one of the AppKit and Foundation team at Apple. Based on an older implementation of toll-free bridging, specifically on OS X, but it is very likely that the current iOS implementation is similar.

`ridiculousfish.com/blog/posts/bridge.html`

Core Foundation Source Code (as of 10.8.3)

`http://opensource.apple.com/source/CF/CF-744.18/`

## Chapter 11

Behind the Scenes: Background Processing

iOS devices have limited resources. There isn't enough memory to keep dozens of apps running at the same time. Battery life is at a premium, so it is critical that a device spend energy only on activities the user really cares about. One "busy waiting" app running in the background could drain the battery in a very short time. But users also want apps to be ready at a moment's notice. They want to switch between apps quickly and seamlessly. Good iOS apps give users the illusion that they are always running, while preserving resources by running as little as possible.

Many things users want can't be achieved with illusions, though. Users want files to download while they're running other apps. They want their news feeds to be up to date when they launch their reader. And they don't want to wait around while their latest pictures upload to the server.

With each version of iOS, apps have gained more access to working in the background. iOS 7 adds important new features that allow you to download files in the background and use silent notifications to wake up your app at arbitrary times. In this chapter, you learn to use these new features and also important older features such as state restoration.

In this chapter, I assume that you understand the basics of running tasks in the background and that you're familiar with using `beginBackgroundTaskWithExpirationHandler:`, registering an app as location aware, and handling similar background issues. If you need information about the fundamental technologies, see "Executing Code in the Background" in the *iOS Application Programming Guide* (`developer.apple.com`).

Best Practices for Backgrounding: With Great Power Comes Great Responsibility

In iPhoneOS 3, only one third-party application could run at a time. When the user left your application, it was terminated. This ensured that third-party background applications couldn't waste resources such as memory or battery. Apple wanted to make sure that the iPhone didn't suffer the same performance and stability problems of earlier mobile platforms, most pointedly Windows Mobile.

Starting with iOS 4, Apple began to permit third-party applications to run in the background, but only in limited ways. This continued Apple's focus on not allowing third-party applications to destabilize the platform or waste resources. It can be very frustrating, but the policy has generally met its goal. iOS remains focused on the user, not the developer.

Your application should give the illusion that it's always running, even though it isn't. Although your application may be terminated without warning any time it's suspended, it should give the impression that nothing has changed when it launches again. This means that you need to avoid displaying a splash screen during loading and that you need to save sufficient state when you enter the background to resume seamlessly if terminated. `NSUserDefaults` is a good place to stash small amounts of data during `applicationWillResignActive:`. Larger data structures need to be written to files, usually in `~/Library/Caches`. See the section "When We Left Our Heroes: State Restoration," later in this chapter, to find out how to do much of this work automatically.

Reducing your app's memory footprint is important when going into the background, and so is minimizing the time required to resume. If throwing away your cached information makes resuming from the background as expensive as launching from scratch, there isn't any point to suspending. Be thoughtful about what you throw away and how long it will take you to re-create it. Everything you do drains the battery, so always look to avoid wasteful processing, even if it doesn't visibly delay your app.

When your application is suspended, it doesn't receive memory warnings. If its memory footprint is very large, your application is likely to be terminated when there is memory pressure, and you won't have an opportunity to do anything about it. `NSCache` and `NSPurgeableData` are invaluable in addressing this issue. `NSPurgeableData` is an `NSData` object that you can flag as currently in use or currently purgeable. If you store it in an `NSCache` object and mark it as purgeable with `endContentAccess`, the OS saves it until there is memory pressure. At that point, it discards that data, even if your app is suspended at the time. This saves you the cost of throwing away this object and re-creating it every time the user leaves your app briefly, while ensuring that it can be thrown away if needed.

A lot of framework data is automatically managed for you when your app goes into the background. The data for images loaded with `imageNamed:` are discarded automatically and reread from disk when needed again. Views automatically throw away their backing stores (their bitmap cache). You should expect your `drawRect:` methods to be called when you resume. There is a major exception to this rule: `UIImageView` doesn't discard its data, and this can be quite a lot of memory. If you have a large image shown in a `UIImageView`, you should generally remove it before going into the background. However, decompressing images can be very expensive, so you don't throw them away too often. There's no one right answer. You need to profile your application, as discussed in Chapter 18.

In Instruments, the VM Tracker is useful for determining how much memory you're using while the application is in the background. It's part of the Allocations template. First, create a "memory pressure" app that displays a massive image. Then run your program with the VM Tracker. Note the amount of memory you're using. Press the Home button and note the amount of memory you're using now. This is what you're releasing when you go into the background. Now, launch the memory pressure app. Note how much memory you release. Ideally, your background usage should be less than your normal usage, without being so low that you delay resuming. Your usage under memory pressure needs to be as low as possible.

In Instruments, you see two kinds of memory: dirty memory and resident memory. *Dirty memory* is the memory that iOS can't automatically reclaim when you go into the background. *Resident memory* is your total current memory utilization. Both are important for different reasons. Minimizing dirty memory reduces the likelihood that you'll be terminated in the background. Reducing it should be your primary focus. Your application needs to consume as few resources as possible when it's not the foreground application. `NSCache` and `NSPurgeableData` are excellent tools for reducing dirty memory. Resident memory is your entire memory footprint. Minimizing it helps prevent low memory warnings while you're in the foreground.

> In the Instruments VM Tracker, you may see references to *Memory Tag 70*. That refers to memory for decompressed images and is primarily caused by `UIImage`.

Memory is important, but it's not the only resource. Avoid excessive network activity, disk access, or anything else that will waste battery life.

Some actions are forbidden while in the background. The most significant ones are OpenGL calls. You must stop updating OpenGL views when you go into the background. A subtle issue here is application termination. The application is allowed to run for a brief time after `applicationWillTerminate:` is called. During that time, the application is "in the background" and must not make OpenGL calls. If it does, it's killed immediately, which could prevent it from finishing other application-termination logic.

> Be sure to shut down your OpenGL updates when the application is terminating as well as when going into the background. `GLKViewController` **automatically handles all of this for you. The OpenGL Game template in Xcode uses** `GLKViewController`.

Running in the background creates new challenges for developers, but users expect this key feature. Just be sure to keep the user as your top priority, test heavily, and watch your resource utilization. Your application should delight, even when it's not onscreen.

Important Backgrounding Changes in iOS 7

iOS 7 offers some new, powerful mechanisms for running in the background, but it also makes one major change that may break existing code. Applications no longer can execute their long-running tasks while the device is locked, and background tasks can no longer keep the device awake.

If you are using `beginBackgroundTaskWithExpirationHandler:` to perform a network operation, that operation may not occur for a very long time if the user locks the device. You should use the new `NSURLSession` for this operation instead, as discussed in the section "Network Access with NSURLSession."

It is also possible that your background operation will be paused if the device goes to sleep. You must be able to deal with an arbitrary amount of time passing at any point during your background operation. Although you will still typically get about 10 minutes of processing time, that 10 minutes may not be continuous.

Network Access with NSURLSession

One of the most common background tasks is network requests. A common case is ensuring that user data is saved to a server even if the user leaves your app in the middle of your save action. In iOS 6, you would wrap your network request with `beginBackgroundTaskWithExpirationHandler:` so you would have time to finish the request. This approach still works for small requests, but for longer requests, you may be suspended before it completes. The new solution is to use NSURLSession.

I wish I could tell you that NSURLSession made background transfers simple. It doesn't. Handling them well is complicated, and can be more complicated in iOS 7 because background processes are suspended more quickly and more often than they were in iOS 6. In this section, you learn the basics of using NSURLSession and tips on how to develop robust systems with it. Unfortunately, there is no simple, universal way to implement background transfers with NSURLSession. You need to adapt it heavily to your specific situation.

In its simplest form, NSURLSession is a replacement for NSURLConnection. If you were previously using [NSURLConnection sendAsynchronousRequest:queue:completionHandler:], you can switch easily to [NSURLSession dataTaskWithURL:completionHandler:]. For example, if you currently have this code:

```
[NSURLConnection sendAsynchronousRequest:request
                                   queue:[NSOperationQueue mainQueue]
                       completionHandler:handler];
```

you can replace it with this:

```
NSURLSessionTask *task = [[NSURLSession sharedSession]
                          dataTaskWithRequest:request
                          completionHandler:handler];
[task resume];
```

After you do this, you get a few features. For instance, you can query the task for information such as its originalRequest and its countOfBytesExpectedToReceive. This information can be handy, but the real benefits come when you start configuring the session.

[NSURLSession sharedSession] is a basic session object that is essentially the same as NSURLConnection. But you can create new session objects with their own configuration. For example, you can configure a session that doesn't use cellular data:

```
NSURLSessionConfiguration *configuration =
  [NSURLSessionConfiguration defaultSessionConfiguration];
[configuration setAllowsCellularAccess:NO];
NSURLSession *session = [NSURLSession
sessionWithConfiguration:configuration];
```

Sessions can have their own NSCache, their own NSHTTPCookieStorage, and their own NSURLCredentialStorage. Sessions also can have their own NSURLProtocol registrations, so that you can have special URL schemes that apply only to some sessions. With NSURLConnection, custom caching solutions you put in place for your REST protocol also had to handle UIWebView connections. If you included a third-party library, any NSURLProtocol registrations it included had to impact your entire program. With NSURLSession, you can isolate this kind of customization very easily. Configuration that used to be handled in NSURLRequest, such as cellular access and caching policy, can now be managed more centrally in the session.

Session Configuration

`NSURLSession` has three built-in configurations that you can use as a basis for customization:

- `defaultSessionConfiguration`—This configuration matches the behavior of `NSURLConnection`. It uses the shared cache, cookie storage, etc.

- `ephemeralSessionConfiguration`—This configuration uses only in-memory storage. It writes nothing to disk. It is primarily designed for private browsing, but is also good for situations in which you don't want to waste disk space with long-term caching.

- `backgroundSessionConfiguration:`—This configuration takes an identifier and is configured to perform network transfers while your application is in the background, or even when it is terminated. This is the most interesting configuration, but also the most complex to use correctly.

When you request one of these configurations, you are returned a copy, so you are free to modify your configuration object without worrying about affecting other parts of the system.

Tasks

The three kinds of tasks are data, upload, and download. Data tasks behave the most like `NSURLConnection`. They either call a delegate with `NSData` as it is downloaded, or they execute a completion handler with the `NSData`.

If you are working with large transfers, you should be careful how you handle the `NSData` objects. If possible, you should retain them rather than copy them, and you should avoid calling `bytes` if possible. `NSURLSession` may create noncontiguous data objects to avoid copying. If you call `bytes`, then `NSData` has to merge all its data ranges together to provide you a contiguous memory range. Rather than call `bytes`, you can use the new `enumerateByteRangesUsingBlock:` method, which provides you access to the data without forcing an expensive copy step.

Upload and download tasks are new features, and greatly simplify uploading and downloading files. Using them is the preferred way to deal with large data transfers, and is the only way to perform background transfers. To upload files while in the foreground, you generally use `uploadTaskWithRequest:fromFile:completionHandler:`. To download files in the foreground, you generally use `downloadTaskWithURL:completionHandler:`. Note that downloading allows only a source URL, not a destination URL. The completion handler includes the URL of the downloaded file in the `Caches` directory. You must read or move this file before returning. It can be deleted at any time after the completion handler returns. This example shows how to move the files into your document directory:

```
- (NSURL *)documentURLForPath:(NSString *)path {
  static NSURL *documentDirectoryURL;
  static dispatch_once_t onceToken;
  dispatch_once(&onceToken, ^{
    NSString *docPath =
      [NSSearchPathForDirectoriesInDomains(NSDocumentDirectory,
                                           NSUserDomainMask,
                                           YES) firstObject];
    documentDirectoryURL = [NSURL fileURLWithPath:docPath];
  });
```

```
      return [documentDirectoryURL URLByAppendingPathComponent:path];
}

...

NSURLSessionDownloadTask *download =
   [session downloadTaskWithURL:URL
            completionHandler:^(NSURL *location,
                                 NSURLResponse *response,
                                 NSError *error) {
             NSString *filename = [location lastPathComponent];
             NSURL *dest = [self documentURLForPath:filename];
             if ([[NSFileManager defaultManager] moveItemAtURL:location
                                                 toURL:dest
                                                 error:&error]) {

               // TODO: Notify interested parties
             }
             else {
               // TODO: Error handling
             }
          }];
[download resume];
```

Most nonbackground `NSURLSession` calls are fairly obvious if you understand `NSURLConnection`, so I don't go into all the use cases here. You should be aware of a few tips, however:

▓ Tasks are always created in a suspended state. You must call `resume` to be able to start them. This step is easy to forget, and is a common cause of "nothing happens" bugs.

▓ Tasks and sessions are retained by the system, just like `NSURLConnection`, so you don't technically have to retain them yourself. I recommend you keep track of them with ivars, however, so that you can cancel them if needed.

▓ Callback handlers are generally called on background queues. Remember to use `dispatch_async` to move back to the main queue before updating UI elements. You should also be thoughtful about modifying your own properties in nonthreadsafe ways.

▓ Sessions are reusable and you are encouraged to reuse them. Because they are retained by the system, you will leak them unless you call either `finishTasksAndInvalidate` or `invalidateAndCancel`.

Background Transfers

At this point, you are ready to address background transfers. This is the most interesting use of `NSURLSession`, but also the most complex. It is difficult to handle in a general way. Each app has different issues to address. The following sections provide the basics of the technology, an introduction to the problems, and some tips for how to attack them. A sample project called PicDownloader is available in the sample code.

You configure a background transfer using `[NSSessionConfiguration backgroundSessionConfiguration:]`. The identifier you pass must be unique and must be consistent between runs of your program. When you have this configuration, you can create a session and generally perform upload and download tasks normally, except that you must use the delegate-based calls. You cannot use completion handlers with background transfers.

So the basics of background transfers are simple:

```
NSURLSessionConfiguration *configuration =
   [NSURLSessionConfiguration backgroundSessionConfiguration:identifier];
NSURLSession *session = [NSURLSession sessionWithConfiguration:
                          configuration
                                                delegate:self
                                          delegateQueue:nil];
NSURLSessionDownloadTask *task = [session downloadTaskWithURL:sourceURL];
[task resume];
```

If your app stays in the foreground, it behaves like a normal download task. If your app terminates before the download completes, when the download does finally complete, your app is re-launched, and the app delegate is called with `application:handleEventsForBackgroundURLSession:completionHandler:`. This method passes a block that you need to call when you're done handling the event. Typically, that means copying the completion handler for later, re-creating the `NSURLSession` using the same identifier, and waiting for `URLSession:task:didCompleteWithError:`. At that point, if all the tasks are completed, you can call the completion handler, as shown here:

PTLDownloadManager.m (PicDownloader)

```
- (void)URLSession:(NSURLSession *)session
              task:(NSURLSessionTask *)task
didCompleteWithError:(NSError *)error {
  NSLog(@"%s", __PRETTY_FUNCTION__);
  [self.backgroundSession
    getTasksWithCompletionHandler:^(NSArray *dataTasks,
                                    NSArray *uploadTasks,
                                    NSArray *downloadTasks) {
      NSUInteger count = [dataTasks count] +
                         [uploadTasks count] +
                         [downloadTasks count];
      if (count == 0) {
        if (self.backgroundSessionCompletionHandler) {
          void (^completionHandler)() =
            self.backgroundSessionCompletionHandler;
          self.backgroundSessionCompletionHandler = nil;
          completionHandler();
        }
      }
    }];
}
```

In principle, this approach should work well. In practice, it is actually quite difficult to implement correctly, harder to test, and frustrating to debug. That leads to some current recommendations:

- `NSURLSession` is a useful class that can replace `NSURLConnection` in most cases.

- Using `NSURLSession` to manage background downloads is poorly supported in the initial version of iOS 7. This feature should be approached with caution and used as simply as possible.

- Carefully test how your application and server will behave if transfers are interrupted. Because of the more aggressive suspending behavior, it is more likely that a network transfer will be paused, perhaps for a very long time. Your server may need changes to support this approach cleanly.

- Make sure to use `URLSession:task:didCompleteWithError:` only for final cleanup. It seems natural to use `URLSession:downloadTask:didFinishDownloadToURL:` to clean up the task and session, but doing so can lead to crashes. You need to let the entire task complete before calling the app delegate completion handler or methods such as `finishTasksAndInvalidate`.

- Generally speaking, you should approach `NSURLSession` with some caution. It has a great deal of potential, but it doesn't have the many years of hardening that `NSURLConnection` does. You should experiment with simple problems, such as file download, before exploring complex problems, such as background transfers.

Periodic Fetching and Adaptive Multitasking

Adaptive multitasking is a mostly transparent feature. In iOS 7, the OS monitors the user's behavior, looking for patterns when apps are launched. If the OS determines that your app is consistently launched every day at 7 a.m., the OS may begin launching the app automatically a few minutes before that so that it can have the freshest data.

You can assist the OS in this task by letting it know how often it is useful to let the app update and by implementing a protocol to manage periodic update requests. First, enable the Background fetch background mode in the target settings, as shown in Figure 11-1.

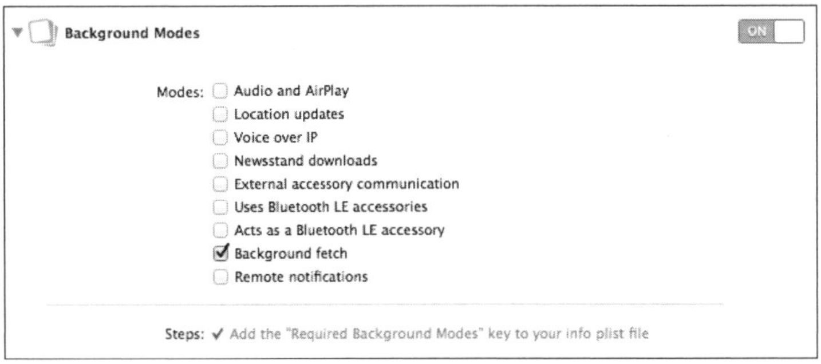

Figure 11-1 Background modes

You then use `[UIApplication setMinimumBackgroundFetchInterval:]` to set the minimum time between updates. The default is "never," so be sure to set this option. It is often useful to set this time to `UIApplicationBackgroundFetchIntervalMinimum`, which leaves it to the system's discretion, but if you know that your data will never update more than once a day or once a week, it is useful to set a longer minimum.

Finally, you implement the application delegate method `application:performFetchWithCompletionHandler:`. You can do whatever you need to do to update your data. When you're done, call the completion handler. The time and energy you spend before calling the completion handler is considered by the OS to determine how often to call your app. If you fail to call the completion handler, the OS may kill your app and may schedule it less frequently for updates.

Waking Up in the Background

Perhaps the most interesting enhancement in background processing is the ability to wake up your app in response to remote notifications. This works like regular remote notifications, except your app is launched automatically without displaying an alert. Set the Remote notifications background mode as shown in Figure 11-1. Then implement `application:didReceiveRemoteNotification:fetchCompletionH andler:`. Use this to update any content the user is likely to want. You can then use a local notification to display an alert, badge, or other indication if appropriate. Be sure to call the completion handler within 30 seconds.

Apple encourages you to send these "silent" remote notifications whenever appropriate for your data. Apple manages rate limiting automatically. Your application receives one notification for every one your server sends, but they may be batched together. For example, if you send a message once a minute, but Apple decides to rate limit you to once an hour, then once an hour your application receives 60 notifications. This is not a mechanism for letting your application run arbitrarily in the background.

When We Left Our Heroes: State Restoration

Your application needs to give the illusion that it is always running, even when it's terminated in the background. If the user is in the middle of typing a message and flips over to another application, that message should be waiting when the user returns. Before iOS 6, this often meant storing a lot of values in `NSUserDefaults`. For example, you need to keep track of which tab is visible, the entire navigation controller stack, the position of the scrollbar, and the partially entered text. The work isn't that difficult, but it is very tedious and somewhat error prone.

Since iOS 6, UIKit handles much of this work for you if you request it. The system is called *state restoration*. It's an opt-in, so you have to tell UIKit that you want it. In this example, you start with a simple, existing app called "FavSpots," available in the sample code for this chapter. The final version is in the project "FavSpotsRestore" in the same directory. This application enables you to bookmark places on a map and add a name and some notes, which are saved in Core Data. To implement state restoration, do the following:

1. Opt in by requesting state saving and restoring in the application delegate.

2. Modify the application delegate's startup handling if needed.

3. Assign restoration identifiers to all the view controllers.

4. Add state encoders and decoders to view controllers with state.

5. Implement `UIDataSourceModelAssociation` for the table view controller.

This example represents a fairly common set of problems, but it doesn't cover every situation that state restoration can handle. For full details, see "State Preservation and Restoration" in the *iOS App Programming Guide* and the other references in the "Further Reading" section at the end of the chapter.

While implementing state preservation, always remember that state preservation is a way to store only *state*. You never use it to store *data*. The information stored in state preservation may be thrown away at any time (and the OS is somewhat aggressive in throwing state preservation away if it suspects a problem). Do not encode actual data objects in the state preservation system. You need to encode identifiers that will allow you to later fetch those data objects. I explain this topic more in the section "State Encoders and Decoders," later in this chapter.

Testing State Restoration

Before adding state restoration, you need to understand how programs behave without it and how to test it to make sure it's working. When the user presses the Home button, your app may not immediately terminate. It typically moves to the background state and later may be moved to the suspended state. In either case, however, your program is still in memory. When the user returns to your application, it typically resumes exactly where it left off. The same screen is showing, any text the user has entered is still in the text fields, and so on. The goal of state restoration is to achieve this same user experience even though the program terminates while in the background. A common error while implementing state restoration is mistaking this default behavior for your restoration code.

To properly test state restoration, you need to make sure that the application actually terminates, but you also need to make sure that it terminates in a way that allows state restoration to happen. For state restoration to run, three things must be true:

- **State preservation must have successfully completed.** This means that the application needs to have entered the background before being terminated.

- **The application must not have been forced to quit.** If you double-tap the Home button and swipe the application out of the application list, the state preservation files are deleted. The assumption is that if the user force-quits the app, there may have been a serious bug and state restoration might create an endless loop.

- **The application must not have failed launching since the last state preservation.** This means that if your application terminates during startup, the state preservation information is deleted. Otherwise, a small bug in your state restoration code could lead to an unlaunchable app.

Now that you have these points in mind, here is the correct way to test state restoration:

1. Run the app in Xcode and navigate to whatever situation you want to test.

2. Press the Home button in the simulator or on the device.

3. Stop the app in Xcode.

4. Relaunch the app.

I also recommend adding some logging to `application:willEncodeRestorableStateWithCoder:` and `application:didDecodeRestorableStateWithCoder:` in the application delegate just so you can see when state preservation and restoration are running.

For iOS 7, Apple provides a mobileconfig file that prevents a device from deleting its restoration files when the application is force-quit. It also provides additional state restoration logging and a tool to read state restoration files. At `https://developer.apple.com/downloads`, **search for "restorationArchiveTool."**

Opting In

State preservation doesn't automatically apply to old programs. Requesting it is trivial. Just add the following to your application delegate:

AppDelegate.m (FavSpotsRestore)

```
- (BOOL)application:(UIApplication *)application
shouldSaveApplicationState:(NSCoder *)coder {
  return YES;
}

- (BOOL)application:(UIApplication *)application
shouldRestoreApplicationState:(NSCoder *)coder {
  return YES;
}
```

You are able to add conditional logic here, but the need for this should be rare.

Startup Changes

Some subtle startup changes occur due to state restoration. You may need to perform some initialization work prior to state restoration and some after state restoration. The method `application:willFinishLaunchingWithOptions:` is called prior to state restoration. After state restoration is complete, Cocoa calls `application:didFinishLaunchingWithOptions:`. You need to think carefully about which logic should go in each.

You can save application-level state in `application:willEncodeRestorableStateWithCoder:`. View controllers manage the majority of state preservation, but you may need to store some global state. You can then restore this state in `application:didDecodeRestorableStateWithCoder:`. I explain encoding and decoding state in the section "State Encoders and Decoders."

This change doesn't apply to the FavSpots project.

Restoration Identifiers

Every subclass of `UIView` and `UIViewController` can have a restoration identifier. These identifiers do not have to be unique, but the full restoration path of every restorable object must be unique. The *restoration path* is a sequence of restoration identifiers, starting at the root view controller and separated by slashes like an URL. The restoration path gives the system a way to glue the UI back together, encoding things such as the entire navigation controller stack. For example, you might encounter the path `RootTab/FirstNav/Detail/Edit` indicating a tab bar controller with a navigation controller in one of its tabs. The navigation controller has a root view controller of Detail, and Edit has been pushed on top of that.

For custom, unique view controller classes, I recommend using the name of the custom class. For standard classes such as `UITabBarController`, I recommend using some more descriptive name to make them unique.

If you use storyboards, managing restoration identifiers is simple. For each view controller you want to save, set the Restoration ID in the Identity Inspector.

You need to assign a restoration identifier to every view controller and view that you want to save. This will typically be most of your view controllers. If you have a particular view controller that wouldn't make sense to come back to, you can leave its identifier blank, and the user is returned to the parent.

Most views should be reconfigured by their view controller, but sometimes it's useful to save the view's state directly. Several UIKit views support state restoration, specifically `UICollectionView`, `UIImageView`, `UIScrollView`, `UITableView`, `UITextField`, `UITextView`, and `UIWebView`. These views save their state, but not their contents. So a text field saves the selection range but does not save the text. See the documentation for each view for information on exactly what state it saves.

In FavSpots, add restoration identifiers to all the view controllers, the table view, the text field, and both text views.

State Encoders and Decoders

The heart of state restoration is the encoding and decoding state. Much like `NSKeyedArchiver`, the state restoration system passes an `NSCoder` to each object via `encodeRestorableStateWithCoder:` and `decodeRestorableStateWithCoder:`.

Unlike `NSKeyedArchiver`, the goal of the restoration system is not to store the entire object. You want to store only the information required to reconstruct your state. For example, if you have a `PersonViewController` that displays a `Person` object, don't encode the `Person` object. You only encode a unique identifier that will allow you to look up that object in the future. If you're using iCloud syncing or a network data source of any kind, it's possible that the record you stored will have been modified, moved, or deleted by the time you restore. Your restoration system should deal gracefully with a missing object.

> If you're using Core Data, the best way to save references to your object is to use `[[obj objectID] URIRepresentation]`. After an object is saved, this identifier is guaranteed to be unique and persistent. However, if the object isn't saved yet, the `objectID` may be temporary. You need to check `[objectID isTemporary]` prior to relying on it. Generally, you save prior to state preservation. Luckily, `applicationDidEnterBackground:` is called prior to state preservation, so if you save your context there, all your identifiers are permanent.

It bears repeating that you need to save user-provided data in a persistent store, not in state preservation. If your user has started writing a long review comment, don't rely on the state preservation system to save that draft. State preservation data can be deleted at any time. If you can't put the data into the persistent store yet, at least write it to a file in `Library/Application Support` so you don't lose it.

Here are the changes for FavSpots in `MapViewController`. It needs to save the region being viewed and the tracking mode. The various `ptl_` methods on `NSCoder` are from a category to simplify encoding `MKCoordinateRegion`. Notice the use of `containsValueForKey:` to distinguish between a missing value and 0. You may be restoring state from a previous version of your program, so be tolerant of missing keys.

MapViewController.m (FavSpotsRestore)

```
- (void)encodeRestorableStateWithCoder:(NSCoder *)coder {
    [super encodeRestorableStateWithCoder:coder];
```

```
        [coder ptl_encodeMKCoordinateRegion:self.mapView.region
                                forKey:kRegionKey];
        [coder encodeInteger:self.mapView.userTrackingMode
                  forKey:kUserTrackingKey];
    }

-   (void)decodeRestorableStateWithCoder:(NSCoder *)coder {
        [super decodeRestorableStateWithCoder:coder];

        if ([coder containsValueForKey:kRegionKey]) {
            self.mapView.region =
            [coder ptl_decodeMKCoordinateRegionForKey:kRegionKey];
        }

        self.mapView.userTrackingMode =
        [coder decodeIntegerForKey:kUserTrackingKey];
    }
```

`TextEditViewController` just needs a `Spot` object. To protect the user's notes (which could be lengthy), `TextEditViewController` automatically saves them when the application resigns active. This happens prior to state preservation, which makes state preservation very simple. It just stores the URI for the object ID. Notice that in `decodeRestoreableStateWithCoder:`, you don't use an accessor. This method is similar to `init` in that you need to be very careful of side effects. As a general rule, add this method to the short list of places where direct ivar access is appropriate. Here are the additions to `TextEditViewController` and the helper methods to encode and decode `Spot`:

TextEditViewController.m (FavSpotsRestore)

```
-   (void)encodeRestorableStateWithCoder:(NSCoder *)coder {
        [super encodeRestorableStateWithCoder:coder];
        [coder ptl_encodeSpot:self.spot forKey:kSpotKey];
    }

-   (void)decodeRestorableStateWithCoder:(NSCoder *)coder {
        [super decodeRestorableStateWithCoder:coder];
        _spot = [coder ptl_decodeSpotForKey:kSpotKey];
    }
```

NSCoder+FavSpots.m (FavSpotsRestore)

```
-   (void)ptl_encodeSpot:(Spot *)spot forKey:(NSString *)key {
        NSManagedObjectID *spotID = spot.objectID;
        NSAssert(! [spotID isTemporaryID],
                @"Spot must not be temporary during state saving. %@",
                spot);

        [self encodeObject:[spotID URIRepresentation] forKey:key];
    }

-   (Spot *)ptl_decodeSpotForKey:(NSString *)key {
        Spot *spot = nil;
```

(continued)

```
    NSURL *spotURI = [self decodeObjectForKey:key];

    NSManagedObjectContext *
    context = [[ModelController sharedController]
                managedObjectContext];
    NSManagedObjectID *
    spotID = [[context persistentStoreCoordinator]
                managedObjectIDForURIRepresentation:spotURI];
    if (spotID) {
      spot = (Spot *)[context objectWithID:spotID];
    }

    return spot;
}
```

DetailViewController is more complicated. It includes an editable text field. Usually, I recommend saving any changes to the persistent store as I did in TextEditViewController, in which case saving state is simple. The name is very short, however, and losing it would not be a great burden on the user, so I demonstrate here how you can store this kind of data in the preservation system.

This demonstration is complicated because the value of the text field can be set two ways. If you're restoring, you use the restored value; if you're not restoring, you use the value from Core Data. This means you have to keep track of whether you're restoring, as shown here:

DetailViewController.m (FavSpotsRestore)

```
@property (nonatomic, readwrite, assign, getter = isRestoring) BOOL
 restoring;
...
- (void)encodeRestorableStateWithCoder:(NSCoder *)coder {
    [super encodeRestorableStateWithCoder:coder];
    [coder ptl_encodeSpot:self.spot forKey:kSpotKey];
    [coder ptl_encodeMKCoordinateRegion:self.mapView.region
                                forKey:kRegionKey];
    [coder encodeObject:self.nameTextField.text forKey:kNameKey];
}

- (void)decodeRestorableStateWithCoder:(NSCoder *)coder {
    [super decodeRestorableStateWithCoder:coder];
    _spot = [coder ptl_decodeSpotForKey:kSpotKey];

    if ([coder containsValueForKey:kRegionKey]) {
      _mapView.region =
       [coder ptl_decodeMKCoordinateRegionForKey:kRegionKey];
    }

    _nameTextField.text = [coder decodeObjectForKey:kNameKey];
    _restoring = YES;
}
...
- (void)configureView {
```

```
Spot *spot = self.spot;

if (! self.isRestoring || self.nameTextField.text.length == 0) {
  self.nameTextField.text = spot.name;
}

if (! self.isRestoring ||
      self.mapView.region.span.latitudeDelta == 0 ||
      self.mapView.region.span.longitudeDelta == 0) {
  CLLocationCoordinate2D center =
  CLLocationCoordinate2DMake(spot.latitude, spot.longitude);
  self.mapView.region =
  MKCoordinateRegionMakeWithDistance(center, 500, 500);
}

self.locationLabel.text =
[NSString stringWithFormat:@"(%.3f, %.3f)",
 spot.latitude, spot.longitude];
self.noteTextView.text = spot.notes;

[self.mapView removeAnnotations:self.mapView.annotations];
[self.mapView addAnnotation:
 [[MapViewAnnotation alloc] initWithSpot:spot]];

self.restoring = NO;
}
```

In general, saving the data directly to the persistent store is easier and less error-prone, but if you need the ability to cancel edits of very small amounts of data, the preceding approach is useful.

Table Views and Collection Views

`UITableView` and `UICollectionView` have a special problem for state restoration. The user basically wants the view to return to the same state it was left in, but the underlying data may have changed. The specific records being displayed may be in a different order, or records may have been added or removed. In all but the simplest cases, restoring a table view to its exact previous state may be impossible.

To make restoration predictable, UIKit restores table and collection views such that the first index shown represents the same record. So if a record representing "Bob Jones" was at the top of the screen when your application quit, that's the record that should be at the top of the screen when you resume. All the other records on the screen may be different, and even the "Bob Jones" record might display a different name, but it should be the same record.

To achieve this result, UIKit uses the protocol `UIDataSourceModelAssociation`. This protocol maps an index path to a string identifier. UIKit doesn't care how you perform this mapping. During state preservation, it asks you for the identifier for an index path. During state restoration, it asks you for the index path for an identifier. The index paths do not need to match, but the referenced object should. The easiest way to implement this capability is to use Core Data's `objectID` and take the `URIRepresentation`, which returns a string. Without Core Data, you need to devise your own mapping that is guaranteed to be consistent and unique.

MasterViewController.m (FavSpotsRestore)

```objc
- (NSString *)modelIdentifierForElementAtIndexPath:(NSIndexPath *)idx
                                            inView:(UIView *)view {
  if (idx && view) {
    Spot *spot = [self.fetchedResultsController objectAtIndexPath:idx];
    return [[[spot objectID] URIRepresentation] absoluteString];
  }
  else {
    return nil;
  }
}

- (NSIndexPath *)
indexPathForElementWithModelIdentifier:(NSString *)identifier
                                inView:(UIView *)view {
  if (identifier && view) {
    NSUInteger numberOfRows =
    [self tableView:self.tableView numberOfRowsInSection:0];
    for (NSUInteger index = 0; index < numberOfRows; ++index) {
      NSIndexPath *indexPath = [NSIndexPath indexPathForItem:index
                                                   inSection:0];
      Spot *spot = [self.fetchedResultsController
                    objectAtIndexPath:indexPath];
      if ([spot.objectID.URIRepresentation.absoluteString
           isEqualToString:identifier]) {
        return indexPath;
      }
    }
  }
  return nil;
}
```

> In iOS 6, these methods were a mandatory part of state restoration. In iOS 7, these methods are optional. If you don't implement them, table and collection views restore state using indexes rather than identifiers. This feature is convenient for static tables and other cases in which the indexes would never change unexpectedly.

Advanced Restoration

In the previous sections, I discussed the common problems facing state restoration. I assumed the use of an application using storyboards and with a typical view controller hierarchy. But state restoration can handle more complicated situations.

Although storyboards are definitely the easiest way to manage state restoration, you also can perform state restoration without them. First, you can set restoration identifiers in nib files rather than the storyboard. Alternatively, you can set the restoration identifier at runtime by calling `setRestorationIdentifier:`. Setting this to `nil` prevents preserving its state.

You can also change which class handles restoration. By default, the view controller is created using the storyboard, but you may require different initialization for restoration. You can set the `restorationClass` property to the `Class` of whatever object you want to construct your view controller. This is normally the view controller's class (such that view controllers create themselves). If this property is set, the view controller is re-created by calling `+viewControllerWithRestorationIdentifierPath:coder:` rather than pulling it from the storyboard. You are responsible for creating the view controller in that method and returning it. If you return `nil`, this view controller is skipped. The system still calls `decodeRestorableStateWithCoder:` on the resulting object, unless the object you return isn't of the same class as the object that was encoded (or a subclass of that object). See "Restoring Your View Controllers at Launch Time" in the *iOS App Programming Guide* for full details.

State preservation and restoration is a powerful system for simplifying this complicated problem. It relies heavily on your using UIKit in "normal" ways. If you use storyboards, Core Data, and don't get overly "clever" in how you use view controllers, it does much of the work for you.

Summary

iOS apps can be terminated at any time when they're in the background. It's up to you to create the illusion that your app is always ready and up to date. iOS 7 brings new tools such as background transfers and silent notifications. It also enhances and simplifies existing tools such as state restoration. At the same time, its more aggressive suspend behavior makes many kinds of background operations more challenging. Getting the most out of this new technology requires careful design and extensive testing. Be sure to factor this into your plans.

Further Reading
Apple Documentation

The following documents are available in the iOS Developer Library at `developer.apple.com` or through the Xcode Documentation and API Reference.

iOS App Programming Guide, "Executing Code in the Background"

iOS App Programming Guide, "State Preservation and Restoration"

WWDC Sessions

The following session videos are available at `developer.apple.com`.

WWDC 2011, "Session 320: Adopting Multitasking in Your App"

WWDC 2012, "Session 208: Saving and Restoring Application State on iOS"

WWDC 2013, "Session 204: What's New with Multitasking"

WWDC 2013, "Session 222: What's New in State Restoration"

WWDC 2013, "Session 705: What's New in Foundation Networking"

Chapter 12

REST for the Weary

At some point, most iOS applications have to communicate with a remote web server in one way or another. Some apps can run and be useful without a network connection, and web server communication might be short-lived (or even optional) for the application. Apps that fall into this category are those that sync data with a remote server when a connection is present, such as to-do lists.

Another set of apps needs nearly continuous network connectivity to provide any meaningful value to the user. These apps typically act as a mobile client for a web service. Twitter clients, foursquare, Facebook, and most apps you write fall into this category. This chapter presents some techniques for writing apps the right way for consuming a web service.

The iPhone can connect to the Internet from nearly anywhere. Most iOS apps use this capability, which makes these apps one of the best Internet-powered devices ever made. However, because this device is constantly on the move, connectivity, reception, or both can be poor. This poses a problem for iOS developers, who need to ensure that their apps' perceived response time remains more or less constant, as though the complete content were available locally. You do this by caching your data locally. Caching data means saving it temporarily so that it can be accessed faster than making a round trip to the server. That's easier said than done, and most apps don't get this right. This chapter shows you the caching techniques you can use to solve the problem of slow performance caused by poor or unavailable connectivity.

As of this writing, a quick search for Twitter in the App Store turns up tens of thousands of apps. Today, if you want to create the next Twitter client, you don't have to know anything about web services or the Twitter API. There are more than a dozen implementations of the Twitter API in Objective-C. The same is true for most public services such as Facebook's Graph API or Dropbox. So, instead of explaining how to build your next Twitter client, this chapter provides insights and best practices for designing your next iPhone app that consumes a generic, simple, and hypothetical web service.

Because the ideas and techniques presented here are generic, you can easily apply them to any projects you undertake. If you've been an iOS developer for at least one year, you may have already implemented a project like this, where your customer sends you documentation of his server APIs. You've been introduced to the server developer and probably have some control over negotiating the output format and error-handling stuff. In most cases, both the client and the server code were developed in tandem.

In addition to discussing the REST implementation and caching on iOS, this chapter provides some guidelines for the server that will help you achieve the following:

- Improve the code quality.
- Reduce development time.
- Improve code readability and maintainability.
- Increase the perceived performance of the app.

The Worldwide Web Consortium has identified two major classes of web services (W3C Web Services Architecture 2004): RESTful services that manipulate XML representation of web resources using a uniform set of stateless operations and arbitrary services that might expose any operation. SOAP and WSDL are in the second category. Web services used in 2013 are mostly RESTful, including but not limited to Twitter APIs, foursquare, and Dropbox. This chapter focuses on consuming a RESTful service in your application.

The REST Philosophy

The three most important features of a RESTful server are its *statelessness, uniform resource identification,* and *cacheability.*

A RESTful server is always stateless. This means every API is treated as a new request, and no client context is remembered on the server. Clients do maintain the state of the server, which includes but isn't limited to caching responses and login access tokens.

Resource identification on a RESTful server is done through URLs. For instance, instead of accepting a resource ID as a parameter, a REST server accepts it as a part of the URL. For example, `http://example.com/resource?id=1234` becomes `http://example.com/resources/1234`.

This method of resource identification and the fact that a RESTful server doesn't maintain the state of the client together allow clients to cache responses based on the URL, just as a browser caches web pages.

Response from a RESTful server is usually sent in a uniform, agreed-upon format, usually to decouple the client/server interface. The client iOS app communicates with a RESTful server through this agreed-upon data exchange format. As of today, the most commonly used formats are XML and JSON. The next section discusses the differences among the formats and the ways you can parse them in your app.

Choosing Your Data Exchange Format

Web services traditionally support two major kinds of data exchange format: JSON (JavaScript Object Notation) and XML (eXtensible Markup Language). Microsoft pioneered XML as the default data exchange format for its SOAP services, whereas JSON became an open standard described in RFC 4627. Although there are debates about which is superior, as an iOS developer, you need to be able to handle both kinds of data formats on your app.

Several parsers are available for both XML and JSON for Objective-C. The following sections discuss some of the most commonly used toolkits.

Parsing XML on iOS

You can do XML parsing using two kinds of parsers: a DOM parser or a SAX parser. A SAX parser is a sequential parser and returns parsed data on a callback as it steps through the XML document. Most SAX parsers work by taking in a URL as a parameter and giving you data as it becomes available. For example, the `NSXMLParser` foundation class has a method called `initWithContentsOfURL:`, which looks like this:

```
(id)initWithContentsOfURL:(NSURL *)url;
```

You essentially initialize a parser object with the URL, and the `NSXMLParser` does the rest. Parsed data becomes available through callback via delegate methods defined in `NSXMLParserDelegate`. The most commonly handled methods are

```
parserDidStartDocument:

parserDidEndDocument:

parser:didStartElement:namespaceURI:qualifiedName:attributes:

parser:didEndElement:namespaceURI:qualifiedName:

parser:foundCharacters:
```

Because the parser uses delegation to return data, you need an `NSObject` subclass conforming to `NSXMLParserDelegate` for every object you're handling. This tends to make your code base a bit more verbose compared to a DOM parser.

> **Although you can use the one class (even your controller) to conform to the** `NSXMLParserDelegate`, **your code will be highly unmanageable when the XML format changes. Create a separate subclass of** `NSObject` **and make that class conform to** `NSXMLParserDelegate` **for clarity.**

A DOM parser, in contrast, loads the complete XML before it starts parsing. The advantage of using a DOM parser is its capability to access data at random using XPath queries, and there's no delegation as in the SAX model.

Mac OS X SDK has an Objective-C-based DOM parser, `NSXMLDocument`. iOS doesn't have an Objective-C-based DOM parser built in. You can use libxml2 or third-party Objective-C wrappers such as KissXML, TouchXML, or GDataXML built around libxml2. Some third-party libraries cannot write XML. A web service that responds in XML will mostly expect the post body in XML. In that case, you will require a library that can write XML (for example, from an `NSObject` or `NSDictionary`). KissXML and GDataXML are two good libraries. For a complete comparison, check the link to "How To Choose The Best XML Parser for Your iPhone Project" in the "Further Reading" section.

If you are using KissXML, the most commonly used classes are `DDXMLDocument` and `DDXMLNode`, and the commonly used methods are `initWithXMLString:options:error:` in `DDXMLDocument` and `elementWithName:stringValue:` in `DDXMLNode`.

Using a DOM parser makes your code cleaner and easier to read. Although this comes at the expense of execution time for handling web service requests, the effect is minor because DOM parsers become slower only for documents larger than a megabyte or so. A web service response generally is less than that. Any performance gain you get is negligible compared to the time of the network operation. These performance gains make a lot of sense when you're parsing XML from your resource bundle. For handling web service requests, I always recommend a DOM parser.

To learn more about XML performance, download and test the XML Performance app published by Apple and Ray Wenderlich (see the "Further Reading" section at the end of this chapter).

Parsing JSON on iOS

The second data exchange format is JSON, which is much more commonly used than XML. Although Apple has a JSON processing framework, it was a private API in iOS 4 and the Mac Snow Leopard, and wasn't available for general use. With iOS 5, Apple introduced `NSJSONSerialization` (Apple's JSON parsing and serializing framework) that you can use for parsing.

You also can choose from plenty of third-party JSON processing frameworks. The most commonly used frameworks by far are SBJson, TouchJSON, YAJL, and JSONKit. (See the "Further Reading" section for the links to download these frameworks.) Almost all frameworks have category extensions on `NSString`, `NSArray`, and `NSDictionary` to convert to and from JSON. The code samples in this chapter use Apple's own `NSJSONSerialization`. JSONKit is not recommended if you are planning to deploy on 64-bit architectures like the iPhone 5s. As of this writing, JSONKit, for performance reasons, still accesses the object's `isa` pointer instead of using `objc_getClass`. The `isa` pointer is a tagged pointer in 64-bit runtime. I highly recommend using `NSJSONSerialization`.

> When you're choosing a library for your app, you might have to do some performance evaluation. You can compare the frameworks using the open source test project json-benchmarks on GitHub (see the "Further Reading" section for the link to this tool). Because all five (SBJson, TouchJSON, YAJL, JSONKit, and `NSJSONSerialization`) are actively developed, every library is equally good, and there's no one best library as of this writing. Keep a close eye on them and be ready to swap frameworks if one seems superior to another. Usually, swapping a JSON library doesn't require monumental refactoring, because in most cases, it involves changing the class category extension methods.

XML Versus JSON

Source code fragments in this chapter are based on using JSON. You'll discover how to design your classes to make it easy to add XML support without affecting the rest of the code base. In every case, JSON processing on iOS is an order of magnitude easier than using XML. So if your server supports both XML and JSON formats, choosing JSON is a wise decision. If your back-end code is not yet written, start by supporting JSON initially.

Designing the Data Exchange Format

It's essential to keep in mind that I am talking about data exchange between client and server. The most common mistake iOS developers make is to think of JSON as some arbitrary data sent by the server in response to an API call. Although that's true to some extent, a quick look at what happens on the server will give you a better picture of what JSON actually is.

Internally, most servers are coded using some object-oriented programming language. Whether it's Java, Scala, Ruby, or C# (even PHP and Python support objects to some extent), any data you need on your iOS app will likely be an object on the server as well. Whether the object is an ORM-mapped entity (ORM stands for *object relational mapping*) or a business object is of little importance. Just call them model objects, and these objects are serialized to JSON only at the transport level. Most object-oriented languages provide interfaces to serialize objects, and developers usually harness this capability to convert their objects to JSON. This means the JSON you see on the response is just a different representation of the objects (or object list) on the server.

Keep this concept in mind while writing your code, and you will probably create model objects for every equivalent server model object. When you do that, you need not worry about changes affecting your code later. Refactoring will be far easier.

Rather than thinking in terms of JSON strings, it makes more sense to think in terms of objects and (RESTful) resources. Design and develop your code such that you always reconstruct model objects for every object on the server. When the reconstructed objects on your iOS app match 100% with the objects on the server, the goal of data exchange is attained, and your app will be easier to maintain and update.

In short, think of JSON as a data exchange format instead of a language with a bunch of syntax. Consider documenting the data exchange contracts on an object/resource basis rather than as primitive data types. foursquare's developer documentation is a good example. In fact, I recommend using foursquare's documentation as the starting point. These objects in turn become the model objects for your app. You see this in detail in "Key Value Coding JSONs," later in this chapter, and you look at how to convert JSON dictionaries into models by using Objective-C's key-value coding/observing (KVC/KVO) mechanism.

Model Versioning

In the past, at least from the late 1990s or early 2000s until the first iPhone was launched in 2007, most client/server development happened in tandem with a web-based interface. Native clients were not commonly used. The client app running on the web browser was always deployed together with the server. As such, you didn't really need to handle versioning in your models. However, on iOS, deploying the client requires that the app be physically installed on your user's device. Deploying it could take days or months, so you should also handle situations when the server is accessed with an older client. How many older versions of the client you want to support depends on your business goals. As an iOS developer, you should probably build support for catering to those business needs. Using class clusters on your iOS app is one way to do that. You find out more about this design in the "Handling Additional Formats Using Categories" later in this chapter.

A Hypothetical Web Service

This chapter describes a hypothetical app concept, the iHotelApp, and you develop the iOS code for it. Later, in the "Caching" section, you revisit this app and add a caching layer.

Assume that you're in charge of developing an iOS app for a restaurant. The restaurant uses iPads to take orders. Orders can be placed directly with waiters who enter the orders into their iPads. Customers can also directly place orders using the kiosks (dedicated iPads running your app) at their tables. Here's a brief description of the top-level functionalities of the app:

- Customer orders are sent to the remote servers based on the customers' table numbers, whereas waiters pick a table number along with every order they send through their own login accounts. So, it's clear that there are two kinds of login/authentication mechanisms. One is the traditional username/password-based type, and the other is based on customer table numbers. In all cases, the server exchanges an access token for given authentication information. The important point is that you need to develop one code base that caters to both types of login. After you log in, every web service requires you to send an access token with every subsequent call you make.

This requirement translates to the `/loginwaiter` and `/logintable` web service endpoints.

Both of these endpoints return an access token. In the iOS client implementation, I show you how to "remember" this access token and send it along with every request.

■ Customers should be able to see the menu, along with the details of every menu item including the photos/videos of the food and ratings left by other customers.

This requirement translates to a web service `/menuitems` endpoint and a `/menuitem/<itemid>` that returns a JSON object that will be modeled as a `MenuItem` object.

In the iOS implementation, you discover how to map the JSON keys to your model object with as little code as possible by making use of Objective-C's most powerful technique: key-value coding (KVC).

■ Customers should be able to submit reviews of an item.

This requirement translates to a web service endpoint `/menuitem/<itemid>/review`.

In these cases, some iOS apps show a floating heads-up display (commonly known as HUD) that prevents users from doing any operation until the review is posted. This approach is clearly bad from a user experience perspective. You see how to post reviews in the background without showing a modal HUD.

Although this app has other requirements, these three cover the most commonly used patterns when talking to a web service.

Important Reminders

Keep these essential points in mind as you build your app:

■ **Never make a synchronous network call.** Even if they're on a background thread, synchronous calls don't report progress. Another reason is that to cancel a synchronous request running on a background thread, you have to kill the thread, which is again not a good idea. Additionally, you will not be able to control the number of network calls in your app, which is critical to the performance of your app. You find more tips to improve performance on iOS later in this chapter.

■ **Avoid using NSThread or GCD-based threading directly for network operations (unless your project is small and has just a couple of API calls).** Running your own threads or using GCD has some caveats, as just explained.

■ **Use `NSOperationQueue`-based threading instead.** `NSOperationQueue` helps with controlling the queue length and the number of concurrent network operations. GCD-based threads also cannot be cancelled after the block has been dispatched.

Now, it's time to start designing the iOS app's web service architecture.

RESTfulEngine Architecture (iHotelApp Sample Code)

iOS apps traditionally use model-view-controller (MVC) as the primary design pattern. When you're developing a REST client in your app, you need to isolate the REST calls to their own class. The stateless nature of REST and its cacheable nature can be best applied when it's written in its own class. Moreover, it also provides a layer of isolation (which is also good for unit testing) and helps keep your controller code cleaner.

To get started, you choose a network management framework.

NSURLConnection Versus Third-Party Frameworks

Apple provides classes in `CFNetwork.framework` and provides a Foundation-based `NSURLConnection` for making asynchronous requests. However, for developing RESTful services, you need to customize those classes by subclassing them. Rather than reinvent what's already available for the development of web services, I recommend using `MKNetworkKit` (see the "Further Reading" section at the end of this chapter). `MKNetworkKit` encapsulates many often-used features such as basic or digest authentication, form posts, and uploading or downloading files. Another important feature it provides is an `NSOperationQueue` encapsulation, which you can use to queue network requests.

> **My advice is generally to refrain from using third-party code when you're developing for iOS. However, some components and frameworks are worth using. As far as possible, avoid third-party code that is heavily interdependent.** `MKNetworkKit` **is a block-based wrapper around** `NSURLConnection` **that doesn't bloat your code base while providing powerful features, most importantly, caching. You learn the benefits of caching your responses later in section Caching. Other similar frameworks that you can consider are AFNetworking or RestKit.**

The code sample provided in the download files for this chapter uses `MKNetworkKit`. You can find this code in the Chapter 12 directory (iHotelApp) on the book's website.

> **Note that the code download for this chapter is quite large. The chapter provides important code snippets, and you should look at the corresponding files. Open the project in Xcode to better understand the code and the architecture.**

> **The necessary server component for the RESTfulEngine is hosted on the book's website at** `iosptl.com`**.**

Creating the RESTfulEngine

The `RESTfulEngine` is the heart of the iHotelApp. This class encapsulates every call to the web service standalone class, which handles your network calls. Data should be passed from `RESTfulEngine` to view controllers only as `Model` objects instead of JSON or `NSDictionary` objects. (I discuss the process of creating model classes in the next section.) Now, what should happen when there's a back-end-related error? Communicating errors from `RESTfulEngine` to the view controller is covered in the subsequent section. Here are the first two important steps:

1. Add `MKNetworkKit` code to your project. You can add it as a submodule and drag the relevant files or add a cocoapod dependency to your project.

2. Create a RESTfulEngine object of type MKNetworkEngine in your project's AppDelegate. For a demo implementation, refer to this chapter's source code on the book's website.

This `RESTfulEngine` encapsulates most of the network-related mundane tasks, such as managing the concurrent queue, showing the activity indicator, and so on.

The `RESTfulEngine` object automatically changes the maximum number of concurrent operations to 6 when on WiFi and to 2 when on carrier data. This change improves the performance of your REST client considerably. I explain this later in the section "Tips to Improve Performance on iOS."

Refrain from writing or storing state variables in your application delegate. Doing so is almost as bad as using global state variables. However, pointers to commonly used modules such as `managedObjectContext` or `persistentStoreCoordinator` or your own `networkEngine` are fine.

Adding Authentication to the RESTfulEngine

Now that the class is ready, you can add methods to handle web service calls: first and foremost, authentication. `MKNetworkKit` provides wrapper methods for a variety of authentication schemes, including, but not limited to, HTTP Basic Authentication, HTTP Digest Authentication scheme, NIL Authentication, and so on. I don't go through the details of the authentication mechanisms in this book, but for the sake of simplicity, assume that you exchange an access token with the server by sending the username and password to the /loginwaiter request or to the /logintable request. You need to define macros for these URL endpoints. Add the following code to the RESTfulEngine class header file:

The Constants in RESTfulEngine.h

```
#define LOGIN_URL @"loginwaiter"
#define MENU_ITEMS_URL @"menuitem"
```

Next, create a property in RESTfulEngine to hold the access token and then create a new method, `loginWithName:password:onSucceeded:onError:`, as in the following code:

The init Method (and Property Declaration) in RESTfulEngine.h

```
@property (nonatomic, strong) NSString *accessToken;

-(id) loginWithName:(NSString*) loginName
          password:(NSString*) password
       onSucceeded:(VoidBlock) succeededBlock
           onError:(ErrorBlock) errorBlock;
```

The init Method (and Property Declaration) in RESTfulEngine.m

```
-(RESTfulOperation*) loginWithName:(NSString*) loginName
          password:(NSString*) password
       onSucceeded:(VoidBlock) succeededBlock
           onError:(ErrorBlock) errorBlock
{
```

```
RESTfulOperation *op = [self operationWithPath:LOGIN_URL];

[op setUsername:loginName password:password basicAuth:YES];
[op onCompletion:^(MKNetworkOperation *completedOperation) {

  NSDictionary *responseDict = [completedOperation responseJSON];
  self.accessToken = [responseDict objectForKey:@"accessToken"];
  succeededBlock();
} onError:^(NSError *error) {

  self.accessToken = nil;
  errorBlock(error);
}];

[self enqueueOperation:op];
return op;
}
```

This code completes your web service call. Now you need to notify the caller (which is usually the view controller) about the outcome of the web service call. You use blocks for this task.

Adding Blocks to the RESTfulEngine

Every web service call in this `RESTfulEngine` mandates the caller to implement two block methods: one for notifying a successful call and another for error notification.

Block Definitions

```
typedef void (^VoidBlock)(void);
typedef void (^ModelBlock)(JSONModel* aModelBaseObject);
typedef void (^ArrayBlock)(NSMutableArray* listOfModelBaseObjects);
typedef void (^ErrorBlock)(NSError* engineError);
```

You use the first block type to notify success without passing additional information. The second block type is to notify success and send a model object.

The `RESTfulEngine` class implementation is now complete for the first method, `loginWithName:pass word:onSucceeded:onFailure`. You can call this method from the view controller (which is usually the login page that shows the username and password fields):

Login Button Event Handling in iHotelAppViewController.m

```
-(IBAction) loginButtonTapped:(id) sender
{
  [AppDelegate.engine loginWithName:@"mugunth"
                           password:@"abracadabra"
                        onSucceeded:^{

                            [[[UIAlertView alloc]
                                initWithTitle:
                                NSLocalizedString(@"Success", @"")
```

(continued)

```
                    message:NSLocalizedString(@"Login successful", @"")
                   delegate:self
        cancelButtonTitle:NSLocalizedString(@"Dismiss", @"")
        otherButtonTitles: nil] show];
        } onError:^(NSError *engineError){
          [UIAlertView showWithError:engineError];
          }];
        }
```

Thus, with just a few lines of code, you're able to implement the login functionality of the web service.

Remember the access token? If your access token is simply a string, you can store it in keychain or in `NSUserDefaults`. Storing it in keychain is more secure than storing it in `NSUserDefaults`. You learn more about security in Chapter 14. The easiest and probably the cleanest way to store your keychain is to write a custom synthesizer for `accessToken` like the one shown here:

Access Token Custom Accessor in RESTfulEngine.m

```
-(NSString*) accessToken
{
    if(!_accessToken)
    {
        _accessToken = [[NSUserDefaults standardUserDefaults]
                          stringForKey:kAccessTokenDefaultsKey];
    }

    return _accessToken return_accessToken;
}
-(void) setAccessToken:(NSString *) aAccessToken
{
    _accessToken = _accessToken = aAccessToken;
    [[NSUserDefaults standardUserDefaults] setObject:self.accessToken
       forKey:kAccessTokenDefaultsKey];
    [[NSUserDefaults standardUserDefaults] synchronize];
}
```

If your web server sends user profile information at login, to cache the data, you may need a mechanism that is bit more sophisticated than `NSUserDefaults`. You can use `Keyed Archiving` or Core Data for that purpose.

Whew! That completes your first endpoint, but you're not done yet! Next, you create a second endpoint, `/menuitems`, which is used to download a list of menu items from the server.

Authenticating Your API Calls with Access Tokens

In most web services, every call after login is probably protected and can be accessed only by passing the access token. Instead of sending the access token in every method, you can write a cleaner factory method in your `RestfulEngine` that creates a request object. This request object can then be filled with parameters specific to the call.

Overriding Methods to Add Custom Authentication Headers in RESTfulEngine.m

MKNetworkKit (and most other third-party networking frameworks) already provides a factory method, operationWithURLString:params:httpMethod:, that internally calls prepareHeaders: and allows you to pass additional headers to a request. You override the method prepareHeaders: and add your custom HTTP Header fields including the Authorization header to add your access token to the request.

Every network operation created on an engine calls this method. Be sure to call the superclass implementation after you're done with it. With techniques like this, you will never again have a buggy API call because you accidentally forgot to set the access token.

```
-(void) prepareHeaders:(MKNetworkOperation *)operation {
if(self.accessToken)
    [operation setAuthorizationHeaderValue:self.accessToken
    forAuthType:@"Token"];
[super prepareHeaders:operation];
}
```

Note that this prepareHeaders: method can also have additional headers set depending on your web service requirements. If your web service requires you to turn on gzip encoding for all calls or to send the application version number and the device-related information on the HTTP header, this method is the best place to add code for adding these additional header parameters.

Now, it's time to add a method to the RESTfulEngine class for fetching menu items from the server:

Method to Fetch the List of Menu Items in RESTfulEngine.m

```
-(RESTfulOperation*) fetchMenuItemsOnSucceeded:(ArrayBlock) succeededBlock
                                       onError:(ErrorBlock) errorBlock
{
  RESTfulOperation *op = (RESTfulOperation*) [self
  operationWithPath:MENU_ITEMS_URL];
[op onCompletion:^(MKNetworkOperation *completedOperation) {
    // convert the response to model objects and invoke succeeded block
} onError:errorBlock];
  [self enqueueOperation:op];
  return op;
}
```

Custom parameters passed to your API should be added to this method. Your view controller code remains clean of unnecessary strings/dictionaries and URLs.

Canceling Requests

View controllers that need to display the information from your web service call methods like fetchMenuItems: on the RESTfulEngine. To ensure that the view controllers play nicely with system resources, the view controller is responsible for canceling any network operation it creates when the user navigates out of the view. For example, tapping the Back button means that even if the request returns, the response is not used. Canceling the request at this point means that other requests queued in the RESTfulEngine get a chance to run, and your subsequent views' requests are executed faster.

To enable this behavior, every method that is written on your `RESTfulEngine` class should return the operation object back to the view controller. Canceling a running operation speeds up the execution waiting time for the request submitted by the next view. A good example of this scenario on the foursquare app is the user tapping on a profile view and then tapping on the Mayorship button. In this case, the profile view submits a request to fetch the user's profile, but the user has already navigated to the Mayorship view without viewing the profile. It's now the responsibility of the profile view to cancel its request. Canceling the profile fetch request naturally speeds up the Mayorship fetch request by freeing up the bandwidth. This is applicable not only to foursquare but also to most web service apps you develop.

Request Responses

When you call the `fetchMenuItems:` method, the server's response is a list of menu items. In the previous web service call example, the response was an access token, a simple string, so you didn't need to design a model. In this case, you create a model class. Assume that the JSON returned by the server is in the following format:

```
{
"menuitems" : [{
  "id": "JAP122",
  "image": "http://d1.myhotel.com/food_image1.jpg",
  "name": "Teriyaki Bento",
  "spicyLevel": 2,
  "rating" : 4,
  "description" : "Teriyaki Bento is one of the best lorem ipsum dolor sit",
  "waitingTime" : "930",
  "reviewCount" : 4
}]
```

One easy way to create a model from a JSON is to write verbose code to fill in your model class with the JSON. The other, much more elegant way is to piggyback on Objective-C's arguably most important feature: key-value coding. The JSONKit classes (or any other JSON parsing framework, including Apple's `NSJSONSerialization`), which were discussed earlier, convert a JSON-formatted string into an `NSMutableDictionary` (or an `NSMutableArray`). In this case, you get a dictionary with an entry, `menuitems`. The call shown in the following code can extract the menu items' dictionary from the response:

```
NSMutableDictionary *responseDict = [[request responseString]
                                   mutableObjectFromJSONString];
NSMutableArray *menuItems = [responseDict objectForKey:@_menuitems_];
```

Now that you have an array of menu items, you can iterate through them, extract the JSON dictionary of every `menuitem`, and use KVO to convert them into model objects. This process is covered in the next section. The server sends another dictionary entry called `status`. You learn more about that entry in the "Error Handling" section later in this chapter.

Key Value Coding JSONs

Before you start writing your first model class, you need to know a bit about the model class inheritance architecture. Any web service-based app includes more than one model. In fact, a count of ten models for a single app is not uncommon. Instead of writing the KVC code in ten different classes, you write a base class that does the bulk of KVC and delegates very little work to the subclasses. Call this base class `JSONModel`. Any model class in the app that models a JSON and needs JSON observing will inherit from this `JSONModel`.

> Because you are making copies and/or mutable copies of your model classes in this example, implement `NSCopying` and `NSMutableCopying` in this base class. Derived classes must override this base class implementation and provide their own deep copy methods.

To start, add a method called `initWithDictionary:` to the base class. Your `JSONModel.h` should look similar to the following:

JSONModel.h

```
@interface JSONModel : NSObject <NSCopying, NSMutableCopying>
- (id) initWithDictionary:(NSMutableDictionary*) jsonDictionary;
@end
```

Then implement the `initWithDictionary:` method:

JSONModel.m

```
- (id) initWithDictionary:(NSMutableDictionary*) jsonObject
{
    if((self = [super init]))
    {
        [self init];
        [self setValuesForKeysWithDictionary:jsonObject];
    }
    return self;
}
```

The important part of this procedure is the method `setValuesForKeysWithDictionary:`. This method is a part of Objective-C KVC that matches each property in the class that has the same name as a key in the dictionary and sets the property's value to the value of that entry. Most importantly, if `self` is a derived class, it automatically matches the derived class properties and sets their values. There are some exception cases to be handled; they are covered shortly.

Voila! With just one line of code, you've "mapped" the JSON into your model class. But will everything work automatically when you have a derived class? Isn't there a catch here? Before going into the details, you need to understand how the method `setValuesForKeysWithDictionary:` works. Your `MenuItem` dictionary looks like this:

```
"id": "JAP122",
"image": "http://d1.myhotel.com/food_image1.jpg",
"name": "Teriyaki Bento",
"spicyLevel": 2,
```

```
"rating" : 4,
"description" : "Teriyaki Bento is one of the best lorem ipsum dolor sit",
"waitingTime" : "930",
"reviewCount" : 4
```

When you pass this dictionary to the `setValuesForKeysWithDictionary:` method, it sends the following messages along with their corresponding values: `setId`, `setImage`, `setName`, `setSpicyLevel`, `setRating`, `setDescription`, `setWaitingTime`, and `setReviewCount`. So a class modeling this JSON will implement these methods. The easiest way to implement this is to use Objective-C's built-in `@property`, so your `MenuItem.h` model class looks like the following:

MenuItem.h

```
@interface MenuItem : JSONModel
@property (nonatomic, strong) NSString *itemId;
@property (nonatomic, strong) NSString *image;
@property (nonatomic, strong) NSString *name;
@property (nonatomic, strong) NSString *spicyLevel;
@property (nonatomic, strong) NSString *rating;
@property (nonatomic, strong) NSString *itemDescription;
@property (nonatomic, strong) NSString *waitingTime;
@property (nonatomic, strong) NSString *reviewCount;

@end
```

Note that the property names for `id` and `description` have been changed to `itemId` and `itemDescription`. The reason is that `id` is a reserved keyword and `description` is a method in `NSObject` that prints out the address of the object. To avoid conflicts, you must rename them. However, you need to handle these exception cases because the default implementation of the `setValuesForKeysWithDictionary:` method crashes with a familiar error message stating "This class is not key value coding-compliant for the key: id." To handle this case, KVC provides a method called `setValue:forUndefinedKey:`.

In fact, the default implementation of this method raises the `NSUndefinedKeyException`. Override this method in your derived class and set the values accordingly.

Your `MenuItem.m` now looks like this:

MenuItem.m

```
- (void)setValue:(id)value forUndefinedKey:(NSString *)key
{
    if([key isEqualToString:@"id"])
        self.itemId = value;
    if([key isEqualToString:@"description"])
        self.itemDescription = value;
    else
        [super setValue:value forKey:key];
}
```

To avoid crashes in the future because of spurious keys in JSON and be a bit more defensive in your programming style, you can override this setValue:forUndefinedKey: method in the base class, JSONModel.m, like this:

```
- (void) setValue: (id) value forUndefinedKey: (NSString *) key {
    NSLog (@"Undefined Key: %@", key);
}
```

Now, in the fetchMenuItems:onSuccceeded:onError: method of the RESTfulEngine, add the following code to convert the JSON responses to MenuItem model objects:

RESTfulEngine.m

```
NSMutableDictionary *responseDictionary = [completedOperation
responseJSON];
    NSMutableArray *menuItemsJson = [responseDictionary
                                    objectForKey:@"menuitems"];
    NSMutableArray *menuItems = [NSMutableArray array];
    [menuItemsJson enumerateObjectsUsingBlock:^(id obj, NSUInteger idx, BOOL
                                    *stop) {
        [menuItems addObject: [[MenuItem alloc] initWithDictionary:obj]];
    }];
    succeededBlock (menuItems);
```

As you see, you call the MenuItem init method with a JSON dictionary to initialize itself from the dictionary keys. In short, by overriding a method only for special cases, you have successfully mapped a JSON dictionary to your custom model, and this model is clean of any JSON key strings! That's the power of KVC. The code is also inherently defensive, in the sense that whenever a change occurs in JSON keys that the server sends (probably arising from a bug on the server side), you see NSLog statements displaying the wrong undefined key on the console, and you can probably notify the server developers or make changes to your client to support the new keys.

It's also a good idea to add methods for performing deep copy to your derived classes. Just override methods in NSCopying and NSMutableCopying, and you're done. Tools like Accessorizer available from the App Store and full-fledged code editors like AppCode by JetBrains can help you with that. (See the "Further Reading" section for a link to the app.)

> Github open sourced a model mapping library called Mantle that is quite feature rich and powerful; it is built on the techniques explained here.

List Versus Detail JSON Objects

A JSON object is a payload that gets transferred from the server to the client. To improve performance and reduce payload size, server developers commonly use two kinds of payload for the same object. One is a large payload format that contains all information about the object; the second is a small payload that contains information needed just to display the information on a list. For the example in this chapter, a minimum amount of information about the menu item is displayed on the listing page, and most of the other content, including images, photos, and reviews, is displayed on the detail page.

This technique goes a long way toward improving an iOS app's perceived performance. On the implementation side, the iOS app doesn't have to be changed for mapping two kinds of JSON. You get either a complete JSON or a JSON that fills your object partially. The code written to map the detailed JSON should work without any modification in this scenario. For example, the server can send the small payload JSON for /menuitems and a detailed payload for /menuitems/<menuitemid>. The detailed payload will contain exactly the same data plus the first page of reviews and links to the photos of the dishes, and so on.

Nested JSON Objects

In the example, every menu item will have an array of reviews left by the user. If you depend on the default implementation of KVC and declare an NSMutableArray property on your model, the KVC binding will set the array's value to an array of NSMutableDictionary. But what you actually want is to map it as an array of models, which means you have to map the dictionaries as well in a recursive fashion. You can handle this case by overriding the setValue:forKey: method.

Assume that the following represents the format of JSON sent by the /menuitems/<itemid> method:

```
{
"menuitems" : [{
  "id": "JAP122",
  "image": "http://d1.myhotel.com/food_image1.jpg",
  "name": "Teriyaki Bento",
  "spicyLevel": 2,
  "rating" : 4,
  "description" : "Teriyaki Bento is one of the best lorem ipsum dolor sit",
  "waitingTime" : "930",
  "reviewCount" : 4,
  "reviews": [{
    "id": "rev1",
    "reviewText": "This is an awesome place to eat",
      "reviewerName": "Awesome Man",
    "reviewedDate": "10229274633",
    "rating": "5"
  }]
}],
"status" : "OK"
}
```

This code is similar to what you've already seen, but it has one additional payload: an array of reviews. In a real-life scenario, there might be multiple such additions, such as a list of photos, a list of "likes," and so on. But for the sake of simplicity, just assume that the detailed listing of a menu item has only one additional piece of information, which is the array of reviews. Now, before overriding the setValue:forKey: method, create a model object for a review entry. This class's header file will look similar to the following one. The implementation contains nothing but synthesizers and overridden NSCopying and NSMutableCopying (deep copy) methods.

Review.m

```
@property (nonatomic, strong) NSString *rating;
@property (nonatomic, strong) NSString *reviewDate;
@property (nonatomic, strong) NSString *reviewerName;
@property (nonatomic, strong) NSString *reviewId;
@property (nonatomic, strong) NSString *reviewText;
```

Again, you can generate these accessors using tools such as Accessorizer and Objectify, both available on the App Store site. Because the JSON data for Review doesn't have special keys that might be in conflict with Objective-C's reserved list, you don't even have to write explicit code for converting JSON to a review model. The initialization code is in the base class, and the KVC-compliant code is generated by the properties. That's the power of KVC.

Next, override the `setValue:forKey:` method in the `MenuItem` model to convert review dictionaries to `Review` models:

Custom Handling of KVC's setValue:forKey: Method in MenuItem.m

```
-(void) setValue:(id)value forKey:(NSString *)key
{
  if([key isEqualToString:@"reviews"])
  {
    for(NSMutableDictionary *reviewArrayDict in value)
    {
      Review *thisReview = [[[Review alloc]
  initWithDictionary:reviewArrayDict]
                              autorelease];
      [self.reviews addObject:thisReview];
    }
  }
  else
    [super setValue:value forKey:key];
}
```

The idea behind this code is to handle the `reviews` key of the JSON in a specialized way and to let the default superclass implementation handle the other keys.

Less Is More

You may have heard how great KVC and KVO are from various iOS veterans. Now that you understand them, you can put these concepts to use in your next app. You'll realize how powerful they are and how easily they enable you to write less code more efficiently.

Next, you learn how to handle server-side errors gracefully on the iOS client.

Error Handling

Recall that you saw a key called `status` in the JSON payload. Every web service has some way to communicate error messages to the client. In some cases, they're sent through a special key, such as `status`. In other cases, the web server sends an error key with more information about the actual error, and no such error key is sent when the API call is successful. This section shows you how to model both of these scenarios on iOS so that you write as little code as possible yet write it in a way that's easy to read and understand.

The first issue to understand is that not all API errors can be mapped to a custom HTTP error code. In fact, a server might throw errors even when everything is perfectly fine, but the user input is wrong. A website registration web service might throw an error if the user tries to register with an e-mail address that's already taken. This is just one example, and in most cases, you need specialized error handling for handling your own internal business logic errors. In this example, for instance, a missing menu item results in a 404 error. Most web services send a custom error message along with the 404 notice so that clients can understand what caused that 404.

A client implementation should not only report the HTTP error as an error message to the user, but also understand the internal business logic error for elegant error reporting and do proper error reporting. Otherwise, the only error you can ever show is, "Sorry, something bad happened; please try again later" and no one, especially your customer, is interested in seeing that kind of vague message. This section shows you how to handle such cases elegantly.

In the following steps, you subclass MKNetworkOperation to handle custom API errors:

1. Create a subclass of MKNetworkOperation. Name it RESTfulOperation. This subclass will have a property to store the business logic errors thrown from the server.

2. Create an NSError* property called restError in the subclass.

3. Override two methods to handle error conditions. The first method to override is the operationFailedWithError:.

Code in RESTfulOperation.m That Illustrates Error Handling

```
-(void) operationFailedWithError:(NSError *)theError
{
  NSMutableDictionary *errorDict = [[self responseJSON]
  objectForKey:@"error"];

  if(errorDict == nil)
  {
    self.restError = [[RESTError alloc] initWithDomain:kRequestErrorDomain
            code:[theError code]
        userInfo:[theError userInfo]];
  }
  else
  {
    self.restError = [[RESTError alloc] initWithDomain:kBusinessErrorDomain
                          code:[[errorDict
                          objectForKey:@"code"] intValue]
  userInfo:errorDict];
  }

  [super operationFailedWithError:theError];
}
```

Using this class, you check for the presence of the `"error"` JSON key and process it appropriately. The `failWithError` method will be called when there is an HTTP error. You handle non-HTTP, business logic errors in the same manner. As you saw earlier, not every business logic error can be mapped to an equivalent HTTP error code. Moreover, in some cases, a benign error may be sent along with your response, and the server may delegate the responsibility of treating that as an error or normal condition to the client. For handling both of these cases, you have to override another method, `operationSucceeded:`, as shown in the following code:

Code in RESTRequest.m That Illustrates Request Handling for Successful Conditions and Reports Business Logic Errors, if Any

```
- (void)operationSucceeded
{
  // even when request completes without a HTTP Status code, it might be a
     benign error

  NSMutableDictionary *errorDict = [[self responseJSON]
   objectForKey:@"error"];

  if(errorDict)
  {
    self.restError = [[RESTError alloc] initWithDomain:kBusinessErrorDomain
                                        code:[[errorDict
                           objectForKey:@"code"] intValue]
                                    userInfo:errorDict];
    [super operationFailedWithError:self.restError];
  }
  else
  {
    [super operationSucceeded];
  }
}
```

Both of these methods remember the business logic errors in the `restError` property of your subclassed request object. This enables the client to know both the HTTP error (by accessing the `RESTfulOperation`'s superclass's error object) and the business layer error, from the local property `restError`.

Because this handling is done on a subclass, the class `RESTfulEngine` doesn't have to do any additional error handling. All it gets is a nicely wrapped `NSError` object for both kinds of errors, HTTP or business logic. The view controller implementation will now be as simple as checking whether the error is `nil`; if it's not `nil`, show the message inside the `[[request restError] userInfo]`.

With that, you're ready to move on to a discussion of localization.

Localization

This section is about localizing web service-related error messages and not localizing your app. Adding internationalization and localization support to your app is explained in detail in Chapter 16.

Some implementations require you to localize error messages in multiple languages. For errors generated within the app, localizing is simple and can be handled using the Foundation classes and macros. For server-related errors, the previous implementation just showed the server errors on the UI. The best way to show localized errors is for the server to return errors in agreed-upon codes. The iOS client can then look into a localized string table and show the correct error for a given code.

RESTError.m

```
+ (void) initialize
{
  NSString *fileName = [NSString stringWithFormat:@"Errors_%@", [[NSLocale
                         currentLocale] localeIdentifier]];
  NSString *filePath = [[NSBundle mainBundle] pathForResource:fileName
                                                ofType:@"plist"];

  if(filePath != nil)
  {
    errorCodes = [[NSMutableDictionary alloc]
      initWithContentsOfFile:filePath];
  }
  else
  {
    // fall back to English for unsupported languages
    NSString *filePath = [[NSBundle mainBundle]
     pathForResource:@"Errors_en_US"
                        ofType:@"plist"];
    errorCodes = [[NSMutableDictionary alloc]
      initWithContentsOfFile:filePath];
  }
}
```

This `RESTError` class can again be initialized with the error dictionary you get from the server using the KVC technique covered earlier in this chapter. Override the `localizedDescription` and `localizedRecoverySuggestion` methods of `NSError` to provide proper user-readable error methods. If your web service provides error codes along with error messages, handling and showing error messages this way is better than showing the server error from the `userInfo` dictionary. Now replace the `NSError` you created in RESTRequest with this `RESTError`. Doing so ensures that for custom error codes sent from the server, the `localizedDescription` and `localizedRecoverySuggestion` come from the `Errors_en_US.plist` file.

Handling Additional Formats Using Categories

Assume that you've written and delivered your app, and for some reason, your client wants to move the server implementation to a Windows-based system, and the server now sends you XML data instead of JSON. With this current architecture in place, you can easily add an additional format parsing to your model. The recommended way to do so is to write a category extension on your model that has a method to convert XML to dictionaries. In short, write a method in your category extension to convert an XML tree into an `NSMutableDictionary` and pass this dictionary to the `initWithDictionary:` method, which you previously wrote. Category classes like this provide a powerful way to extend and add features to your existing implementation without creating unwanted side effects.

Tips to Improve Performance on iOS

The best tip for improving performance for a web service-based app is to avoid sending data that's not immediately necessary. Unlike a web-based app, an iPhone app has limited bandwidth, and in most cases, it is connected to a 3G network. Trying to implement techniques such as prefetching contents for what could be the user's next page only slows down your app.

Avoid multiple small AJAX-like API calls. In the "Creating the RESTfulEngine" section, earlier in this chapter, you learned that `MKNetworkEngine` sets the `networkQueue` to run six concurrent operations because most servers don't allow more than six parallel HTTP connections from a single IP address. Running more than six operations results in only the seventh and subsequent operations timing out.

On a 3G network, at least at the time of this writing, most network operators throttle the bandwidth and limit the number of outbound connections from a mobile device to two, usually one on an EDGE connection. `MKNetworkKit` automatically changes the number of concurrent operations based on the network that the device is connected to. In case you aren't using `MKNetworkKit`, you should check for reachability notifications using the Reachability classes provided by Apple and change the queue size dynamically as and when the connectivity changes. Again, this count of two on 3G and one on EDGE is not absolute, and you need to test your customer base's network and use the results accordingly.

If you have control over the server development, the following tips can help you get the best out of the iOS app you develop:

- A server that caters to a web-based client should almost always have multiple small web service calls that are usually performed using AJAX. On iOS, it's best to avoid these APIs and possibly use or develop a custom API that gives more customized data per call.

- Unlike a browser, most carrier networks throttle the number of parallel data connections. Again, it's safe to assume that you run not more than one network operation on an EDGE connection, not more than two parallel network operations on a 3G network, and not more than six on a WiFi connection.

Caching

Now you've learned how to write a web service app the right way. The next important features any web service app should have is effective caching.

Reasons for Going Offline

The main reason your app might need to work offline is to improve the perceived performance of the app. You go offline by caching your app's content. You can use two kinds of caching to make your app work offline. The first is on-demand caching, where the app caches request responses as and when they're made, much as your web browser does. The second is precaching, where you cache your contents completely (or a recent "n" items) for offline access.

Web service apps like the one you developed earlier in this chapter use on-demand caching techniques to improve the perceived performance of the app rather than to provide offline access. Offline access just happens to be an added advantage. Twitter and foursquare are great examples of this. The data that these apps bring in often becomes stale quickly. How often are you interested in a tweet that was posted a couple of days ago or in knowing where a friend was last week? Generally, the relevance of a tweet or a check-in is important only for a couple of hours but loses some or all of its importance after 24 hours. Nevertheless, most Twitter clients cache tweets, and the official foursquare client shows you the last state of the app when you open it without an active Internet connection.

You can even try this on your favorite Twitter client, Twitter for iPhone, Tweetbot, or whatever you prefer: Open a friend's profile and view his timeline. The app fetches the timeline and populates the page. While it loads the timeline, you see a loading spinner. Now go to a different page and come back again and open the timeline. You will see that it's loaded instantly. The app still refreshes the content in the background (based on when you previously opened it), but instead of showing a rather uninteresting spinner, it shows previously cached content, thereby making it appear fast. Without this caching, users would see the spinner for every single page, which would slowly frustrate them. Whether the Internet connection is fast or slow, you, as an iOS developer, are responsible for mitigating this effect and providing the perception that the app is loading fast. This effort goes a long way toward improving your customers' satisfaction and thereby boosting your app's ratings on the App Store.

The other kind of caching gives more importance to the data being cached and the ability to edit the cached items on the fly without connecting to the server. Examples include apps such as Google Reader clients, read-later apps such as Instapaper, and so on.

Strategies for Caching

The two caching techniques discussed in the previous section—on-demand caching and precaching—are quite different when it comes to design and implementation. With on-demand caching, you store the content fetched from the web service locally on the file system (in some format), and then for every request, you check for the presence of this data in the cache and perform a fetch from the server only if the data isn't available (or is stale). Hence, your cache layer will behave more or less like cache memory on your processor. The speed of fetching the data is more important than the data itself. On the other hand, when you precache, you save content locally for future access. With precaching, a loss of data or a cache-miss is not acceptable. For example, consider a scenario in which the user has downloaded articles to read while on the subway, only to find that they're no longer present on her device.

Apps such as Twitter, Facebook, and foursquare fall into the on-demand category, whereas apps such as Instapaper and Google Reader clients fall into the precaching category.

To implement precaching, you'll probably use a background thread that accesses data and stores it locally in a meaningful representation so that local cache can be edited without reconnecting to the server. Editing can be either "marking items as read" or "favoriting items" or a similar operation on the item. By *meaningful representation,* I mean you save your contents in a way that allows you to make these kinds of modifications locally without talking to the server and then are able to send the changes back when you're connected again. This capability is in contrast to apps such as foursquare where you cannot become a Mayor of a place without an Internet connection, although you can see the list of your Mayorships without an Internet connection (if it is cached). Core Data (or any structured storage) is one way to do this.

On-demand caching works like your browser cache. It allows you to view content that you've viewed or visited before. You can implement on-demand caching by caching your data models (to create a data model cache) when you open a view controller on demand rather than on a background thread. You can also implement on-demand caching when a URL request returns a successful (200 OK) response (to create a URL cache). There are advantages and disadvantages to the two methods, and I show you the pros and cons of each of them in the "Creating a Data Model Cache" and "Creating a URL Cache" sections later in this chapter.

A quick-and-dirty way of deciding whether to go for on-demand caching or precaching is to determine whether you might ever post-process any data after downloading it. Post-processing could be in the form of either user-generated edits or updates made to downloaded data, such as rewriting image links in an HTML page to point to locally cached images. If your app requires any of the aforementioned post-processing, you must implement precaching.

Storing the Cache

Third-party apps can save information only to the application's sandbox. Because cache data is not user-created, it should be saved to the `NSCachesDirectory` instead of the `NSDocumentsDirectory`. A good practice is to create a self-contained directory for all your cached data. In this example, you create a directory named `MyAppCache` in the `Library/caches` folder. You can create this directory by using the following code:

```
NSArray *paths = NSSearchPathForDirectoriesInDomains(NSCachesDirectory,
    NSUserDomainMask, YES);
NSString *cachesDirectory = [paths objectAtIndex:0];
cachesDirectory = [cachesDirectory
    stringByAppendingPathComponent:@"MyAppCache"];
```

The reason for storing the cache in the caches folder is that iCloud (and iTunes) backups exclude this directory. If you create large cache files in the Documents directory, they get uploaded to iCloud during backup and use up the limited space (about 5GB at the time of this writing) fairly quickly. You don't want to do that. You want to be a good citizen on your user's iPhone, right? `NSCachesDirectory` is meant for that purpose.

Precaching is implemented using a higher-level database such as raw SQLite or an object serialization framework such as Core Data. You need to carefully choose the technology based on your requirements. I offer suggestions on when to use a URL cache or data model cache and when to use Core Data in the "Which Caching Technique Should You Use?" section later in this chapter. Next, I show you the implementation-level details of a data model cache.

Implementing Data Model Caching

You implement data model caching using the `NSKeyedArchiver` class. To archive your model objects using `NSKeyedArchiver`, the model class must conform to the `NSCoding` protocol, as shown here:

NSCoding Protocol Methods
```
- (void)encodeWithCoder:(NSCoder *)aCoder;
- (id)initWithCoder:(NSCoder *)aDecoder;
```

When your models conform to `NSCoding`, archiving them is as easy as calling one of the following methods:

```
[NSKeyedArchiver archiveRootObject:objectForArchiving toFile:
 archiveFilePath];

[NSKeyedArchiver archivedDataWithRootObject:objectForArchiving];
```

The first method creates an archive file specified at the path `archiveFilePath`. The second method returns an `NSData` object. `NSData` is usually faster because it doesn't have any file-access overhead, but it is stored in your application's memory and will soon use up memory if it's not checked periodically. Periodic caching to flash memory on the iPhone is also not advisable because cache memory, unlike hard drives, comes with limited read/write cycles. You need to balance both in the best possible way. You learn in detail about caching using archives in the "Creating a Data Model Cache" section later in this chapter.

The `NSKeyedUnarchiver` class is used to unarchive your models from a file (or an `NSData` pointer). You can use either one of the following class methods, depending on where you have to unarchive from:

```
[NSKeyedUnarchiver unarchiveObjectWithData:data];

[NSKeyedUnarchiver unarchiveObjectWithFile:archiveFilePath];
```

These four methods come in handy when you are converting to and from serialized data.

Use of any of the `NSKeyedArchiver`/`NSKeyedUnarchiver` methods requires that your models implement the `NSCoding` protocol. However, doing so is so easy that you can automate implementing the `NSCoding` protocol using tools such as Accessorizer. (See the "Further Reading" section at the end of this chapter for a link to Accessorizer at the App Store.)

The next section explains strategies that can be used for precaching. You learned previously that precaching requires you to use a more structured data format. I introduce both Core Data and SQLite here.

Core Data

Core Data, as Marcus Zarra says, is more of an object serialization framework than just a database API:

> It's a common misconception that Core Data is a database API for Cocoa. . . . It's an object framework that can be persisted to disk (Zarra, 2009; Pg: 9).

> For a good, in-depth explanation of Core Data, read *Core Data: Apple's API for Persisting Data on Mac OS X* by Marcus S. Zarra (Pragmatic Bookshelf, 2009. ISBN 9781934356326).

To store data in Core Data, first create a Core Data model file and create your Entities and Relationships; then write methods to save and retrieve data. Using Core Data, you get true offline access for your app, such as Apple's built-in Mail and Calendar apps. When you implement precaching, you must periodically delete data that's no longer needed (stale); otherwise, your cache will start growing and hurt the app's performance. You synchronize local changes by keeping track of changesets and sending them back to the server. Many algorithms can be used for changeset tracking, and the one I recommend is the one used by the Git version control system. I don't cover syncing your cache with a remote server because that topic is beyond the scope of this book.

Using Core Data for On-Demand Caching

Although technically you can use Core Data for on-demand caching, I advise against using it this way. The benefit Core Data offers is individual access to the models' properties without unarchiving the complete data. However, the complexity of implementing Core Data in your app defeats the benefits; moreover, for on-demand cache implementation, you probably wouldn't require individual access to the models' properties.

Raw SQLite

SQLite can be embedded into your app by linking against the libsqlite3 libraries, but doing so has significant drawbacks. All sqlite3 libraries and object relational mapping (ORM) mechanisms are almost always slower than Core Data. In addition, although sqlite3 is thread-safe, the binary bundled with iOS is not. So unless you ship a custom-built sqlite3 library (compiled with the thread-safe flag), it becomes your responsibility to ensure that data access to and from the sqlite3 database is thread-safe. Because Core Data has so much more to offer and has thread-safety built in, I suggest avoiding native SQLite as much as possible on iOS.

> **The only exception for using Raw SQLite over using Core Data in your iOS app is when you have application-specific data in the resource bundle that is shared by all other third-party platforms your app supports—for example, a location database for an app that runs on iPhone, Android, BlackBerry, and, say, Windows Phone. But again, that's not caching, either.**

Which Caching Technique Should You Use?

Of the different techniques available to save data locally, three of them stand out: URL cache, data model cache (using NSKeyedArchiver), and Core Data.

On one hand, if you're developing an app that needs to cache data to improve perceived performance, you should implement on-demand caching (using a data model cache or URL cache). On the other hand, if you need your data to be available offline and in a more meaningful way so that editing offline data is possible, use a higher-level serialization such as Core Data.

Data Model Cache Versus URL Cache

On-demand caching can be implemented using either a data model cache or a URL cache. Both have advantages and disadvantages, and choosing which to use depends on the server implementation. A URL cache is implemented like a browser's cache or a proxy server's cache. It works best when your server is correctly designed and conforms to the HTTP 1.1 caching specifications. If your server is a SOAP server (or servers implemented like RPC servers, other than a RESTful server), you need to use data model caching. If your server adheres to the HTTP 1.1 caching specification, use URL caching. Data model caching allows the client (iOS app) to have control over cache invalidation, whereas when you implement a URL cache, the server dictates invalidation through HTTP 1.1 cache control headers. Although some programmers find this approach counterintuitive and complicated to implement, especially on the server, it's probably the right way to do caching. As a matter of fact, MKNetworkKit provides native support for the HTTP 1.1 caching standard.

Cache Versioning and Invalidation

When you cache data, you need to decide whether to support version migration. If you're using an on-demand caching technique, version migration might be necessary if you use a data model cache. But the easiest way is to delete the cache when the user downloads the new version because old data is not important. If, however, you have implemented precaching, chances are that you have cached multiple megabytes of data, and it only makes sense to migrate them to the new version. With Core Data, data migration across versions is easy (at least compared to raw SQLite).

When you're using URL cache-based on-demand caching, URL responses are stored against the URLs as raw data. Versioning never becomes a problem. A change in version is either reflected by a URL change or invalidated from the server through cache-control headers.

In the following sections, I show you how to implement the two different types of on-demand caching: data model caching (using `AppCache`) and URL caching (using `MKNetworkKit`). You can download the complete source code located in this chapter's files on the book's website.

Creating a Data Model Cache

In this section, you add on-demand caching to the iHotelApp that you created earlier by implementing a data model cache. You implement this capability using a data model cache. On-demand caching is done as and when the view disappears from the hierarchy (technically, in your `viewWillDisappear:` method). The basic construct of the view controller that supports caching is shown in Figure 12-1. You can get the complete code for AppCache Architecture from the downloaded source code for this chapter. From this point on, I assume that you've downloaded the code and have it available to use.

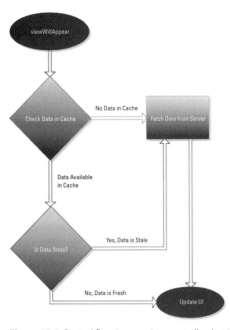

Figure 12-1 Control flow in your view controller that implements on-demand caching

In your `viewWillAppear` method, check your cache for the data necessary to display this view. If it's available, fetch it and update your user interface with cached data. Now, check whether your data from the cache is old. Your business rules should dictate what's new and what's old. If you decide that the content is old, show the data on the UI, and in the background, fetch data from the server and update the UI again. If the data is not available in the cache, fetch the data from the server while showing a loading spinner. After fetching data, update your UI.

The flowchart in Figure 12-1 assumes that what you show on the UI are models that can be archived. Implement the NSCoding protocol in the MenuItem model in the iHotelApp. The NSKeyedArchiver mandates that this protocol be implemented, as illustrated by the following code snippets:

NSCoding encodeWithCoder Method for the MenuItem Class (MenuItem.m)

```
- (void)encodeWithCoder:(NSCoder *)encoder
{
    [encoder encodeObject:self.itemId forKey:@"ItemId"];
    [encoder encodeObject:self.image forKey:@"Image"];
    [encoder encodeObject:self.name forKey:@"Name"];
    [encoder encodeObject:self.spicyLevel forKey:@"SpicyLevel"];
    [encoder encodeObject:self.rating forKey:@"Rating"];
    [encoder encodeObject:self.itemDescription forKey:@"ItemDescription"];
    [encoder encodeObject:self.waitingTime forKey:@"WaitingTime"];
    [encoder encodeObject:self.reviewCount forKey:@"ReviewCount"];
}
```

initWithCoder Method for the MenuItem Class (MenuItem.m)

```
- (id)initWithCoder:(NSCoder *)decoder
{
    if ((self = [super init])) {
        self.itemId = [decoder decodeObjectForKey:@"ItemId"];
        self.image = [decoder decodeObjectForKey:@"Image"];
        self.name = [decoder decodeObjectForKey:@"Name"];
        self.spicyLevel = [decoder decodeObjectForKey:@"SpicyLevel"];
        self.rating = [decoder decodeObjectForKey:@"Rating"];
        self.itemDescription = [decoder
            decodeObjectForKey:@"ItemDescription"];
        self.waitingTime = [decoder decodeObjectForKey:@"WaitingTime"];
        self.reviewCount = [decoder decodeObjectForKey:@"ReviewCount"];
    }
    return self;
}
```

As mentioned previously, you can generate the NSCoding protocol implementation using Accessorizer.

Based on the cache flow flowchart you saw in Figure 12-1, you have to implement the actual caching logic in the viewWillAppear: method. The following code added to viewWillAppear: implements that logic:

Code Snippet in the viewWillAppear: of Your View Controller That Restores Your Model Objects from Cache

```
NSArray *paths = NSSearchPathForDirectoriesInDomains(NSCachesDirectory,
    NSUserDomainMask, YES);
NSString *cachesDirectory = [paths objectAtIndex:0];
NSString *archivePath = [cachesDirectory
    stringByAppendingPathComponent:@"AppCache/MenuItems.archive"];

NSMutableArray *cachedItems = [NSKeyedUnarchiver
    unarchiveObjectWithFile:archivePath];
```

(continued)

```
  if(cachedItems == nil)
    self.menuItems = [AppDelegate.engine localMenuItems];
  else
    self.menuItems = cachedItems;
  NSTimeInterval stalenessLevel = [[[[NSFileManager defaultManager]
      attributesOfItemAtPath:archivePath error:nil]
  fileModificationDate] timeIntervalSinceNow];

  if(stalenessLevel > THRESHOLD)
    self.menuItems = [AppDelegate.engine localMenuItems];

  [self updateUI];
```

The logical flow of the caching mechanism is as follows:

1. The view controller checks for previously cached items in the archive file `MenuItems.archive` and unarchives it.

2. If `MenuItems.archive` is not present, the view controller makes a call to fetch data from the server.

3. If `MenuItems.archive` is present, the view controller checks the archive file modification date to determine how stale this cached data is. If it's old (as determined by your business requirements), fetch the data again from the server. Otherwise, display the cached data.

Next, the following code added to the `viewDidDisappear` method saves (as `NSKeyedArchiver` archives) your models to the `Library/Caches` directory:

Code Snippet in the viewWillDisappear: of Your View Controller That Caches Your Models

```
NSArray *paths = NSSearchPathForDirectoriesInDomains(NSCachesDirectory,
    NSUserDomainMask, YES);
NSString *cachesDirectory = [paths objectAtIndex:0];
NSString *archivePath = [cachesDirectory
stringByAppendingPathComponent:@"AppCache/MenuItems.archive"];

  [NSKeyedArchiver archiveRootObject:self.menuItems toFile:archivePath];
```

As the view disappears, you save the contents of the `menuItems` array to an archive file. Take care not to cache this if you didn't fetch from the server in `viewWillAppear:`.

So, just by adding fewer than ten lines in the view controller (and a bunch of Accessorizer-generated lines in the model), you add caching support to your app.

Refactoring

When you have multiple view controllers, the preceding code will probably get duplicated. You can avoid that duplication by abstracting out the common code and moving it to a new class called `AppCache`. This `AppCache` is the heart of the application that handles caching. By abstracting out common code to `AppCache`, you can avoid duplicated code in `viewWillAppear:` and `viewWillDisappear:`.

Refactor this code so that your view controller's `viewWillAppear/viewWillDisappear` block looks like the following code. The lines in bold show the changes made while refactoring, and I explain them after the code.

Refactored Code Snippet in the viewWillAppear: of Your View Controller That Caches Your Models Using the AppCache Class (MenuItemsViewController.m)

```objc
-(void) viewWillAppear:(BOOL)animated {

  self.menuItems = [AppCache getCachedMenuItems];
  [self.tableView reloadData];

  if([AppCache isMenuItemsStale] || !self.menuItems) {

    [AppDelegate.engine fetchMenuItemsOnSucceeded:^(NSMutableArray
     *listOfModelBaseObjects) {

      self.menuItems = listOfModelBaseObjects;
      [self.tableView reloadData];
    } onError:^(NSError *engineError) {
      [UIAlertView showWithError:engineError];
    }];
  }

  [super viewWillAppear:animated];
}

-(void) viewWillDisappear:(BOOL)animated {

  [AppCache cacheMenuItems:self.menuItems];
  [super viewWillDisappear:animated];
}
```

The `AppCache` class abstracts the knowledge of staleness from the view controller. It also abstracts exactly where the cache is stored. Later in this chapter, you modify this `AppCache` to introduce another layer of cache where the content is stored in memory.

Because the `AppCache` class abstracts out exactly where the cache is stored, you don't have to worry about copying and pasting code that gets the application's cache directory. In case your app is like the iHotelApp example, you also can easily add security to the cached data by creating subdirectories for every user. The helper method in `AppCache` then returns the cache directory that can be modified to return the correct subdirectory for the currently logged-in user. This way, data cached by user 1 will not be visible to user 2, who logs in later.

The complete code listing is available from the source code download for this chapter on the book's website.

Cache Versioning

The `AppCache` class you wrote in the preceding section abstracted out on-demand caching from your view controllers. When the view appears and disappears, caching happens behind the scenes. However, when you update the app, you might change your model classes, which means that any previously archived data will no longer be restored on your new models. As you learned earlier, in on-demand caching, your data is not that important, and you can delete it when you update the app. Next, I show you a code snippet that can be used to delete the cache directory when version upgrades are performed.

Invalidating the Cache

First, save the application's current version somewhere. NSUserDefaults is a candidate. To detect version updates, every time the app launches, check whether the previously saved version is older than the app's current version. If it is, delete the cache folder and resave the new version to NSUserDefaults. Following is the code for doing so. Add it to your AppCacheinit method.

Code Snippet in the AppCache Initialize Method That Handles Cache Versioning (AppCache.m)

```
+(void) initialize
{
  NSString *cacheDirectory = [AppCache cacheDirectory];
  if(![[NSFileManager defaultManager] fileExistsAtPath:cacheDirectory])
  {
    [[NSFileManager defaultManager] createDirectoryAtPath:cacheDirectory
    withIntermediateDirectories:YES
    attributes:nil
    error:nil];
  }

  double lastSavedCacheVersion = [[NSUserDefaults standardUserDefaults]
    doubleForKey:@"CACHE_VERSION"];
  double currentAppVersion = [[AppCache appVersion] doubleValue];

  if( lastSavedCacheVersion == 0.0f || lastSavedCacheVersion <
    currentAppVersion)
  {
    [AppCache clearCache];
    // assigning current version to preference
    [[NSUserDefaults standardUserDefaults] setDouble:currentAppVersion
      forKey:@"CACHE_VERSION"];
    [[NSUserDefaults standardUserDefaults] synchronize];
  }
}
```

Note that this code depends on a helper method that gets the application's current version. You can read the version from your app's Info.plist file using this block of code:

Code to Get the Current App Version from the Info.plist File (AppCache.m)

```
+(NSString*) appVersion
{
  CFStringRef versStr =
  (CFStringRef)CFBundleGetValueForInfoDictionaryKey
    (CFBundleGetMainBundle(), kCFBundleVersionKey);
  NSString *version = [NSString stringWithUTF8String:CFStringGetCStringPtr
                      (versStr,kCFStringEncodingMacRoman)];

  return version;
}
```

The preceding code calls a method to clear the cache directory. The following snippet illustrates that:

Code Snippet That Clears All Cached Files from the Cache Directory (AppCache.m)

```
+(void) clearCache
{
  NSArray *cachedItems = [[NSFileManager defaultManager]
                          contentsOfDirectoryAtPath:[AppCache
                          cacheDirectory] error:nil];

  for(NSString *path in cachedItems)
    [[NSFileManager defaultManager] removeItemAtPath:path error:nil];
}
```

Again, the cache invalidation and versioning issue is also abstracted out of the view controllers using the `AppCache` architecture. Now, go ahead and create an in-memory cache for the `AppCache` class. An in-memory cache improves the performance of caching drastically, but at the expense of memory. However, because on iOS, only one app runs in the foreground, this shouldn't be a problem.

Creating an In-Memory Cache

Every iOS device shipped so far has included flash memory, and this flash memory has one little problem: It has limited read-write cycles before it wears out. Although this limit is generally very high compared to the device's life span, it's still important to avoid writing to and reading from flash memory too often. In the previous example, you were caching directly to disk when the view was hidden and reading directly from disk whenever the view was shown. This behavior could tax the flash memory on users' devices. To avoid this problem, you can introduce another cache layer, which uses the device RAM instead of flash (`NSMutableDictionary`).

In the "Implementing Data Model Caching" section, you read about two methods for creating archives: one for saving them to a file and one for saving them as `NSData` objects. Here, you use the second method, which gives you an `NSData` pointer that you can store in an `NSMutableDictionary` rather than as flat files in the file system. The other advantage you get by introducing an in-memory cache layer is slightly higher performance when you archive and unarchive contents. Although this sounds complicated, it isn't really. In this section, you look at how to add a transparent in-memory cache to the `AppCache` class. (In-memory cache is transparent in the sense that the calling code—the ViewController—doesn't even know about its presence and doesn't need any code changes.) You also design a Least Recently Used (LRU) algorithm to save the cached data back to disk.

You follow the steps shown here to create the in-memory cache. These steps are explained in more detail in the following sections.

1. Add variables to hold your cached data in memory.

2. Limit the size of the in-memory cache and write the least recently used items to a file and remove it from in-memory cache. RAM is limited, and when you hit the limit, you get a memory warning. Failing to release memory when you receive this warning crashes your app. You obviously don't want that to happen, right? So you set a maximum threshold for the memory cache. When anything is added to the cache after it's full, the last used object (least recently used) should be saved to file (flash memory).

3. Handle memory warnings and write the in-memory cache to flash memory (as files).

4. Write all in-memory cache to flash memory (files) when the app is closed or quit or when it enters the background.

Designing the In-Memory Cache for AppCache

You start designing the `AppCache` class by adding the variables to hold the cache data. Add an `NSMutableDictionary` for storing your cache data, an `NSMutableArray` to keep track of recently used items, in chronological order, and an integer that limits the maximum size of this cache, as shown in the following code:

Variables in AppCache

```
static NSMutableDictionary *memoryCache;
static NSMutableArray *recentlyAccessedKeys;
static int kCacheMemoryLimit;
```

Now you have to make changes to the `cacheMenuItems:` and `getCachedMenuItems` methods in `AppCache` to save the model objects transparently to this in-memory cache, as shown here:

```
+(void) cacheMenuItems:(NSMutableArray*) menuItems
{
  [self cacheData:[NSKeyedArchiver archivedDataWithRootObject:menuItems]
          toFile:@"MenuItems.archive"];
}
+(NSMutableArray*) getCachedMenuItems
{
   return [NSKeyedUnarchiver unarchiveObjectWithData:[self
     dataForFile:@"MenuItems.archive"]];
}
```

Instead of writing directly to the file, the preceding code calls a helper method, `cacheData:toFile:`. This method saves the `NSData` from the `NSKeyedArchiver` to the in-memory cache. It also checks and removes the least recently accessed data and saves it to file when the prefixed memory limit for the number of in-memory items is reached. The implementation for this is shown in the following code:

Helper Method That Transparently Caches Data to In-Memory Cache (AppCache.m)

```
+(void) cacheData:(NSData*) data toFile:(NSString*) fileName
{
   [memoryCache setObject:data forKey:fileName];
   if([recentlyAccessedKeys containsObject:fileName])
   {
     [recentlyAccessedKeys removeObject:fileName];
   }

   [recentlyAccessedKeys insertObject:fileName atIndex:0];

   if([recentlyAccessedKeys count] > kCacheMemoryLimit)
   {
```

```
    NSString *leastRecentlyUsedDataFilename = [recentlyAccessedKeys
                                                lastObject];
    NSData *leastRecentlyUsedCacheData =
      [memoryCache objectForKey:leastRecentlyUsedDataFilename];
    NSString *archivePath = [[AppCache cacheDirectory]
                               stringByAppendingPathComponent:fileName];
    [leastRecentlyUsedCacheData writeToFile:archivePath atomically:YES];

    [recentlyAccessedKeys removeLastObject];
    [memoryCache removeObjectForKey:leastRecentlyUsedDataFilename];
  }
}
```

Similarly to the preceding code, which caches data (`cacheData:toFile:`), in the following code, you need to write a method that checks the in-memory cache and returns this data, instead of directly reading from a file. The method should access the file only if it isn't present in the in-memory cache.

Helper Method That Transparently Retrieves the Cached Data from In-Memory Cache (AppCache.m)

```
+(NSData*) dataForFile:(NSString*) fileName
{
  NSData *data = [memoryCache objectForKey:fileName];
  if(data) return data; // data is present in memory cache

  NSString *archivePath = [[AppCache cacheDirectory]
                             stringByAppendingPathComponent:fileName];
  data = [NSData dataWithContentsOfFile:archivePath];

  if(data)
    [self cacheData:data toFile:fileName]; // put the recently accessed
                                            data to memory cache
  return data;
}
```

This method also saves the data read from flash memory back to in-memory cache, which is just the way a Least Recently Used caching algorithm works.

Handling Memory Warnings

For the most part, the `AppCache` is now complete, and you've added a transparent in-memory cache without modifying the calling code. However, you need to do one more important thing. Because you're retaining data used by views in `AppCache`, the memory consumption of your app continues to grow, and the chances of receiving a memory warning become very high. To avoid this situation, you handle the memory warning notifications in `AppCache`. In the static initialize method, add a notification observer to `UIApplicationDidReceiveMemoryWarningNotification`:

```
[[NSNotificationCenter defaultCenter] addObserver:self
  selector:@selector(saveMemoryCacheToDisk:)
 name:UIApplicationDidReceiveMemoryWarningNotification object:nil];
```

Now write a method to save the in-memory cache items to files:

```
+(void) saveMemoryCacheToDisk:(NSNotification *)notification
{
  for(NSString *filename in [memoryCache allKeys])
  {
    NSString *archivePath = [[AppCache cacheDirectory]
                                stringByAppendingPathComponent:filename];
    NSData *cacheData = [memoryCache objectForKey:filename];
    [cacheData writeToFile:archivePath atomically:YES];
  }

  [memoryCache removeAllObjects];
}
```

This method ensures that your `AppCache` doesn't eat up the available system memory and is faster than writing directly to files from your view controller.

Handling Termination and Entering Background Notifications

You also need to ensure that your in-memory cache is saved when the app quits or enters the background. Doing so gives an added advantage to your on-demand caching: offline access.

Now, you add the third and final step, which is to watch for the app's resigning active or closing notifications and handle memory warnings as you did in the previous section. No extra methods are needed; just add observers in the initialize method for `UIApplicationDidEnterBackgroundNotification` and `UIApplicationWillTerminateNotification`. This way, you ensure that your in-memory cache is saved to the file system.

Observing Notifications and Saving In-Memory Cache to Disk (AppCache.m)

```
[[NSNotificationCenter defaultCenter] addObserver:self
  selector:@selector(saveMemoryCacheToDisk:)
name: UIApplicationDidEnterBackgroundNotification object:nil];

[[NSNotificationCenter defaultCenter] addObserver:self
  selector:@selector(saveMemoryCacheToDisk:)
name: UIApplicationWillTerminateNotification object:nil];
```

Remember to call `removeObserver` in `dealloc` as well. For the complete AppCache code, download the code sample from the book's website.

Whew! That was a bit of overload. But it's not finished yet. I told you that on-demand caching can be implemented using a data model cache or a URL cache. The `AppCache` implementation you learned about earlier is a data model cache. Next, I show you how to implement a URL cache. But don't worry; implementing a URL cache is much simpler on the client side. Most of the heavy lifting and cache invalidation is done remotely on the server, and the server dictates cache invalidation through cache control headers.

Creating a URL Cache

You implement a URL cache by caching the responses for every URL request made by the app. This cache is similar to the `AppCache` you implemented in the previous section. The difference is that the key and value you stored in the cache will differ. A data model cache uses a filename as the key and the archives of the data models as values. A URL cache uses the URL as the key and the response data as values. Most of the implementation is similar to the `AppCache` you wrote earlier, except for caching invalidation.

A URL cache works like a proxy server in the way it handles caching. As a matter of fact, `MKNetworkKit` handles the HTTP 1.1 caching standard transparently for you. But you still need to understand how it works under the hood.

Earlier, I told you that the server dictates caching invalidation for a URL cache. The HTTP 1.1 (RFC 2616 Section 13) specification explains the different cache control headers that a server might send. The RFC specifies two models: an expiration model and a validation model.

Expiration Model

The expiration model allows the server to set an expiry date after which the resource (your image or response) is assumed to be stale. Intermediate proxy servers or browsers are expected to invalidate or expire the resource after this set amount of time.

Validation Model

The second model is a validation model where the server usually sends a checksum (Etag). All subsequent requests that get fulfilled from the cache should be revalidated with the server using this checksum. If the checksum matches, the server returns an HTTP 304 Not Modified status.

Example

Here's an example of both the expiration model and the validation model, along with tips on when to use which model on your server. Although this information may be beyond the usual scope of iOS developers, understanding how caching works will help you become a better developer. If you've ever configured your server, you've already written something like this to your `nginx.conf` file.

nginx Configuration Setting for Specifying an Expiration Header

```
location ~ \.(jpg|gif|png|ico|jpeg|css|swf)$ {
            expires 7d;
        }
```

This setting tells `nginx` to emit a cache control header (`Expires` or `Cache-Control: max-age=n`) that instructs the intermediate proxy servers to cache files ending with jpg and others for seven days. On API servers, the sysadmin probably does this, and on all your image requests, you see the cache control header. Caching your images on your device by respecting these headers is probably the right way to go, rather than invalidating all images in your local cache every few days.

Whereas static responses, images, and thumbnails use the expiration model, dynamic responses mostly use the validation model for controlling cache invalidation. This requires computing the checksum of the response objects and sending it along with the Etag header. The client (your iOS app) is expected to send the Etag on the IF-NONE-MATCH header for every subsequent request (revalidating its cache). The server, after computing the checksum, checks whether the checksum matches the IF-NONE-MATCH header. If yes, the server sends a 304 Not Modified message, and the client serves the cached content. If the checksum is different, the content has indeed changed. The server can be programmed either to return the complete content or a changeset explaining what was exactly changed.

Let me give you an example to make this clear. Assume that you're writing an API that returns the list of outlets in the iHotelApp. The endpoint /outlets return the list of outlets. On a subsequent request, the iOS client sends the Etag. The server fetches the list of outlets again (probably from the database) and computes the checksum of the response. If the checksum matches no new outlets were added, and the server should send a 304 Not Modified message. If the checksum is different, a couple of things could have happened. A new outlet may have been added, or an existing outlet may have been closed. The server may choose to send you the complete list of outlets again or send you a changeset that contains the list of new outlets and the list of closed outlets. The latter is more efficient (although it's slightly more complicated in terms of implementation) if you don't foresee frequent changes to outlets.

The World Wide Web increased from a dozen computers in 1990 to billions of computers worldwide within a decade. The one driving factor was the HTTP protocol being built with scalability and caching, and the fact that every major browser adhered to these HTTP standards. With the computing world becoming more and more app-centric, your web service application needs to reimplement what browsers did to help the Web grow. Adhering to some of these simple standards will go a long way toward improving the performance of your application and thereby pushing the limits of what it can do.

Caching Images with a URL Cache

A URL cache like the preceding one is transparent to the data it caches. This means that it's immaterial to a URL whether you cache images or music or videos or URL responses. Just remember that when you use a URL cache for caching images, the URL of the image becomes the key and the image data becomes the cached object. The advantage you get here is superior performance because a URL cache doesn't post-process responses (such as converting them to JPEGs or PNGs). This fact alone is a good reason to use URL caching for caching your images.

Summary

In this chapter, you learned how to architect an iOS application that uses a web service. The chapter also presented the different data exchange formats and ways to parse them in Objective-C, and you learned a powerful method of processing responses from a RESTful service using Objective-C's KVC method. You then learned about using queues for handing concurrent requests and maximizing performance by altering the maximum concurrent operations on the queue-based available network. You also learned about the various types of caching and the pros and cons of different methods of caching. Caching goes a long way toward improving the performance of an app, yet a majority of the apps on the App Store don't implement it properly. The techniques you discovered (for both iOS and your API server) in this chapter can help you push the limits and take your app to the next level.

Further Reading
Apple Documentation

The following documents are available in the iOS Developer Library at developer.apple.com or through the Xcode Documentation and API Reference.

Reachability

Apple XMLPerformance Sample Code

NSXMLDocument Class Reference

Archives and Serializations Programming Guide

iCloud

Books

The Pragmatic Bookshelf, Core Data.

```
http://pragprog.com/titles/mzcd/core-data
```

Other Resources

Callahan, Kevin. Accessorizer. *Mac App Store.* (2011)

```
http://itunes.apple.com/gb/app/accessorizer/id402866670?mt=12
```

Kumar, Mugunth. "MKNetworkKit Documentation"

```
http://mknetworkkit.com
```

Callahan, Kevin. "Mac App Store" (2011)

```
http://itunes.apple.com/gb/app/accessorizer/id402866670?mt=12
```

Cocoanetics. "JSON versus Plist, the Ultimate Showdown" (2011)

```
www.cocoanetics.com/2011/03/json-versus-plist-the-ultimate-showdown/
```

Wenderlich, Ray. "How To Choose The Best XML Parser for Your iPhone Project"

```
http://www.raywenderlich.com/553/how-to-chose-the-best-xml-parser-for-
your-iphone-project
```

Crockford, Douglas. "RFC 4627. 07 01" (2006)

```
http://tools.ietf.org/html/rfc4627
```

W3C. "Web Services Architecture. 2 11" (2004)

```
www.w3.org/TR/ws-arch/#relwwwrest
```

Brautaset, Stig. "JSON Framework 1 1" (2011)

http://stig.github.com/json-framework/

Wight, Jonathan. "TouchCode/TouchJSON. 1 1" (2011)

https://github.com/TouchCode/TouchJSON

Gabriel. "YAJL-ObjC" (2011)

https://github.com/gabriel/yajl-objc

Johnezang. "JSONKit" (2011)

https://github.com/johnezang/JSONKit

"mbrugger json-benchmarks" on GitHub

https://github.com/mbrugger/json-benchmarks/

## Chapter 13

Getting More Out of Your Bluetooth Devices

Building software that work closely with hardware is always fun. In fact, one of the key strengths of Apple devices prior to the iPhone was the suite of hardware that worked together with iPods connecting through a 30-pin connector. After the iPhone was introduced, the same suite of hardware (mostly speaker docks and car audio interfaces) that worked with the iPod (the clickwheel iPod and iPod nano) worked just as well with the iPhone. One of the key strengths of Apple is its capability to build and nurture an ecosystem of hardware devices working closely with its suite of iDevices.

In the past, most hardware devices that worked with an iPhone, iPod, or iPod touch connected through the 30-pin accessory (on iPods and older iPhones) and the lightning connector on new devices. Hardware programming can be an expensive undertaking (at least compared to making apps). This added cost comes in the form of royalties because, as was true in the past, almost every piece of hardware that works closely with Apple's iDevices has to be *Made For iPhone/iPod,* or *MFi,* licensed. However, MFi licensing is only for devices that connect to the iPhone or iPod touch through the lightning (or the old 30-pin) connector.

More recently, iDevices (iPhone 4S onwards, iPad 3G onwards, and iPad mini) started shipping with Bluetooth low energy (LE) chips, and with the advent of Bluetooth low energy (also called as Bluetooth LE or Bluetooth Smart), a independent software vendor can also sell hardware that works with his app without having MFi licensing.

This chapter is about Bluetooth low energy. In the course of this chapter, I show you how to harness the power of iOS Bluetooth hardware.

History of Bluetooth

Bluetooth has been around for two decades. Today, every iOS device comes with a Bluetooth chip. Even feature phones and not-so-smart phones from 2000 to 2007 came with Bluetooth chips. In fact, Bluetooth was first developed by Swedish company Ericsson in 1994. Early versions of Bluetooth (1 and 1.1) were predominantly used (by early iPhones and older feature phones) for wireless communication between a headset and a device. Bluetooth 2.0 and 2.1+EDR (Enhanced Data Rate) increased data rate, thus allowing stereo music from your phone's music player to be streamed to wireless headphones. Bluetooth 3.0 allowed even higher data rates and made it possible to transfer larger files wirelessly between devices.

Before Bluetooth 4.0, to use Bluetooth, a specific device had to implement certain required Bluetooth profiles. Profiles exist for common use cases such as Bluetooth audio headsets, Bluetooth stereo headsets, Bluetooth remotes, and SIM card access. However, building a custom profile for your own unique hardware is quite a tedious job. Moreover, building a custom profile requires you to install the necessary drivers. Although you can accomplish this task, it's hard, and you still have to get MFi licensed if your app uses a custom profile. But that was the past.

Now, say hello to Bluetooth 4.0 (also known as Bluetooth LE or Bluetooth Smart).

Why Bluetooth Low Energy?

Bluetooth low energy makes it much easier to implement custom protocols and custom data transfers that, you can fully implement in the user space, without requiring a system level driver. That means, your app is the driver and it is the software. You don't need a system-level pairing (although you can still pair with a device for encrypted communication), and that means you don't require MFi licensing.

Easier implementation and lower energy requirements have made Bluetooth LE more important than ever before. So far, more than a billion Bluetooth LE devices have been shipped. These devices touch our lives in many areas, including health care, sports/fitness, security systems such as car locks, home automation that ranges from opening your door to controlling your thermostat, peer-to-peer gaming, point-of-sales terminals, and much more. The range of devices that use Bluetooth LE is overwhelming.

For iOS developers, the ease of building apps coupled with the lack of strict licensing requirements such as the Apple's MFi makes becoming an independent hardware vendor an attractive option for independent software vendors.

Bluetooth SDK in iOS

The Bluetooth SDK in iOS is provided through the `CoreBluetooth.framework`. CoreBluetooth has been available since iOS 5 and is specifically designed for Bluetooth LE devices.

> Prior to iOS 5, there wasn't a native way to access a Bluetooth device. iOS handles getting remote control events from a Bluetooth remote (AVRCP profiles) internally and sends them as normal remote control messages through `UIResponder`. The `GameKit.Framework` also provided ways to create a peer-to-peer game using Bonjour over Bluetooth, without exposing Bluetooth-specific functionality.

`Bluetooth.framework` has seen remarkable changes since 2011. Before delving into details, you need to understand some core concepts behind Bluetooth. Bluetooth LE is based on a peer-to-peer communication system in which one device acts as a server and another acts as a client. The device that has data acts as the server, and the device that consumes data acts as the client.

Server

In any Bluetooth LE network, a server generates data and a client consumes data. Usually, the devices that generate data act as servers. They could be heart rate monitors, thermostats, joysticks, or just about anything. All those tiny Bluetooth gadgets lying around you—your Fitbit tracker, the Jawbone UP—are servers advertising that they can track your activity or fitness. They are responsible for advertising the kind of data they generate and serving them to connected clients.

Client

The device that is interested in data (this is normally your iPhone) must discover for an interested service. It's the client's responsibility to initiate the connection to the server and start reading the data.

Classes and Protocols

In iOS parlance, the server is called a *peripheral,* and the client is called a *central*. iOS 5 allowed the iOS device to act as a client and read data from a Bluetooth device. The class in `CoreBluetooth.framework` that you use for this is `CBCentralManager`. Peripherals are modeled using the class `CBPeripheral`.

In iOS 6, the SDK added a few more classes that allowed an iOS device to act as a peripheral that can send data from the iPhone. This data can be a list of notifications, the currently playing music track, a photo, or just about anything that your app wants to send to a device. There are even apps that download firmware updates for the peripheral and send the new firmware by acting as a server. You use the `CBPeripheralManager`, and the `CBCentral` class models a client device.

iOS 7 adds methods to scan and retrieve peripherals synchronously. A minor change is replacing the `Central` and `Peripheral` identifiers from the `CoreFoundation`-based `CFUUIDRef` with `NSUUID`. A major addition to `CoreBluetooth.framework` in iOS 7 is State Preservation and Restoration for apps that communicate and send data over Bluetooth in the background. You learn more on that later in the "State Preservation and Restoration" section of this chapter.

> With iOS 5 and iOS 6, the iOS Simulator supported Bluetooth. You were able to use the iOS Simulator to discover, connect, and serve or read data from another peripheral/central. iOS 7 drops support for Bluetooth simulation, so the only way to accomplish those tasks during development is to use your iOS device.

Working with a Bluetooth Device

There are plenty of Bluetooth LE devices laying around for you to choose from, but in this chapter, I use the Texas Instruments SensorTag Bluetooth LE device and write a demo application that connects to the temperature sensor of that device. After you are familiar with writing a Bluetooth LE app that consumes data, I show you how to make a Bluetooth LE server app that serves data.

> The reason for choosing the SensorTag is that it is one of the cheapest Bluetooth LE devices you can buy. As of this writing, it costs $25 including shipping to most of the world. In the "Further Reading" section at the end of this chapter, you will find a link to the Texas Instruments website where you can place an order to ship one.

If you are planning to write a Bluetooth LE app for your own device, don't worry Before I show you the code, you need to understand some concepts behind Bluetooth. Bluetooth LE networking works similarly to Apple's

Bonjour Service. If you know Bonjour, you will feel at home. If you don't know Bonjour, don't worry, you don't need to know Bonjour to understand what follows.

Scanning for a Service

Bluetooth devices or peripherals are the servers that serve data. Before a server can serve data, it has to *advertise* the services it offers. A Bluetooth peripheral advertises itself by sending out advertisement packets. A Bluetooth central (client) scans for these packets to detect a nearby peripheral. A service is identified using a universally unique identifier (UUID).

In the next section, you learn how to connect to a peripheral. The first step toward doing that is to start scanning for the peripheral, as shown in the following code. Later in this chapter, you learn how to create your own peripherals.

Scanning for a Peripheral (SCTViewController.m)

```
self.centralManager = [[CBCentralManager alloc]
  initWithDelegate:self queue:nil options:nil];

if(self.centralManager.state == CBCentralManagerStatePoweredOn) {
  [self.centralManager scanForPeripheralsWithServices:nil options:nil];
}
```

You learned about `CBCentralManager` earlier in the "Classes and Protocol" section. You use this class to act as a central. The `CBCentralManager` class has a delegate called `CBCentralManagerDelegate`. The delegate notifies you of discovered peripherals, services, characteristics of services, and value updates.

If the Bluetooth on your iPhone is not powered on, you get a callback on the `CBCentralManagerDelegate` method `didUpdateState:`. You should start scanning for peripherals on that method as well.

Universally Unique Identifiers (UUIDs)

The Bluetooth SIG consortium maintains a list of commonly used Bluetooth LE services. These adopted services have a 16-bit UUID. A heart rate monitor's UUID is 0x180D (0001 1000 0000 1101), a glucose meter uses 0x1808, and a time service uses 0x1805. If your hardware is doing something that is not available in the specification, you are free to use your own UUIDs. Be aware that such UUIDs should be 128 bits long. The TI SensorTag advertises multiple services; the service of interest for this example, the ambient temperature service, uses f000aa00-0451-4000-b000-000000000000, which is one such 128-bit UUID.

CBCentralManagerDelegate Method (SCTViewController.m)

```
- (void)centralManagerDidUpdateState:(CBCentralManager *)central {

  if(central.state == CBCentralManagerStatePoweredOn) {

    [self.centralManager scanForPeripheralsWithServices:nil
                                              options:nil];
  }
}
```

> You also have to handle error conditions such as Bluetooth LE not being available on a given device (iPhone 4 and older devices) and Bluetooth being turned off by the user in Settings. These conditions are shown in the sample code for this chapter on the book's website but are omitted here for brevity.

The `scanForPeripheralsWithServices:` method takes an array of services as a parameter and scans the surrounding area for peripherals advertising these services. If you pass `nil` to this parameter, it scans for all available peripherals, but it is slow.

> The TI SensorTag, for unexplained reasons, doesn't advertise the service UUID in the advertisement packet. Hence, you are forced to use `nil` as the first parameter. Texas Instruments might fix this issue in a firmware update later.

When a peripheral is found, you get details of the peripheral in the CBCentralManagerDelegate method centralManager:didDiscoverPeripheral:advertisementData:RSSI:.

Connecting to a Device

When a peripheral is found, the next step is to connect to the peripheral and discover the services it provides.

Connecting to the Device

```
- (void)centralManager:(CBCentralManager *)central
 didDiscoverPeripheral:(CBPeripheral *)peripheral
     advertisementData:(NSDictionary *)advertisementData
                  RSSI:(NSNumber *)RSSI {

  // optionally stop scanning for more peripherals
  // [self.centralManager stopScan];
  if(![self.peripherals containsObject:peripheral]) {

    NSLog(@"Connecting to Peripheral: %@", peripheral);
    peripheral.delegate = self;
    [self.peripherals addObject:peripheral];
```

(continued)

```
    [self.centralManager connectPeripheral:peripheral options:nil];
    }
}
```

You should retain the peripheral before you connect to it; otherwise, the ARC compiler will deallocate the peripheral object and you will not be able to connect to it. In the preceding example, you retain the peripheral by adding it to an array that maintains the list of peripherals.

Retrieving Peripherals Directly

If you know the identifier of the peripheral, you can use the `retrievePeripheralsWithIdentifiers:` method instead of scanning for it. This method was added in iOS 7.

Retrieving Peripherals

```
NSArray *peripherals = [self.centralManager
    retrievePeripheralsWithIdentifiers:
    @[[CBUUID UUIDWithString:@"7BDDC62C-D916-7E4B-4E09-285E11164936"]]];
```

In the preceding example, you add the peripheral list to an array. It's a good practice to store this array and attempt a connection to the known peripheral every time instead of scanning. Scanning taxes the battery and should be avoided if possible.

When you have the pointer to the peripheral, you can connect to it directly, just as in the preceding example.

Discovering Services

Your attempt to make a connection may or may not succeed. You know if the peripheral is connected on the delegate method, `didConnectPeripheral`. The next step is to discover services offered by the peripheral.

didConnectPeripheral Delegate Method

```
-(void) centralManager:(CBCentralManager *)central
    didConnectPeripheral:(CBPeripheral *)peripheral {

[peripheral discoverServices:nil];
}
```

The list of services offered is notified via another delegate method in `CBCentralManagerDelegate`; this method is `peripheral:didDiscoverServices:`.

You retrieve the list of services from the `services` property of the peripheral. In the case of the SensorTag, the peripheral has about ten services, and you are interested in the Infrared Room Temperature service. This service is identified by the UUID F000AA00-0451-4000-B000-000000000000.

Discovering Characteristics

After you get the service, you can then discover its characteristics. A service usually has one or more characteristics, such as the capability to turn a service on and off or the capability to read the current value of the service. For the SensorTag, the Room Temperature service has those two characteristics: The first allows the room temperature sensor to be turned on or off, and the second characteristic gives you the actual reading of the room temperature sensor.

Discovering Characteristics

```
-(void) peripheral:(CBPeripheral *)peripheral
  didDiscoverServices:(NSError *)error {

  [peripheral.services enumerateObjectsUsingBlock:^(id obj,
  NSUInteger idx, BOOL *stop) {
    CBService *service = obj;

    if([service.UUID isEqual:[CBUUID UUIDWithString:
  @"F000AA00-0451-4000-B000-000000000000"]])
      [peripheral discoverCharacteristics:nil forService:service];
  }];
}
```

You get the list of characteristics for a given service in the `didDiscoverCharacteristcsForService:` `error:` delegate callback. As previously mentioned, the Room Temperature service has two characteristics. The first one, identified by the UUID F000AA02-0451-4000-B000-000000000000, is the on/off characteristic; and the second one, identified by the UUID F000AA01-0451-4000-B000-000000000000, gives the sensor reading.

Some characteristics are read-only, some are read/write, and some even support update notifications. For this example, the temperature sensor can automatically update you when the temperature changes. But before you can get notified, you should turn on the sensor. You do this by writing a value of `1` to the On/Off characteristic.

The next step is to enable notifications for the characteristic that gives you the sensor reading. This step is illustrated in the following code snippet.

Delegate Method Returning the Discovered Characteristics

```
- (void)peripheral:(CBPeripheral *)peripheral
  didDiscoverCharacteristicsForService:(CBService *)
  service error:(NSError *)error {

  [service.characteristics enumerateObjectsUsingBlock:^(id obj, NSUInteger
  idx, BOOL *stop) {

    CBCharacteristic *ch = obj;
    if([ch.UUID isEqual:[CBUUID UUIDWithString:
  @"F000AA02-0451-4000-B000-000000000000"]]) {
      uint8_t data = 0x01;
      [peripheral writeValue:[NSData dataWithBytes:&data length:1]
          forCharacteristic:ch
                    type:CBCharacteristicWriteWithResponse];
    }

    if([ch.UUID isEqual:[CBUUID UUIDWithString:
  @"F000AA01-0451-4000-B000-000000000000"]]) {

      [peripheral setNotifyValue:YES forCharacteristic:ch];
    }
  }];
}
```

Getting Update Values

At this point, you can sit back and relax for a bit. You will start receiving updates about temperature on the delegate method `peripheral:didUpdateValueForCharacteristic:error:`, as shown in the following code.

Getting Value Updates on the Delegate

```objc
- (void)peripheral:(CBPeripheral *)peripheral
  didUpdateValueForCharacteristic:(CBCharacteristic *)characteristic
  error:(NSError *)error {

    float temp = [self temperatureFromData:characteristic.value];
    NSLog(@"Room temperature: %f", temp);
}

-(float) temperatureFromData:(NSData *)data {

    char scratchVal[data.length];
    int16_t ambTemp;
    [data getBytes:&scratchVal length:data.length];
    ambTemp = ((scratchVal[2] & 0xff)| ((scratchVal[3] << 8) & 0xff00));

    return (float)((float)ambTemp / (float)128);
}
```

The value update is sent to you as an `NSData`. You should convert it to the actual temperature value by meaningfully interpreting the values. In the case of SensorTag, the second and third bytes give you the temperature. The helper method `temperatureFromData:` extracts the actual temperature from these two bytes and swaps the bytes around to change the byte order. In the sample code, you are just logging the temperature and displaying it on a `UILabel`.

The complete code listing for this example is available from the book's website.

The process of connecting to and reading data from most other Bluetooth LE devices is similar to what you have just learned. If you have a heart rate monitor, try reading your heart rate using your app. The service UUID for heart rate monitors is standardized by Bluetooth SIG; it is 0x180D.

Creating Your Own Peripherals

A peripheral, as you know by now, is a Bluetooth LE device that "has data." Data can be anything. Data normally comes from a sensor (as in the case of a heart rate monitor or the SensorTag), and it's usually a small hardware device. You can also emulate a peripheral by writing an app. The iPhone hardware has some interesting sensors that you can use. For example, you can read the accelerometer or gyroscope data and send it through Bluetooth by writing an app that acts as a peripheral. Most people would not be interested in doing that because such an app will only be a thousand dollar SensorTag. Most people would instead be interested in reading data from a remote server. For example, you can write an iOS app that can read Yahoo! weather and advertise it as a peripheral that knows local weather. Similarly, you can write a peripheral that reads your Twitter or Facebook notifications and sends it off to your wearable device (something similar to Pebble watch).

The first version of `CoreBluetooth.framework` did not support creating peripherals. iOS 6 added the `CBPeripheralManager` and `CBCentral` classes that allow you to create a peripheral and manage centrals that connect to your peripheral. In the next few sections, I show you how to create a barebones Bluetooth server that advertises and serves data to a central.

Advertising Your Service

You already know that a central works by scanning for peripherals that advertise their services. Now, when your app is the peripheral, you should advertise your service (and the service characteristics). You already learned that services and characteristics are identified using UUIDs and that there are two kinds of UUIDs: 16 bit and 128 bit. Because your peripheral is a nonstandard peripheral, you can use a 128-bit UUID.

> You can create a unique identifier on your Mac by typing `uuidgen` on your terminal.

Now you are going to make an app that sets up a peripheral server and advertises local weather. The sample code shows you the Bluetooth aspects of the app. I leave getting the current location of the user and fetching local weather to you.

The first step is to initialize your peripheral manager and wait for the callback.

Initializing Your Peripheral Manager (SCTViewController.m)

```
self.manager = [[CBPeripheralManager alloc] initWithDelegate:self
   queue:nil];
```

Peripheral Manager Delegate (SCTViewController.m)

```
- (void)peripheralManagerDidUpdateState:(CBPeripheralManager *)manager {

   if(manager.state == CBPeripheralManagerStatePoweredOn) {
     [self createService];
   }
}
```

When you know that Bluetooth is powered on, you can create your service and characteristic, as shown in the following code. You create a service using the class `CBMutableService` and a UUID. You create a characteristic using the class `CBMutableCharacteristic`. You should always use unique identifiers generated using `uuidgen` for every UUID you use in your application.

Creating Your Peripheral Service (SCTViewController.m)

```
- (void)createService {

    CBUUID *serviceUUID = [CBUUID UUIDWithString:kServiceUUID];
    self.service = [[CBMutableService alloc]
                    initWithType:serviceUUID primary:YES];

    CBUUID *characteristicUUID = [CBUUID UUIDWithString:kCharacteristicUU
ID];
    self.characteristic = [[CBMutableCharacteristic alloc]
    initWithType:kCharacteristicUUID
      properties:CBCharacteristicPropertyNotify
           value:nil
     permissions:CBAttributePermissionsReadable];
    [self.service setCharacteristics:@[self.characteristic]];
    [self.manager addService:self.service];
}
```

When you create a characteristic, you need to specify its properties and the permission levels that the client has on it. The value for the property is one of the following in the enum `CBCharacteristicProperties`. The most commonly used ones are `CBCharacteristicPropertyRead`, `CBCharacteristicPropertyWrite`, and `CBCharacteristicPropertyNotify`:

- `CBCharacteristicPropertyRead` denotes that the characteristic value can be read by the central.
- `CBCharacteristicPropertyWrite` denotes that the central can write to the characteristic value.
- `CBCharacteristicPropertyNotify` denotes that the central will be updated by the peripheral whenever the value of the characteristic changes.

Read and notify properties are almost always used together, and in some cases, you just use notify.

In case of a thermostat, the "required room temperature" would be a `CBCharacteristicProperty Write` characteristic (that the central would set), and the "ambient room temperature" would be a `CBCharacteristicPropertyRead` and `CBCharacteristicPropertyNotify` characteristic, meaning that a central can read the current temperature and can also ask the peripheral to notify it when the temperature changes. For this example, you have to periodically (or whenever the location changes) read the weather data and notify the central of the new temperature. In the preceding code, the parameter following `CBCharacteristicPropertyNotify` allows you to set permissions. You use this parameter if you are planning to encrypt your data.

> At the time this chapter was written, encryption and/or other permissions are allowed only for reading a value of a characterstic. Encryption for service and/or characteristic discovery requires system-level pairing and therefore MFi licensing.

After you add the service, the last step is to start advertising. You start the advertising in the delegate callback that notifies you of successful service addition.

Peripheral didAddService Delegate (SCTViewController.m)

```objc
- (void)peripheralManager:(CBPeripheralManager *)peripheral
  didAddService:(CBService *)service error:(NSError *)error {

  if (!error) {
    [self.manager startAdvertising:
    @{ CBAdvertisementDataLocalNameKey : @"LocalWeather",
    CBAdvertisementDataServiceUUIDsKey :
    @[[CBUUID UUIDWithString:kServiceUUID]]}];
  }
}
```

Your peripheral is now set up. The next step is to wait for the central to connect and maintain a list of connected centrals. Note that this step is required only when the peripheral is responsible for notifying the central of value changes.

Central Subscription Delegate Callback (SCTViewController.m)

```objc
- (void)peripheralManager:(CBPeripheralManager *)peripheral
  central:(CBCentral *)central didSubscribeToCharacteristic:
  (CBCharacteristic *)characteristic {

  if([characteristic isEqual:self.characteristic])
    [self.subscribedCentrals addObject:central];
}
```

It's the peripheral's responsibility to maintain a list of subscribed centrals and add the new central when it subscribes and remove it when it unsubscribes.

> The preceding code snippet shows only the subscription part. You can download the complete code for this example from the book's website to learn how to remove the central when it unsubscribes itself.

Now that you have a peripheral set up and a list of subscribed centrals, the final step is to notify the centrals of the new value. For illustration purposes, you can use a text field, and whenever the text field changes, you can update the subscribed centrals of the new value.

Updating Subscribed Centrals (SCTViewController.m)

```objc
- (BOOL)textFieldShouldReturn:(UITextField *)textField {

  if(self.subscribedCentrals.count > 0) {

    NSData *data = [textField.text dataUsingEncoding:NSUTF8StringEncoding];
    [self.manager updateValue:data
            forCharacteristic:self.characteristic
         onSubscribedCentrals:self.subscribedCentrals];
  }

  [textField resignFirstResponder];
  return YES;
}
```

Now make another app that acts as a client and scan for this service. You can clone the SensorTag app that you made earlier and change the UUIDs. When the central and peripheral are up and running, you can see that data you typed into the text field gets displayed on the label. The complete code is available from the book's website. Download both BluetoothServer and BluetoothClient and test it out.

> **If you are making your own Bluetooth server based on this code, don't forget to replace the UUIDs. You don't want your server to clash with that of someone else who happens to use the UUIDs shown in the sample code.**

Common Scenarios

In the preceding example, you created a Bluetooth server that "served" data in a text field. You can easily modify the sample code to serve data from a weather service, Twitter feeds, Facebook notifications, e-mail, or anything that you could think of. Some apps even take it to the next level by serving software updates for the connected centrals. You can do almost anything you want; you're limited only by the power of your imagination. I am excited to see what you will come up with!

Running in the Background

The next few sections show you how to optimize power when running in the background. The Bluetooth chip on your iPhone may not be as power hungry as the GPS chip. However, keeping the Bluetooth service turned on can still cause a significant battery drain. First, I show you some common ways to optimize battery usage.

Background Modes

Apps that use `CoreBluetooth.framework` can be made to run in the background. The way backgrounding is handled in iOS 7 is slightly different from previous iOS versions, as explained in the "State Preservation and Restoration" section later in the chapter. To make an app run in the background, you add the `bluetooth-central` and/or `bluetooth-peripheral` keys to your `Info.plist`.

> **Xcode 5 has a new target editor that allows you to edit the capabilities without editing the `Info. plist` directly. Just open the target and go to the Capabilities tab. Expand the Background Modes section and select Uses Bluetooth LE accessories and/or Acts as a Bluetooth LE accessory accordingly.**

When you use one of these background modes, calls to `scanForPeripheralsWithServices:options:` continue to scan and return discovered peripherals even after your app enters the background. However, if your app was terminated by the OS, it will not be launched in the background again. In iOS 7, Apple added methods to `CoreBluetooth.framework` for supporting State Preservation and Restoration so that when using these APIs, your app will be launched in the background when one of your connected devices is discovered.

Power Considerations

If you are a hardware (Bluetooth peripheral) manufacturer, Apple recommends that you send the advertisement packets from your peripheral every 20 ms for the first 30 seconds. After the initial 30 seconds, you might want to save power and increase the delay. In that case, use 645 ms, 768 ms, 961 ms, 1065 ms, or 1294 ms to increase the chance of getting discovered by the iOS device. This is even more critical when your app is backgrounded. When a Bluetooth app is backgrounded, the scanning frequency enters a lower power mode, so getting discovered and connected to the iOS device will be faster if your advertising interval is one of the Apple recommended values. If your app is the peripheral, you don't have to worry about these values. The `CoreBluetooth.framework` internally uses these values for advertising.

State Preservation and Restoration

Prior to iOS 7, when your app was terminated by the system due to memory pressure or terminated forcefully by the user, your app would not be launched again automatically. iOS 7 introduces a feature called State Preservation and Restoration. It is similar to UIKit's feature by the same name. If you implement the State Preservation and Restoration API in your app, when an app enters background mode and gets terminated because of memory pressure, the system starts acting like a proxy and launches your application when needed. A few events that can trigger your app to be launched are when a peripheral connects, when a central subscribes, or when a notification is received from a central you were connected to. For example, if your car supports Bluetooth and your app that communicates with the car is not running, it will be launched automatically when you move near to the car and a "peripheral connect" event happens.

You have to make a few changes to support State Preservation and Restoration. First, you should use the new initializer method, `initWithDelegate:queue:options:`, to initialize your `CBCentralManager`/`CBPeripheralManager`. The options parameter is a dictionary that takes a restore identifier in the `CBCentralManagerOptionRestoreIdentifierKey`.

The next step is to implement the delegate method `centralManager:willRestoreState:` (or `peripheralManager:willRestoreState:`). The `willRestoreState:` method is called before the `didUpdateState:` method if your app was preserved and restored. The delegate also provides a dictionary that contains one or more of the following keys: `CBCentralManagerRestoredStatePeripheralsKey`, `CBCentralManagerRestoredStateScanServicesKey`, or `CBCentralManagerRestoredStateScanOptionsKey`. The last two keys merely tell you the scan services and options that you set before. This information might not be useful for most apps. The `CBCentralManagerRestoredStatePeripheralsKey` provides the list of peripherals that connect, which in turn trigger the app launch. From this key, you can get the list of peripherals, set its delegate, and start discovering the peripherals' services and characteristics.

On the peripheral side, the `didUpdateState:` method's key would be `CBPeripheralManagerRestoredStateServicesKey` and/or `CBPeripheralManagerRestoredStateAdvertisementDataKey`. As their names imply, the first key gives you a list of services advertised by your peripheral, and the second gives you the data that was advertised.

Summary

Bluetooth LE is an exciting new technology that has taken Bluetooth to whole new levels. Bluetooth was invented in 1994, but after the introduction of Bluetooth LE in 2010, the number of devices using Bluetooth skyrocketed. Today, in 2013, you can connect to more than a billion devices! Interesting devices such as the Pebble Smart Watch even have an iOS SDK that you can integrate within your app to send notifications from your app to the watch. If the past decade was the decade of web services, this decade is going to be the decade of connected devices. Of all competing technologies such as WiFi Direct or Near Field Communication (NFC), Bluetooth LE is gaining traction, and I believe that knowledge of the Bluetooth architecture and `CoreBluetooth.framework` will help you push the limits of what you can do with your connected app.

Further Reading

Apple Documentation

The following document is available in the iOS Developer Library at `developer.apple.com` or through the Xcode Documentation and API Reference.

> *CoreBluetooth—What's new in iOS 7*
>
> *CoreBluetooth Framework reference*

WWDC Sessions

The following session videos are available at `developer.apple.com`.

> WWDC 2012, "Core Bluetooth 101"
>
> WWDC 2012, "Advanced Core Bluetooth"
>
> WWDC 2013, "Core Bluetooth"

Other Resources

> Bluetooth Low Energy SensorTag, TI.com
>
> `http://www.ti.com/ww/en/wireless_connectivity/sensortag/index.shtml?DCMP=sensortag&HQS=sensortag-bn`
>
> Services, Bluetooth Low Energy Portal
>
> `http://developer.bluetooth.org/gatt/services/Pages/ServicesHome.aspx`
>
> PebbleKit iOS Reference, getpebble.com
>
> `http://developer.getpebble.com/iossdkref/index.html`

## Chapter 14

Batten the Hatches with Security Services

iOS is likely the first platform most developers encounter that employs a true least-privilege security model. Most modern operating systems employ some kind of privilege separation, allowing different processes to run with different permissions, but it is almost always used in a very rough way. Most applications on UNIX, OS X, and Windows either run as the current user or as an administrative user that can do nearly anything. Attempts to segment privileges further, whether with Security Enhanced Linux (SELinux) or Windows User Account Control (UAC), have generally led developers to revolt. The most common questions about SELinux are not how to best develop for it, but how to turn it off.

> With the Mac App Store, and particularly OS X 10.8, Apple has expanded some of iOS's least-privilege approach to the desktop. Time will tell whether it's successful.

Coming from these backgrounds, developers tend to be shocked when encountering the iOS security model. Rather than ensure maximal flexibility, Apple's approach has been to give developers the fewest privileges it can and see what software developers are *incapable* of making with those privileges. Then Apple provides the fewest additional privileges that allow the kinds of software it wants for the platform. This approach can be very restrictive on developers, but it's also kept iOS quite stable and free of malware. Apple is unlikely to change its stance on this approach, so understanding and dealing with the security model are critical to iOS development.

This chapter shows the way around the iOS security model, dives into the numerous security services that iOS offers, and provides the fundamentals you need to really understand Apple's security documentation. Along the way, you'll gain a deeper understanding of how certificates and encryption work in practice so that you can leverage these features to really improve the security of your products.

The code for this chapter is available in the online sample code. A clear-cut project called `FileExplorer` is also available; it enables you to investigate the public parts of the file system.

Understanding the iOS Sandbox

The heart of the iOS security model is the *sandbox*. When an application is installed, it's given its own home directory in the file system, readable only by that application. This makes it difficult to share information between applications, but also makes it difficult for malicious or poorly written software to read or modify your data.

Applications are not separated from each other using standard UNIX file permissions. All applications run as the same user ID (501, `mobile`). Calling `stat` on another application's home directory fails, however, because of operating system restrictions. Similar restrictions prevent your application from reading `/var/log` while allowing access to `/System/Library/Frameworks`.

Within your sandbox, there are four important top-level directories: your `.app` bundle, `Documents`, `Library`, and `tmp`. Although you can create new directories within your sandbox, how iTunes will deal with them is not well defined. I recommend keeping everything in one of these top-level directories. You can always create subdirectories under `Library` if you need more organization.

Your `.app` bundle is the package built by Xcode and copied to the device. Everything within it is digitally signed, so you can't modify it. In particular, this includes your `Resources` directory. If you want to modify files that you install as part of your bundle, you need to copy them elsewhere first, usually somewhere in `Library`.

The `Documents` directory is the place where you store user-visible data, particularly files such as word-processing documents or drawings that the user assigns a filename. These files can be made available to the desktop through file sharing if `UIFileSharingEnabled` is turned on in `Info.plist`.

The `Library` directory stores files that shouldn't be directly visible to users. Most files should go into `Library/Application Support`. These files are backed up, so if you want to avoid that, you can attach the attribute `NSURLIsExcludedFromBackupKey` to the files using the method `NSURL setResource Value:forKey:error:`.

> The addition of `setResourceValues:error:` to `NSURL` was a very strange move by Apple in iOS 5. `NSURL` represents a URL, not the resource at that URL. This method makes more sense as part of `NSFileManager`. I've opened a radar on this, but Apple has indicated that it does not plan to change this method.

The `Library/Caches` directory is special because it isn't backed up, but it is preserved between application upgrades. This is the place where you want to put most things you don't want copied to the desktop.

The `tmp` directory is special because it's neither backed up nor preserved between application upgrades, which makes it ideal for temporary files, as the name implies. The system may also delete files in a program's `tmp` directory when that program isn't running.

When you are considering the security of the user's data, backups are an important consideration. Users may choose whether to encrypt the iTunes backup with a password. If you have data that shouldn't be stored unencrypted on the desktop machine, you store it in the keychain (see the "Using Keychains" section, later in this chapter). iTunes backs up the keychain only if backup encryption is enabled.

If you have information you don't want the user to have access to, you can store it in the keychain or in `Library/Caches` because they are not backed up. This is weak protection, however, because the user can always jailbreak the phone to read any file or the keychain. There is no particular way to prevent the owner of a device from reading data on that device. iOS security is about protecting the user from attackers, not about protecting the application from the user.

Securing Network Communications

The greatest risk to most systems is their network communication. Attackers don't need access to the device, only to the device's network. The most dangerous areas are generally coffee shops, airports, and other public WiFi networks. It's your responsibility to make sure that the user's information is safe, even on hostile networks.

The first and easiest solution is to use Hypertext Transfer Protocol Secure (HTTPS) for your network communication. Most iOS network APIs automatically handle HTTPS, and the protocol eliminates many of the easiest attacks. In the simplest deployment, you put a self-signed certificate on the web server, turn on HTTPS, and configure `NSURLConnection` to accept untrusted certificates, as discussed shortly. This solution is still vulnerable to several kinds of attacks, but it's easy to deploy and addresses the most basic attacks.

How Certificates Work

Hopefully, you have encountered public-private key infrastructure (PKI) systems before. This section gives a quick overview of the technology and then discusses how it affects the security of your application.

Asymmetric cryptography is based on the mathematical fact that you can find two very large numbers (call them A and B) that are related in such a way that anything encrypted with one can be decrypted with the other, and vice versa. Key A cannot decrypt things that key A encrypted, nor can key B decrypt things that key B encrypted. Each can decrypt only the other's encrypted data (called *ciphertext*). There is no real difference between key A and key B, but for the purposes of public key cryptography, one is called the *public key*, which generally everyone is allowed to know, and the other is called the *private key*, which is secret. You can use a public key to encrypt data such that only a computer with the private key can decrypt it. This is an important property that is used repeatedly in public key systems. If you want to prove that some entity (person or machine) has the private key, you make up a random number, encrypt it with the entity's public key, and send it. That entity decrypts the message with the entity's private key and sends it back to you. Because only the private key could have decrypted the message, the entity you're communicating with must have the private key.

This property also allows you to *digitally sign* data. Given some data, you first hash it with some well-known hashing algorithm and then encrypt it with your private key. The resulting ciphertext is the signature. To validate the signature, you hash the data again with the same algorithm, decrypt the signature using the public key, and compare the hashes. If they match, you know the signature was created by some entity that had access to the private key.

Just because an entity has access to the private key doesn't prove it's who it says it is. You need to ask two questions. First, how well is the private key protected? Anyone with access to the private key can forge a signature. Second, how do you know that the public key you have is related to the entity you care about? If I approach you on the street and hand you a business card that says I'm the President of the United States, it hardly proves anything. I'm the one who handed you the business card. Similarly, if a server presents you with a public key that claims to be for `www.apple.com`, why should you believe it? This is where a *certificate chain* comes in, and it's relevant to both questions.

A certificate is made up of a public key, some metadata about the certificate (more on that later), and a collection of signatures from other certificates. In most cases, there is a short chain of certificates, each signing the one below it. In very rare cases, one certificate may have multiple signatures. An example of a certificate chain is shown in Figure 14-1.

Figure 14-1 The certificate chain for daw.apple.com

In this example, the server `daw.apple.com` presents a certificate that includes its own public key, signed by an intermediate certificate from VeriSign, which is signed by a root certificate from VeriSign. Mathematically, you can determine that the controllers of each of these certificates did sign the next certificate in the chain, but why would you trust any of them? You trust them because Apple trusts the VeriSign root certificate, which has signed the intermediate certificate, which has signed the Apple certificate. Apple ships the VeriSign root certificate in the trusted root store of every iOS device, along with more than a hundred other trusted root certificates. The *trusted root store* is a list of certificates that is treated as explicitly trustworthy. Explicitly trusted certificates are called *anchors*. You can programmatically set your own anchors if you don't want to trust Apple's list.

It's now time to address the much-misused term *self-signed certificate*. For cryptographic reasons, every certificate includes a signature from itself. A certificate that has only this signature is called self-signed. Often, when people talk about a self-signed certificate, they mean a certificate that you shouldn't trust. But the VeriSign root certificate is a self-signed certificate, and it's one of the most trusted certificates in the world. Every root certificate, by definition, is a self-signed certificate. What's the difference? It isn't how many signatures a certificate has in its chain that matters, but how well all the private keys in the chain are protected and whether the identity of the owner has been authenticated.

If you generate your own self-signed certificate and protect the private key very well, that's more secure than a certificate that VeriSign issues you. In both cases, you're dependent on protecting your private key, but in the latter case, you also have to worry about VeriSign protecting *its* private key. VeriSign spends a lot of money and effort doing that, but protecting two keys is always more risky than protecting just one of them.

I'm not saying that commercial certificates from VeriSign, DigiTrust, and other providers are bad. But you don't get a commercial certificate to improve the security of your system. You get one for convenience because the commercial certs are already in the root key store. But remember, you control the root key store in your own application. This leads to a surprising fact: *There is no security reason to purchase a commercial certificate to secure your application's network protocol to your own server.*

Commercial certificates are valuable only for websites visited by browsers or other software you don't control. Taking the step to generate your own certificate and ship the public key in your application is marginally *more* secure than using a commercial certificate. If you already have a commercial certificate for your server, using it for your application's network protocol is somewhat more convenient; it's just not more secure. This is not to say that trusting random certificates is okay (that is, turning off certificate validation). It's to say that trusting only *your* certificates is slightly better than trusting commercial certificates.

Certificates can be corrupt or not, valid or not, and trusted or not. These are separate attributes that need to be understood individually. The first question is whether a certificate is corrupt. A certificate is *corrupt* if it doesn't conform to the X.509 data format or if its signatures are incorrectly computed. A corrupt certificate should never be used for anything, and the iOS certificate function generally rejects such certificates automatically.

> **X.509 refers to the data format specification and semantics originally defined by ITU-T** (`www.itu.int/ITU-T`). **The current version (v3) is defined by IETF RFC 5280** (`www.ietf.org/rfc/rfc5280.txt`).

Checking Certificate Validity

Given that a certificate is not corrupt, is it valid? Certificates contain a great deal of metadata about the public key they contain. The public key is just a very large number. It doesn't represent anything by itself. It's the metadata that gives that number meaning. A certificate is *valid* if the metadata it presents is consistent and appropriate for the requested use.

The most important piece of metadata is the subject. For servers, this is generally the fully qualified domain name (FQDN), such as `www.example.org`. The first test of validity is a name match. If you walk into a bank and identify yourself as "John Smith," you might be asked for your driver's license. If you hand over a license that says "Susan Jones," that would not help in identifying you no matter how authentic the driver's license. Similarly, if you're visiting a site named `www.example.org` and the site presents a certificate with a common name `www.badguy.com`, you should generally reject it. Unfortunately, the situation is not always that simple.

What if you visit `example.org` and it presents a certificate that says `www.example.org`? Should you accept that certificate? Most people would assume that `example.org` and `www.example.org` refer to the same server (which may or may not be true), but certificates use a simple string match. If the strings don't match, the certificate is invalid. Some servers present wild card certificates with subjects like `*.example.org`, and iOS accepts them, but in some cases it still rejects a certificate because of a name mismatch you believe it should accept. Unfortunately, iOS doesn't make ambiguous name mismatches easy to manage, but it can be done.

The primary tool for determining whether to accept a certificate is the `NSURLConnection` delegate method `connection:willSendRequestForAuthenticationChallenge:`. In this method, you're supposed to determine whether you're willing to authenticate to this server and, if so, to provide the credentials. The following code authenticates to any server that presents a noncorrupt certificate, whether or not the certificate is valid or trusted:

```
- (void)connection:(NSURLConnection *)connection
  willSendRequestForAuthenticationChallenge:
  (NSURLAuthenticationChallenge *)challenge
{
  SecTrustRef trust = challenge.protectionSpace.serverTrust;
  NSURLCredential *cred;
  cred = [NSURLCredential credentialForTrust:trust];
  [challenge.sender useCredential:cred
        forAuthenticationChallenge:challenge];
}
```

This code extracts the trust object, discussed later, and creates a credential object for it. HTTPS connections always require a credential object, even if you're not passing credentials to the server.

In the next example, you're trying to connect to the IP address 72.14.204.113, which is `encrypted.google.com`. The certificate you receive is `*.google.com`, which is a mismatch. The string `72.14.204.113` doesn't include the string `.google.com`. You decide to accept any trusted certificate that includes `google.com` in its subject. To compile this example, you need to link `Security.framework` into your project.

ConnectionViewController.m (Connection)

```
- (void)connection:(NSURLConnection *)connection
  willSendRequestForAuthenticationChallenge:
  (NSURLAuthenticationChallenge *)challenge
{
  NSURLProtectionSpace *protSpace = challenge.protectionSpace;
  SecTrustRef trust = protSpace.serverTrust;
  SecTrustResultType result = kSecTrustResultFatalTrustFailure;

  OSStatus status = SecTrustEvaluate(trust, &result);
  if (status == errSecSuccess &&
      result == kSecTrustResultRecoverableTrustFailure) {
    SecCertificateRef cert = SecTrustGetCertificateAtIndex(trust,
                                                           0);
    CFStringRef subject = SecCertificateCopySubjectSummary(cert);

    NSLog(@"Trying to access %@. Got %@.", protSpace.host,
          subject);
    CFRange range = CFStringFind(subject, CFSTR(".google.com"),
                                 kCFCompareAnchored|
                                 kCFCompareBackwards);
    if (range.location != kCFNotFound) {
      status = RNSecTrustEvaluateAsX509(trust, &result);
    }
    CFRelease(subject);
  }
```

```objc
    if (status == errSecSuccess) {
      switch (result) {
        case kSecTrustResultInvalid:
        case kSecTrustResultDeny:
        case kSecTrustResultFatalTrustFailure:
        case kSecTrustResultOtherError:
// We've tried everything:
        case kSecTrustResultRecoverableTrustFailure:
          NSLog(@"Failing due to result: %u", result);
          [challenge.sender cancelAuthenticationChallenge:challenge];
          break;

        case kSecTrustResultProceed:
        case kSecTrustResultUnspecified: {
          NSLog(@"Success with result: %u", result);
          NSURLCredential *cred;
          cred = [NSURLCredential credentialForTrust:trust];
          [challenge.sender useCredential:cred
                forAuthenticationChallenge:challenge];
        }
          break;

        default:
          NSAssert(NO, @"Unexpected result from trust evaluation:%u",
                   result);
          break;
      }
    }
    else {
      // Something was broken
      NSLog(@"Complete failure with code: %lu", status);
      [challenge.sender cancelAuthenticationChallenge:challenge];
    }
}
```

In this routine, you're passed a challenge object and extract the trust object. You evaluate the trust object (SecTrustEvaluate) and receive a recoverable failure. Typically, a recoverable failure is something like a name mismatch. You fetch the certificate's subject and determine whether it's "close enough" (in this case, checking whether it includes .google.com). If you're okay with the name you were passed, you reevaluate the certificate as a simple X.509 certificate rather than as part of an SSL handshake (that is, you evaluate it while ignoring the hostname). This is done with a custom function RNSecTrustEvaluateAsX509, shown here:

```objc
static OSStatus RNSecTrustEvaluateAsX509(SecTrustRef trust,
                                         SecTrustResultType *result
                                         )
{
  OSStatus status = errSecSuccess;

  SecPolicyRef policy = SecPolicyCreateBasicX509();
  SecTrustRef newTrust;
  CFIndex numberOfCerts = SecTrustGetCertificateCount(trust);
  NSMutableArray *certs = [NSMutableArray new];
```

```
for (NSUInteger index = 0; index < numberOfCerts; ++index) {
  SecCertificateRef cert;
  cert = SecTrustGetCertificateAtIndex(trust, index);
  [certs addObject:(__bridge id)cert];
}

status = SecTrustCreateWithCertificates((__bridge CFArrayRef)certs,
                                        policy,
                                        &newTrust);
if (status == errSecSuccess) {
  status = SecTrustEvaluate(newTrust, result);
}

CFRelease(policy);
CFRelease(newTrust);
return status;
}
```

This function creates a new trust object by copying all the certificates from the original trust object created by the URL loading system. This trust object uses the simpler X.509 policy, which checks only the validity and trust of the certificate itself, without considering the hostname as the original SSL policy does.

A certificate may also be invalid because it has expired. Unfortunately, although you can reevaluate the certificate using any date you want using `SecTrustSetVerifyDate`, there is no easy, public way to determine the validity dates for the certificate. The following private methods allow you to work out the valid range:

```
CFAbsoluteTime SecCertificateNotValidBefore(SecCertificateRef);
CFAbsoluteTime SecCertificateNotValidAfter(SecCertificateRef);
```

As with all private methods, they may change at any time and may be rejected by Apple. The only other practical way to parse the certificate is to export it with `SecCertificateCopyData` and parse it again using OpenSSL. The process of building and using OpenSSL on iOS is beyond the scope of this book. Search the Web for "OpenSSL iOS" for several explanations on how to build this library.

After the trust object is evaluated, the final result is a `SecTrustResultType`. Several results represent "good" or "possibly good" certificates:

- `kSecTrustResultProceed`—The certificate is valid, and the user has explicitly accepted it.

- `kSecTrustResultUnspecified`—The certificate is valid, and the user has not explicitly accepted or rejected it. Generally, you accept it in this case.

- `kSecTrustResultRecoverableTrustFailure`—The certificate is invalid, but in a way that may be acceptable, such as a name mismatch, expiration, or lack of trust (such as a self-signed certificate).

The following results indicate that the certificate should not be accepted:

- `kSecTrustResultDeny`—The certificate is valid, and the user has explicitly rejected it.

- kSecTrustResultInvalid—The validation was unable to complete, likely because of a bug in your code.

- kSecTrustResultFatalTrustFailure—The certificate itself was defective or corrupted.

- kSecTrustResultOtherError—The validation was unable to complete, likely because of a bug in Apple's code. You should never see this error.

Determining Certificate Trust

So far, you've found out how to determine whether a certificate is valid, but that doesn't mean it's trusted. Returning to the example of identifying yourself at the bank, if you present your Metallica fan club membership card, it probably won't be accepted as identification. The bank has no reason to believe that your fan club has done a good job making sure you're who you say you are. That's the same situation an application faces when presented with a certificate signed by an unknown authority.

To be trusted, a certificate must ultimately be signed by one of the certificates in the trust object's list of *anchor certificates*. Anchor certificates are those certificates that are explicitly trusted by the system. iOS ships with more than a hundred of them from companies and government agencies. Some are global names such as VeriSign and DigiTrust; others are more localized such as QuoVadis and Vaestorekisterikeskus. Each of these organizations went through a complex audit process and paid significant amounts of money to be in the root store, but that doesn't mean your application needs to trust them.

If you generate your own certificate, you can embed the public key in your application and configure your trust object to accept only that certificate or certificates signed by it. This gives you greater control over your security and can save you some money.

For this example, you create a self-signed root certificate.

1. Open Keychain Access.

2. Select Keychain Access menu⇨Certificate Assistant⇨Create a Certificate.

3. Enter any name you like, set the Identity Type to Self Signed Root, set the Certificate Type to SSL Client, and create the certificate. You receive a warning that this is a self-signed certificate. That is the intent of this process, so you click Continue. Your newly created certificate displays a warning that "This root certificate is not trusted." That's also as expected because it isn't in the root keychain.

4. Back in the Keychain Access window, select the login keychain and select the category Certificates.

5. Find your certificate and drag it to the desktop to export it. This file includes only the public key. Keychain does not export the private key by default. Drag the public key file into your Xcode project.

You can test that a certificate you've received is signed by your certificate as follows:

```
SecTrustRef trust = ...; // A trust to validate
NSError *error;
NSString *path = [[NSBundle mainBundle] pathForResource:@"MyCert"
                                                 ofType:@"cer"];
NSData *certData = [NSData dataWithContentsOfFile:path
                                          options:0
                                            error:&error];
```

```
SecCertificateRef certificate;
certificate = SecCertificateCreateWithData(NULL,
                                (__bridge CFDataRef)certData);
NSArray *certs = @[ (__bridge id)certificate ];
SecTrustSetAnchorCertificates(trust,
                                (__bridge CFArrayRef)certs);
CFRelease(certificate);
```

You load the certificate from your resource bundle into an `NSData`, convert it into a `SecCertificate`, and set it as the anchor for the trust object. The trust object now accepts only the certificates passed to `SecTrustSetAnchorCertificates` and ignores the system's anchors. If you want to accept both, you can use `SecTrustSetAnchorCertificatesOnly` to reconfigure the trust object.

Using these techniques, you can correctly respond to any certificate in your `connection:willSendRequestForAuthenticationChallenge:` method and control which certificates you accept or reject.

Employing File Protection

iOS provides hardware-level encryption of files. Files marked for protection are encrypted using a per-device key, which is encrypted using the user's password or personal identification number (PIN). Ten seconds after the device is locked, the unencrypted per-device key is removed from memory. When the user unlocks the device, the password or PIN is used to decrypt the per-device key again, which is then used to decrypt the files.

The weakest link in this scheme is the user's password. On an iPhone, users almost exclusively use a 4-digit PIN, which offers only 10,000 combinations (far fewer are used in practice). In May 2011, ElcomSoft Co. Ltd demonstrated that it could brute-force attack a 4-digit PIN in about 20–40 minutes. So a PIN doesn't protect against forensics or device theft, but does protect against attackers who have access to the device only for a few minutes. On the iPad, typing a real password is much more convenient, so the security is similar to file encryption on a laptop.

For a developer, the specifics of the iOS encryption scheme aren't critical. The scheme is effective enough for users to expect it on any application that holds sensitive information.

You can configure the protection of individual files that you create with `NSFileManager` or `NSData`. The options, shown in the following list, have slightly different names. `NSFileManager` applies string attributes to the file, whereas `NSData` uses numeric options during creation, but the meanings are the same. The `FileManager` constants begin with `NSFileProtection...`, and the `NSData` constants begin with `NSDataWritingFileProtection....`

- `...None`—The file is not protected and can be read or written at any time. This is the default value.

- `...Complete`—Any file with this setting is protected 10 seconds after the device is locked. This is the highest level of protection, and the setting you should generally use. Files with this setting may not be available when your program is running in the background. When the device is unlocked, these files are unprotected.

- `...CompleteUnlessOpen`—Files with this setting are protected 10 seconds after the device is locked unless they're currently open. This allows your program to continue accessing the file while running in the background. When the file is closed, it is protected if the device is locked.

■ ...`CompleteUntilFirstUserAuthentication`—Files with this setting are protected only between the time the device boots and the first time the user unlocks the device. The files are unprotected from that point until the device is rebooted. This allows your application to open existing files while running in the background. You can create new files using ...`CompleteUnlessOpen`. This setting is better than the `None` setting but should be avoided if at all possible because it provides very limited protection.

The best way to employ this protection is by enabling Data Protection for your App ID. In the Identifiers section of the Developer Portal, you can configure the default protection level for all files your application creates, as shown in Figure 14-2.

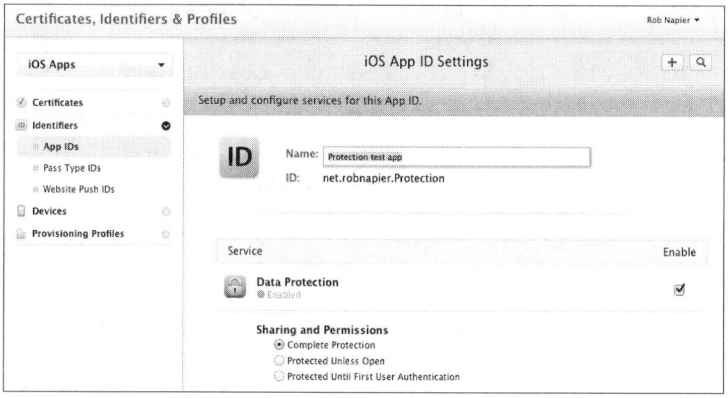

Figure 14-2 iOS App ID Settings

You can also apply protection on a per-file basis. To create a new file with file protection turned on, convert it to an `NSData` and then use `writeToFile:options:error:`. This is preferable to creating the file and then using `NSFileManager` to set its protection attribute.

```
[data writeToFile:dataPath
        options:NSDataWritingFileProtectionComplete
          error:&writeError];
```

To create a protected file in the background, you can apply the option ...`CompleteUnlessOpen`, which allows you to read as long as it's open when the device locks. You should generally avoid this option unless you're actually in the background. The easiest way to create a protected file when you may or may not be in the background is as follows:

```
[data writeToFile:path
        options:NSDataWritingFileProtectionComplete
          error:&error] ||
[data writeToFile:path
        options:NSDataWritingFileProtectionCompleteUnlessOpen
          error:&error];
```

If you use this technique, upgrade your file protection at startup with a routine like this:

```
- (void)upgradeFilesInDirectory:(NSString *)dir
                          error:(NSError **)error {
```

```
NSFileManager *fm = [NSFileManager defaultManager];
NSDirectoryEnumerator *dirEnum = [fm enumeratorAtPath:dir];
for (NSString *path in dirEnum) {
  NSDictionary *attrs = [dirEnum fileAttributes];
  if (![[attrs objectForKey: NSFileProtectionKey]
      isEqual:NSFileProtectionComplete]) {
    attrs = @{NSFileProtectionKey: NSFileProtectionComplete};
    [fm setAttributes:attrs ofItemAtPath:path error:error];
  }
 }
}
```

If your application needs to know whether protected data is available, you can use one of the following:

- Implement the methods `applicationProtectedDataWillBecomeUnavailable:` and `applicationProtectedDataDidBecomeAvailable:` in your application delegate.

- Observe the notifications `UIApplicationProtectedDataWillBecomeUnavailable` and `UIApplicationProtectedDataDidBecomeAvailable` (these constants lack the traditional `Notification` suffix).

- Check `[[UIApplication sharedApplication] protectedDataAvailable]`.

For foreground-only applications, file protection is easy. Because it's so simple and it's hardware-optimized, unless you have a good reason not to, you generally should protect your files. If your application runs in the background, you need to give more careful thought to how to apply file protection but still be sure to protect all sensitive information as well as possible.

Using Keychains

File protection is intended to protect *data*. A keychain is intended to protect *secrets*. In this context, a secret is a small piece of data used to access other data. The most common secrets are passwords and private keys.

The keychain is protected by the operating system and is encrypted when the device is locked. In practice, it works similarly to file protection. Unfortunately, the Keychain API is anything but friendly. Many people have written wrappers around the Keychain API. My current recommendation is SGKeychain by Justin Williams (see the "Further Reading" section for the link). This is what I discuss in this section after a brief introduction to the low-level data structures.

An item in the keychain is called a `SecItem`, but it's stored in a `CFDictionary`. There is no `SecItemRef` type. The `SecItem` has five classes: generic password, Internet password, certificate, key, and identity. In most cases, you want to use a generic password. Many problems come from developers trying to use an Internet password, which is more complicated and provides little benefit unless you're implementing a web browser. SGKeychain uses only generic password items, which is one reason I like it. Storing private keys and identities is rare in iOS applications and isn't discussed in this book. Certificates that contain only public keys should generally be stored in files rather than in the keychain.

Keychain items have several searchable *attributes* and a single encrypted *value*. For a generic password item, some of the more important attributes are the account (`kSecAttrAccount`), service (`kSecAttrService`), and identifier (`kSecAttrGeneric`). The value is generally the password.

So with that background, it's time to see how to use `SGKeychain`. Setting passwords is straightforward:

```
NSError *error;
if (! [SGKeychain setPassword:@"password"
                    username:@"bob"
                 serviceName:@"myservice"
            updatingExisting:YES
                       error:&error]) {
  // handle error
}
```

The `serviceName` can be anything you like. It's useful if you have multiple services. Otherwise, just set it to an arbitrary string.

Reading passwords is simple as well:

```
NSError *error;
NSString *password = [SGKeychain passwordForUsername:@"bob"
                                         serviceName:@"myservice"
                                               error:&error];

if (! password) {
  // handle error
}
```

Reading and writing to the keychain can be expensive, so you don't want to do it too often. The keychain is not a place to store sensitive data that changes often. That data should be written in an encrypted file, as described in the earlier section, "Employing File Protection."

Sharing Data with Access Groups

The iOS sandbox creates a significant headache for application suites. If you have multiple applications that work together, there is no easy way to share information between them. Of course, you can save the data on a server, but the user still needs to enter credentials for each of your applications.

iOS offers a solution to credential sharing with access groups. Every keychain item has an access group identifier associated with it. By default, the access group identifier is your application identifier prefix (also called your "bundle seed") followed by your application bundle identifier. For instance, if my application identifier prefix is `ABCD123456`, and my application identifier is `com.iosptl.AwesomeApp`, any keychain items that I create without specifying an access group have an access group of `ABCD123456.com.iosptl.AwesomeApp`.

To use a given access group, your application's prefix must match the application group, and you must include the access group in your Entitlements file. Because the application prefixes must match, you can share keychain data only between applications signed by the same certificate.

To configure this in Xcode, open the target pane, and in the Capabilities section, enable Keychain Sharing. Then add a new keychain access group, as shown in Figure 14-3.

Xcode is somewhat confusing here. The UI indicates that the access group ID is `net.robnapier.sharedStuff`, but as just discussed, the access group ID is really `$(AppIdentifierPrefix)net.robnapier.sharedStuff`. You'll see this if you look in the Entitlements file.

Figure 14-3 Creating a keychain access group

Keychain items that you create without an explicit group inherit the default group value, which is the prefix followed by the application identifier. So if you just list all your applications' identifiers in the Keychain Groups section, your applications can read each other's items, and you never have to pass an access group. This approach is easy, but I don't recommend it in most cases. The problem is that it is easy to create duplicate items this way. If two applications create the same username and service but use different access groups, you get two records. This can lead to surprising results when you later search for a password.

If you plan to share an access group, I recommend that you create an explicit access group for this purpose. SGKeychain supports passing an access group identifier. There is one problem, however. Your access group must begin with your application identifier prefix, and there is no simple, programmatic way to determine this prefix at runtime. You can hard-code the prefix, but that's not very elegant.

A solution to determining the application prefix (also called the bundle seed ID) at runtime is to create a keychain item and check what access group was assigned. The following code from David H (http://stackoverflow.com/q/11726672/97337) demonstrates this technique:

```objc
- (NSString *)bundleSeedID {
  NSDictionary *query = @{
                (__bridge id)kSecClass : (__bridge
  id)kSecClassGenericPassword,
                (__bridge id)kSecAttrAccount : @"bundleSeedIDQuery",
                (__bridge id)kSecAttrService : @"",
                (__bridge id)kSecReturnAttributes : (id)kCFBooleanTrue
  };
  CFDictionaryRef result = nil;
  OSStatus status = SecItemCopyMatching((__bridge CFTypeRef)query,
                                        (CFTypeRef *)&result);
  if (status == errSecItemNotFound) {
    status = SecItemAdd((__bridge CFTypeRef)query, (CFTypeRef *)&result);
  }
  if (status != errSecSuccess) {
    return nil;
  }
  NSString *accessGroup = CFDictionaryGetValue(result,
  kSecAttrAccessGroup);
```

```
    NSArray *components = [accessGroup componentsSeparatedByString:@"."];
    NSString *bundleSeedID = components[0];
    CFRelease(result);
    return bundleSeedID;
}
```

Using this technique, you can access a shared keychain as follows:

```
NSString *accessGroup = [NSString stringWithFormat:@"%@.%@",
                            [self bundleSeedID], kSharedKeychainName];
 [SGKeychain setPassword:password
              username:username
           serviceName:service
           accessGroup:accessGroup
         updateExisting:YES
                error:&error];
```

Using Encryption

Most of the time, iOS handles all your encryption needs for you. It automatically encrypts and decrypts HTTPS for network traffic and manages encrypted files using file protections. If you have certificates, SecKeyEncrypt and SecKeyDecrypt handle asymmetric (public/private key) encryption for you.

But what about simple, symmetric encryption using a password? iOS has good support for this type of encryption, but limited documentation. The available documentation is in /usr/include/ CommonCrypto. Most of it assumes that you have some background in cryptography and doesn't warn you of common mistakes. This unfortunately has led to a lot of bad AES code examples on the Internet. In an effort to improve things, I developed RNCryptor as an easy-to-use encryption library based on CommonCrypto. In this section, I discuss many of the underlying details of how RNCryptor works and why. This way, you can modify RNCryptor with confidence or build your own solution. If you just want a prebuilt solution, skip this section and read the RNCryptor docs. You can find the source for RNCryptor at github.com/rnapier/ RNCryptor.

> This section intentionally does not include any encryption or decryption code samples. Incorrectly copying crypto code is a major cause of security mistakes. The following section covers the theory you need to use cryptography correctly. The majority of users should just use RNCryptor and should not write their own CommonCrypto calls. It is very easy to make small mistakes that drastically reduce the security of your system.

> You should never copy CommonCrypto examples from the Internet without understanding the materials in this section. For readers who do need to design their own crypto formats, this section will help you understand the principles you need. You should then be able to use the CommonCrypto documentation and RNCryptor as an example to develop your own code.

Overview of AES

The *Advanced Encryption Standard*, or AES, is a symmetric encryption algorithm. Given a key, it converts plaintext into ciphertext. The same key is used to convert ciphertext back into plaintext. Originally named Rijndael, in 2001 the algorithm was selected by the U.S. government as its standard for encryption.

It's a very good algorithm. Unless you need another algorithm for compatibility with an existing system, always use AES for symmetric encryption. The best cryptographers in the world have carefully scrutinized it, and it's hardware-optimized on iOS devices, making it extremely fast.

AES offers three key lengths: 128, 192, and 256 bits. There are slight differences in the algorithm for each length. Unless you have very specialized needs, I recommend AES-128. It offers an excellent tradeoff of security and performance, including time performance and battery life performance.

At its heart, AES is just a mathematical function that takes a fixed-length key and a 16-byte block and returns a different 16-byte block. This has several implications:

- AES uses a fixed-length key, not a variable-length password. You must convert passwords to keys to be able to use them in AES. I explain this further in the next section, "Converting Passwords to Keys with PBKDF2."

- AES can only transform a 16-byte block into another 16-byte block. To work on plaintext that isn't exactly 16 bytes long, you must apply some mechanism outside AES. See the sections "AES Mode and Padding" and "The Initialization Vector (IV)" for how AES works with data that is not exactly 16 bytes long.

- Every possible block of 16 bytes is legal as plaintext or as ciphertext. This means that AES provides no authentication against accidental or intentional corruption of the ciphertext. See the section "Authentication with HMAC" for how to add error detection to AES.

- Every possible key can be applied to every possible block. This means that AES provides no mechanism for detecting that the key was incorrect. See the section "Bad Passwords" for more information on this problem.

- Although AES supports three key lengths, it has only one block size. This sometimes creates confusion over the constant `kCCBlockSizeAES128`. The "128" in this constant refers to the block size, not the key size. So when using AES-192 or AES-256, you still pass `kCCBlockSizeAES128` as the block size.

Converting Passwords to Keys with PBKDF2

A *key* is not the same thing as a *password*. A key is a very large number used to encrypt and decrypt data. All possible keys for an encryption system are called its *key space*. A password is something a human can type. Long passwords that include spaces are sometimes called *passphrases*, but for simplicity, I just use the word "password" no matter the construction.

If you try to use a password directly as an AES key, you significantly shrink the number of available keys. If the user selects a random 16-character password using the 94 characters on a standard keyboard, that only creates about a 104-bit key space, approximately one 10-millionth the size of the full AES key space. Real users select passwords from a much smaller set of characters. Worse yet, if the user has a password longer than 16 bytes (16 single-byte characters or 8 double-byte characters), you throw away part of it when using AES-128. Clearly, using the password directly as the key is not the right approach.

You need a way to convert a password into a usable key that makes it as hard as possible on the attacker to search every possible password. The answer is a *password-based key derivation function*. Specifically, you use PBKDF2, which is defined by RSA Laboratories' Public-Key Cryptography Standards (PKCS) #5. You don't need to know the internals of PBKDF2 or PKCS #5, but it's important to know the names because they show up in the documentation. What is important is that PBKDF2 converts a password into a key.

To use PBKDF2, you need to generate a *salt*, which is just a large random number. The standard recommends at least 64 bits. The salt is combined with the password to prevent identical passwords from generating identical keys. You then iterate through the PBKDF2 function a specific number of times, and the resulting data is your key. This process is called *stretching*. To decrypt the data, you need to preserve the salt and the number of iterations. Typically, the salt is saved with the encrypted data, and the number of iterations is a constant in your source code, but you can also save the number of iterations with the encrypted data.

The important fact here is that the salt, the number of iterations, and the final ciphertext are all public information. Only the key and the original password are secrets.

Generating the salt is easy. It's just a large random number. You can create it with a method like `randomDataOfLength:`, shown in the following code:

RNCryptor.m (RNCryptor)

```
const NSUInteger kPBKDFSaltSize = 8;

+ (NSData *)randomDataOfLength:(size_t)length {
  NSMutableData *data = [NSMutableData dataWithLength:length];

  int result = SecRandomCopyBytes(kSecRandomDefault,
                                  length,
                                  data.mutableBytes);
  NSAssert(result == 0, @"Unable to generate random bytes: %d",
           errno);

 return data;
}
...
NSData *salt = [self randomDataOfLength:kPBKDFSaltSize];
```

Originally, the standard called for 1,000 iterations of PBKDF2, but this has gone up as CPUs have improved. I recommend between 10,000 and 100,000 iterations on an iPhone 4 and 50,000 to 500,000 iterations on a modern MacBook Pro. The reason for the large number is to slow down brute-force attacks. An attacker generally tries passwords rather than raw AES keys because the number of practical passwords is much smaller. When you require 10,000 iterations of the PBKDF2 function, the attacker must waste about 80ms per attempt on an iPhone 4. That adds up to 13 minutes of search time for a 4-digit PIN, and months or years to search for even a very simple password. The extra 80ms for a single key generation is generally negligible. Going up to 100,000 iterations adds nearly a second to key generation on an iPhone but provides much better protection if the password guessing is done on a desktop, even if the password is very weak.

PBKDF2 requires a *pseudorandom function* (PRF), which is just a function that can generate a very long series of statistically random numbers. iOS supports various sizes of SHA-1 and SHA-2 for this. SHA256, which is a specific size of SHA-2, is often a good choice.

> **RNCryptor uses SHA-1 as its PRF for historical compatibility reasons. This will likely change to SHA256 in the future.**

Luckily, it's easier to use PBKDF2 than it is to explain it. The following method accepts a password string and salt data and returns an AES key:

RNCryptor.m (RNCryptor)

```
+ (NSData *)keyForPassword:(NSString *)password salt:(NSData *)salt
                    settings:(RNCryptorKeyDerivationSettings)keySettings
{
  NSMutableData *derivedKey = [NSMutableData
                                 dataWithLength:keySettings.keySize];
  size_t passwordLength = [password
                     lengthOfBytesUsingEncoding:NSUTF8StringEncoding];
  int result = CCKeyDerivationPBKDF(keySettings.PBKDFAlgorithm,
                               password.UTF8String,
                               passwordLength,
                               salt.bytes,
                               salt.length,
                               keySettings.PRF,
                               keySettings.rounds,
                               derivedKey.mutableBytes,
                               derivedKey.length);
  // Do not log password here
  NSAssert(result == kCCSuccess,
    @"Unable to create AES key for password: %d", result);

  return derivedKey;
}
```

The password salt must be stored along with the cipher text.

AES Mode and Padding

AES is a block cipher, which means that it operates on a fixed-sized block of data. AES works on exactly 128 bits (16 bytes) of input at a time. Most things you want to encrypt, however, are not exactly 16 bytes long. Many things you want to encrypt aren't even a multiple of 16 bytes. To address this issue, you need to choose an appropriate mode.

A *mode* is an algorithm for chaining blocks together so that arbitrary data can be encrypted. Modes are applicable to any block cipher, including AES.

> If you don't need the details on how AES modes work, just trust me that the right answer on iOS devices is CBC with PKCS#7 padding and skip to the "The Initialization Vector (IV)" section.

The simplest mode is *electronic codebook* (ECB). *Never use this mode in an iOS app.* ECB is appropriate only for cases when you need to encrypt a huge number of random blocks. This is such a rare situation that you shouldn't even consider it for iOS apps. ECB simply takes each block and applies the block cipher (AES) and outputs the result. It is extremely insecure for encrypting nonrandom data. The problem is that if two plaintext blocks are the same, the ciphertext is the same. This leaks a lot of information about the plaintext and is subject to numerous attacks. Do not use ECB mode.

The most common block cipher mode is *cipher block chaining* (CBC). The ciphertext of each block is XORed with the next plaintext block prior to encryption. This is a very simple mode but very effective in solving the problems with ECB. It has two major problems: Encryption cannot be done in parallel, and the plaintext must be an exact multiple of the block size (16 bytes). Parallel encryption is not a major concern on iOS platforms. iPhone and iPad perform AES encryption on dedicated hardware and don't have sufficient hardware to parallelize the operation. The block size problem is dealt with by padding.

Padding is the practice of expanding plaintext to the correct length prior to encryption. It needs to be done in a way that the padding can be unambiguously removed after decryption. There are several ways to achieve this result, but the only way supported by iOS is called PKCS #7 padding. This is requested with the option `kCCOptionPKCS7Padding`. In PKCS #7 padding, the encryption system appends n copies of the byte n. So, if your last block were 15 bytes long, it would append one 0x01 byte. If your last block were 14 bytes long, it would append two 0x02 bytes. After decryption, the system looks at the last byte and deletes that number of padding bytes. It also performs an error check to make sure that the last n bytes are actually the value n. This padding algorithm has surprising implications that we discuss in the section "Bad Passwords."

There is a special case if the data is exactly the length of the block. To be unambiguous, PKCS #7 requires a full block of 0x10.

PKCS #7 padding means that the ciphertext can be up to a full block longer than the plaintext. Most of the time this is fine, but in some situations, this is a problem. You may need the ciphertext to fit in the same space as the plaintext. There is a CBC-compatible method called *ciphertext stealing* that can do this, but it's not supported on iOS.

CBC is the most common cipher mode, making it the easiest to exchange with other systems. Unless you have a strong reason to use something else, use CBC. This is the mode recommended by Apple's security team, based on my discussions with them.

If padding is impossible in your situation, I recommend cipher feedback (CFB) mode. It's less widely supported than CBC, but it is a fine mode and doesn't require padding. Output feedback (OFB) is also fine for the same reasons. I have no particular advice on choosing one over the other. They are both good.

Counter (CTR) mode is useful if you need to avoid the overhead of an initialization vector but is easy to use insecurely. I discuss CTR, when to use it and how, in the section "The Initialization Vector (IV)."

XTS is a specialized mode for random-access data, particularly encrypted file systems. I don't have enough relevant information to recommend it. FileVault on Mac uses it, so Apple clearly has an interest in it.

All the modes offered by iOS are unauthenticated. This means that modifying the ciphertext does not necessarily create errors. In most cases, it just changes the resulting plaintext. It's possible for attackers who know some of the contents of an encrypted document to modify it in predictable ways without the password. There are encryption modes (called *authenticated modes*) that protect against modification, but none are supported by iOS. See the section "Authentication with HMAC" for more information.

The many encryption modes can be confusing, but in almost all cases, the most common solution is best: CBC with PKCS #7 padding. If padding is problematic for you, I recommend CFB.

The Initialization Vector (IV)

As discussed in the earlier section "AES Mode and Padding," in chaining modes such as CBC, each block influences the encryption of the next block. This ensures that two identical blocks of plaintext will not generate identical blocks of ciphertext.

The first block is a special case because there's no previous block. Chaining modes allow you to define an extra block called the *initialization vector* (IV) to begin the chain. This is often labeled optional, but you need to always provide one. Otherwise, an all-zero block is used, and that leaves your data vulnerable to certain attacks.

As with the salt, the IV is just a random series of bytes that you save with the ciphertext and use during decryption:

```
iv = [self randomDataOfLength:kAlgorithmIVSize];
```

You then store this IV along with the ciphertext and use it during decryption. In some cases, adding this IV creates unacceptable overhead, particularly in high-performance network protocols and disk encryption. In that case, you still cannot use a fixed IV (such as NULL). You must use a mode that doesn't require a random IV. Counter (CTR) mode uses a nonce rather than an IV. A *nonce* is essentially a predictable IV. Whereas chaining modes like CBC require that the IV be random, nonce-based modes require that the nonce be unique. This requirement is very important. In CTR mode, a given nonce/key pair must never, ever be reused. This is usually implemented by making the nonce a monotonically increasing counter starting at 0. If the nonce counter ever needs to be reset, the key must be changed.

> The WEP wireless security protocol is a famous example of a protocol that allowed a nonce/key pair to be reused. This is why WEP is so easy for attackers to break.

The advantage of a nonce is that it doesn't have to be stored with the ciphertext the way an IV must. If a key is chosen randomly at the beginning of a communication session, both sides of the communication can use the current message number as the nonce. As long as the number of messages cannot exceed the largest possible nonce, this way of communicating is a very efficient.

The nonce uniqueness requirement is often more difficult to implement than it seems. Randomly selecting a nonce, even in a large nonce-space, may not be sufficient if the random number generator is not sufficiently random. A nonce-based mode with a nonunique nonce is completely insecure. IV-based modes, on the other hand, are weaker if the IV is reused, but they aren't completely broken.

In most iOS applications, the correct solution is CBC with PKCS #7 padding and a random IV sent along with the data.

Authentication with HMAC

None of the modes discussed so far protect against accidental or malicious modification of the data. AES does not provide any authentication. If an attacker knows that a certain string occurs at a certain location in your data, the attacker can change that data to anything else of the same size. This could change "$100" to "$500" or "bob@example.com" to "sue@example.net."

Many block cipher modes provide authentication. Unfortunately, iOS does not support any of them. To authenticate data, you need to do it yourself. The best way to do this is to add an *authentication code* to the data. The best choice on iOS is a hash-based message authentication code (HMAC).

An HMAC requires its own key, just like encryption. If you're using a password, you can use PBKDF2 to generate the HMAC key using a different salt. If you're using a random encryption key, you need to also generate a random HMAC key. If you use a salt, you include the HMAC salt with the ciphertext, just like the encryption salt. See the section "Converting Passwords to Keys with PBKDF2" for more information.

Because an HMAC can hash anything, it can be computed on the plaintext or on the ciphertext. It is best to encrypt first and then hash the ciphertext. This approach allows you to verify the data prior to decrypting it. It also protects against certain kinds of attacks.

In practice, the hashing code is similar to the encryption code. You either pass all the ciphertext to a single HMAC call, or you call the HMAC routines multiple times with blocks of ciphertext.

Bad Passwords

The lack of authentication has a commonly misunderstood side effect. AES cannot directly detect when you pass the wrong key (or password). Systems that use AES often appear to detect incorrect passwords, but that's actually just luck.

Most AES systems use CBC encryption with PKCS #7 padding. As I explained in the earlier section "AES Mode and Padding," the PKCS #7 padding scheme adds a predictable pattern to the end of the plaintext. When you decrypt, the padding provides a kind of error detection. If the plaintext does not end in a legal padding value, the decryptor returns an error. But if the last decrypted byte happens to be 0x01, this appears to be legal padding, and you aren't able to detect a bad password. For systems without padding (such as CFB or CTR), even this check is not possible.

If you use a password and PBKDF2 to generate your encryption key and HMAC key, the HMAC gives you better password validation. Because the HMAC key is based on the password, and the password is wrong, the HMAC doesn't validate. But if you use a random encryption key and random HMAC key, you still aren't able to detect a bad password.

To some extent, this is a security feature. If the attacker cannot easily detect the difference between a correct and incorrect password, it's harder to break the encryption. In almost all cases in which you want to report a bad password, the reason is that you're using PBKDF2, and HMAC will solve the problem. Otherwise, you may have to add error detection inside your plaintext to be certain.

Note that even with HMAC, you cannot easily distinguish a corrupted ciphertext from a bad password. The best way to check for bad passwords is to encrypt some known piece of data along with the plaintext. Be sure to use a different IV for this or prepend the known data to the plaintext before encrypting. During decryption, you can check the decryption of the known data to verify the password is correct.

Combining Encryption and Compression

Compressing data before encrypting it is sometimes a good idea. There's a theoretical security benefit to doing so, but generally it's just to make the data smaller. The important point to remember is that you must compress before you encrypt. You can't compress encrypted data. If you could, that would suggest patterns in the ciphertext, which would indicate a poor encryption algorithm. In most cases, encrypting and then compressing leads to a larger output than the original plaintext.

Summary

iOS provides a rich collection of security frameworks to make it as easy as possible to secure your users' data. This chapter showed you how to secure network communications, files, and passwords. You also found out how to properly validate certificates so that you can ensure that your application communicates only with trusted sources. Securing your application requires a few extra lines of code, but taking care of the basics is generally not difficult using the code provided in this chapter.

Further Reading

Apple Documentation

The following documents are available in the iOS Developer Library at `developer.apple.com` or through the Xcode Documentation and API Reference.

> *Certificate, Key, and Trust Services Programming Guide*
>
> *Secure Coding Guide (/usr/lib/CommonCrypto)*

WWDC Sessions

The following session videos are available at `developer.apple.com`.

> WWDC 2011, "Session 208: Securing iOS Applications"
>
> WWDC 2012, "Session 704: The Security Framework"
>
> WWDC 2012, "Session 714: Protecting the User's Data"

Other Resources

Aleph One. *Phrack, Volume 7, Issue Forty-Nine*, "Smashing the Stack for Fun and Profit" (1996). More than 15 years later, this is still one of the best introductions to buffer overflows available, with examples.

```
www.phrack.org/issues.html?issue=49&id=14#article
```

Boneh, Dan. Stanford "Cryptography" Course. There is no better free resource for learning how to do cryptography correctly than this course. This is not a quick overview. You'll discover the math behind cryptosystems and how they're put together correctly and incorrectly. You'll learn how to reason about the security of systems and mathematically prove security theorems about them. It's more math than computer programming. I highly recommend it, but it is a significant commitment. It's broken into two six-week sections.

```
https://www.coursera.org/course/crypto
```

Napier, Rob. "RNCryptor." This is my framework for AES encryption, based on CommonCryptor. Its purpose is to make it easy to use AES correctly, and it implements all the features discussed in this chapter.

```
https://github.com/rnapier/RNCryptor
```

Schneier, Bruce. *Applied Cryptography* (John Wiley & Sons, 1996). Anyone interested in the guts of cryptography should read this book. The main problem is that after reading it, you may think you can create your own cryptography implementations. You shouldn't. Read this book as a fascinating, if dated, introduction to cryptography. Then put it down and use a well-established implementation.

Williams, Justin. "SGKeychain." This is my current recommendation for accessing the keychain.

```
https://github.com/secondgear/SGKeychain
```

Chapter 15

Running on Multiple iPlatforms, iDevices, and 64-bit Architectures

The iOS SDK was announced to the public in February 2008. At that time, only two devices were using it: iPhone and iPod touch. Apple has since been innovating vigorously, and in 2010 it introduced a bigger brother to the family, the iPad. Also in 2010, Apple introduced another new device running iOS: the Apple TV. Who knows what the future might hold—Apple might even announce an SDK for Apple TV development and may even enable running games from Apple TV controlled by your iPhone on iPod touch.

Every year, a new version of the SDK comes out along with at least two or three new device updates, and these new devices often come with additional sensors. The GPS sensor debuted with iPhone 3G; the magnetometer—a sensor used to show the direction of magnetic north (more commonly known as a compass)—debuted in iPhone 3GS; and the gyroscope (for life-like gameplay) was unveiled in iPhone 4. The iPad was introduced later with a whole new UI, a far bigger screen than the iPhone, but without a camera. The iPad added a couple of cameras (including a front-facing camera) in the second iteration, iPad 2. The iPad 2 was superseded by the new iPad, which has a better camera and features such as face detection and video stabilization. Recently, Apple announced the iPhone 5s, which runs on 64-bit architecture. I show you how to plan and implement 64-bit migration later in this chapter.

Similarly, every version of the SDK comes with powerful new features: In App Purchases, Push Notification Service, Core Data, and MapKit support in iOS 3; multitasking, blocks, and Grand Central Dispatch in iOS 4; iCloud, Twitter integration, and storyboards in iOS 5; and PassKit in iOS 6, 64-bit architecture in iOS 7 to name a few. When you use one of these features, you might be interested in providing backward compatibility to users running an older version of the operating system. Keep in mind, however, that if you're using a feature available in a newer version of the SDK, you must either forget about old users (not a good idea) or write code that adapts to both users (either by supporting an equivalent feature for older users or by prompting them that additional features are available if they run a newer version).

As a developer, you need to know how to write code that easily adapts to any device (known or unknown) and platform. For that purpose, it's easier to depend on Cocoa framework APIs to detect capabilities than to write code assuming that a certain sensor will be present on some given hardware. In short, developers need to avoid making assumptions about hardware capabilities based on device model strings.

This chapter looks at some strategies that can help you write code that adapts easily to multiple platforms and devices using the various APIs provided by the Cocoa framework. You also learn how to adapt your code to support the new, taller, iPhone 5 and the iPhone 5's new 64-bit arm64 architecture. Finally, you learn about transitioning your application to iOS 7. In the course of this chapter, you also write a category extension on the UIDevice class and add methods that check for features that aren't readily exposed by the framework.

Developing for Multiple Platforms

iOS debuted with a public SDK in version 2.0, and version 7.0 is the sixth iteration that is available for developers. One important advantage of iOS over competing platforms is that users don't have to wait for carriers to "approve" their OS updates, and because the updates are free of charge, most users (more than 75%) get the latest available OS within a month. It's usually fine for iOS developers to support just the two latest iterations of the SDK. That is, in late 2012, it was enough to support iOS 5 and iOS 6. In early 2013 most of us supported only iOS 6 (iOS 5 was used in about 10% of devices). As of this writing, in September 2013, iOS 7 reached an adoption rate of 60% within four days of general availability. For developers who are starting with a brand new app, my recommendation is to forget about iOS 6 completely. If you have an existing iOS 6 app, you should support iOS 6 and iOS 7 for a couple of months.

iOS 7 in itself is very new and different from iOS 6. The process of supporting and transitioning your codebase from iOS 6 is not as straightforward as it was before. You learn about transitioning to iOS 7 (while supporting iOS 6) later in the section titled "Transitioning to iOS 7."

Configurable Target Settings: Base SDK versus Deployment Target

To customize the features your app can use and the devices and OS versions your app can run, Xcode provides two configurable settings for the target you build. The first is your base SDK setting, and the second is the iOS Deployment Target.

Configuring the Base SDK Setting

The first configurable setting is called Base SDK. You can configure this setting by editing your target. To do so, follow these steps:

1. Open your project and select the project file on the project navigator.

2. On the editor pane, select the target and select the Build Settings tab. The Base SDK setting is usually the third option here, but the easiest way to look for a setting in this pane is to search for it in the search bar.

You can change the value to "Latest iOS SDK" or any version of the SDK installed on your development machine. The Base SDK setting instructs the compiler to use that version of SDK to compile and build your app, and this means it directly controls which APIs are available for your app. By default, new projects created with Xcode always use the latest-available SDK, and Apple handles API deprecation. Unless you have very specific reasons not to, stick to this default value.

Configuring the Deployment Target Setting

The second setting is the Deployment Target, which governs the minimum required OS version necessary for using your app. If you set this to a particular version, say 6.0, the AppStore app automatically prevents users running previous operating systems from downloading or installing your app. You can set the Deployment Target on the same Build Settings tab as the Base SDK setting.

When you're using a feature available in the iOS 7 SDK but still want to support older versions, set your Base SDK setting to the latest SDK (or iOS 7) and your Deployment Target to at least iOS 6. However, when your app is running on iOS 6 devices, some frameworks and features may not be available. It's your responsibility as a developer to adapt your app to work properly without crashing.

Considerations for Multiple SDK Support: Frameworks, Classes, and Methods

You need to handle three cases when you support multiple SDKs: frameworks, classes, and methods. In the following sections, you find out about the ways to make this possible.

Framework Availability

Sometimes a new SDK may add a whole new framework, which means that a complete framework is not available on older operating systems. An example from iOS 7 is the `GameController.framework`. This framework is available only to users running iOS 7 and above. You have two choices here. Either set the deployment target to iOS 7 and build your app only for customers running iOS 7 and above, or check whether the given framework is present on the user's operating system and hide necessary UI elements that invoke a call to this framework. Clearly, the second choice is the optimal one.

When you use a symbol that's defined in a framework that is not available on older versions, your app will not load. To avoid this problem and to selectively load a framework, you must *weak-link* it. To weak-link a framework, open the target settings page from the project settings editor. Then open the Build Phases tab and expand the fourth section (Link Binary With Libraries). You will see a list of frameworks that are currently linked to your target. If you haven't yet changed a setting here, all the frameworks are set to Required by default. Click the Required combo box and change it to Optional. This weak-links the said framework.

When you weak-link a framework, missing symbols automatically become null pointers, and you can use this null check to enable or disable UI elements.

An example on iOS 7 is the `GameController.Framework`. When you use the built-in GameController framework, weak-link it and do a runtime check to see if it is available. If not, you have to implement your own methods to mimic that functionality.

> **When you link a framework that is present only on a newer version of the SDK, but you still specify the iOS Deployment target to an SDK older than that, your application will fail to launch and crash almost immediately. This will cause your app to be rejected. When you receive a crash report from the Apple review team stating that the app crashes immediately on launch (mostly without any useful crash dumps), this is what you have to look for. The fix for this crash is to weak-link the framework. To learn more about debugging, read Chapter 17 in this book.**

Class Availability

Sometimes a new SDK might add new classes to an existing framework. This means that even if the framework gets linked, not all symbols are available to you on older operating systems. An example from iOS 7 is the `NSLayoutManager` class defined in `UIKit.Framework`. This framework is linked with every iOS app, so when you're using this class, you need to check for its presence by instantiating an object using the `NSClassFromString` method. If it returns `nil`, that class is not present on the target device.

Another method to check for class availability is to use the `class` method instead of `NSClassFromString`, as shown in the following code:

Checking for Availability of the UIStepper Control

```
if ([NSLayoutManager class])  {
    // Use the new TextKit based layout
} else {
    // Use the old Core Text based layout
}
```

Method Availability

In some cases, new methods are added to an existing class in the new SDK. A classic example from iOS 4 is multitasking support. The class `UIView` has a method called `setTintColor:`. The following code checks for this class:

Code for Checking Whether a Method Is Available in a Class

```
if ([self.view respondsToSelector:@selector(setTintColor:)]) {

  // code to set the view's tint color goes here
}
```

To check whether a method is available in a given class, use the `respondsToSelector:` method. If it returns `YES`, you can use the method you checked for.

If the method you're checking is a global C function, equate it to `NULL` instead, as shown in the following code:

Checking Availability of a C Function

```
if (CFunction != NULL) {
  CFunction(a);
}
```

> You have to equate the function name explicitly to `NULL`. Implicitly assuming pointers as `nil` or `NULL` does not work. Do note that the you should not use `CFunction()`. It's just `CFunction` without the parentheses. Checking the condition does not invoke the method.

Checking the Availability of Frameworks, Classes, and Methods

Although remembering framework availability is quite easy, remembering the availability of every single class and method can be challenging. Equally difficult is reading through the complete iOS documentation to find out which method is available and which method is not. I recommend two different ways to check the availability of a framework, class, or method.

Developer Documentation

The straightforward way to check the availability of symbols or frameworks is to search in the Availability section of the developer documentation. Figure 15-1 is a screenshot from the developer documentation showing how to look for multitasking availability.

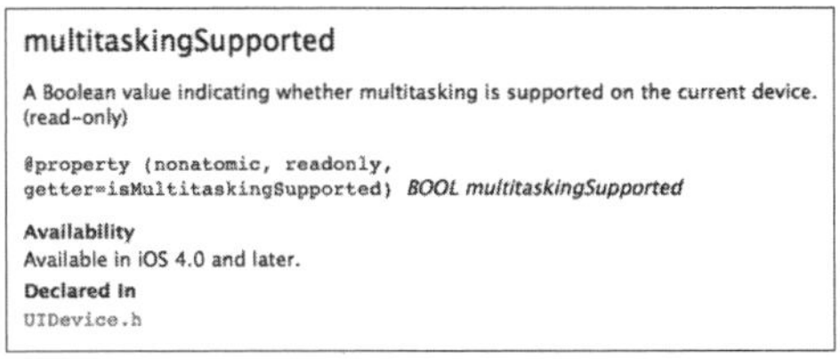

```
multitaskingSupported

A Boolean value indicating whether multitasking is supported on the current device.
(read-only)

@property (nonatomic, readonly,
getter=isMultitaskingSupported) BOOL multitaskingSupported

Availability
Available in iOS 4.0 and later.
Declared In
UIDevice.h
```

Figure 15-1 Multitasking availability in developer documentation

Macros in iOS Header Files

The other method for checking the availability of a method or class is to read through the header files. I find this approach easier than fiddling through the documentation. Just ⌘-click the symbol from your source code, and Xcode opens the header file where it's defined. Most newly added methods have either one of the macro decorations shown in Figure 15-2.

Availability Macros

```
UIKIT_CLASS_AVAILABLE
__OSX_AVAILABLE_STARTING
__OSX_AVAILABLE_BUT_DEPRECATED
```

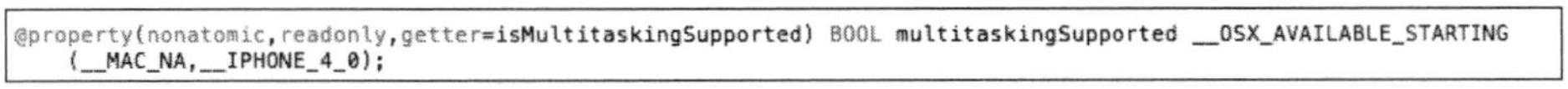

```
@property(nonatomic,readonly,getter=isMultitaskingSupported) BOOL multitaskingSupported __OSX_AVAILABLE_STARTING
    (__MAC_NA,__IPHONE_4_0);
```

Figure 15-2 Multitasking availability in header file

It's usually easier and faster to check availability of a class or method for a given SDK version from the header file. But not all methods have this macro decoration. If your method doesn't, you have to look at the developer documentation.

> If a method doesn't have a macro decoration, it probably means that the method was added ages ago to the SDK, and you normally don't have to worry if you're targeting the two most recent SDKs.

Now that you know how to support multiple SDK versions, it's time to focus on the meat of the chapter: supporting multiple devices. In the next section, you discover the subtle differences between the devices and learn the correct way to check for availability of a particular feature. In parallel, you also write a category extension class on UIDevice that adds methods and properties for checking features not exposed by the framework.

Detecting Device Capabilities

The first and most common mistake that developers made in the past, when there were only two devices (iPod touch and iPhone), was to detect the model name and check whether it was an "iPhone," thereby assuming capabilities. This approach worked well for a year or so. But soon, when new devices with new hardware sensors became available, the method became highly prone to error. For example, the initial version of the iPod touch didn't have a microphone; however, after the iPhone OS 2.2 software update, users could add one by connecting an external microphone/headset. If your code assumes device capabilities based on a model name, it will still work, but it's not correct and not the right thing to do.

Detecting Devices and Assuming Capabilities

Consider the following code fragment, which assumes the capabilities of the iPhone:

Detecting a Microphone the Wrong Way

```
if(![[UIDevice currentDevice].model isEqualToString:@"iPhone"])  {
        UIAlertView *alertView = [[UIAlertView alloc] initWithTitle:@"Error"
message:@"Microphone not present"
delegate:self
        cancelButtonTitle:@"Dismiss"
otherButtonTitles: nil];
        [alertView show];
    }
```

The problem with the preceding code is that the developer has made a broad assumption that only iPhones will ever have microphones. This code worked well initially. But with the iOS software 2.2 update, when Apple added the external microphone capability to the iPod touch, the preceding code prevents users from using the app. Another problem is that this code shows an error for any new device introduced later—for example, iPad.

You should use some other method for detecting hardware or sensor availability rather than assume devices' capabilities. Fortunately or unfortunately, these methods are scattered around on various frameworks.

Now, it's time to start looking at various methods for checking device capabilities the right way and grouping them under a UIDevice category class.

Detecting Hardware and Sensors

The first thing to understand is that instead of assuming capabilities, you need to check for the presence of the exact hardware or sensor you need. For example, instead of assuming that only iPhones have a microphone, use APIs to check for the presence of a microphone. The first advantage of the following code is that it automatically works for new devices to be introduced in the future and for externally connected microphones.

What's the second advantage? The code is a one-liner.

Correct Way to Check for Microphone Availability

```
- (BOOL) microphoneAvailable {
  return [AVAudioSession sharedInstance].inputIsAvailable;
}
```

In the case of a microphone, you also need to consider detecting input device change notifications. That is, enable your Record button on the UI when the user plugs in a microphone, in addition to `viewDidAppear`. Sounds cool, right? Here's how to do that:

Detecting Whether a Microphone Is Being Plugged In

```
void audioInputPropertyListener(void* inClientData,
AudioSessionPropertyID inID, UInt32 inDataSize, const void *inData) {

    UInt32 isAvailable = *(UInt32*)inData;
    BOOL micAvailable = (isAvailable > 0);
    // update your UI here
}

- (void)viewDidLoad {
    [super viewDidLoad];
AudioSessionAddPropertyListener(
kAudioSessionProperty_AudioInputAvailable,
audioInputPropertyListener, nil);
}
```

All you need to do here is to add a property listener for `kAudioSessionProperty_AudioInputAvailable` and on the callback check for the value.

With just a few extra lines of code, you're able to write the correct version of device detection code. Next, you extend this capability for other hardware and sensors.

> `AudioSessionPropertyListeners` **behave much like observing** `NSNotification` **events. When you add a property listener to a class, you are responsible for removing it at the right time. In the preceding example, because you added the property listener in** `viewDidLoad`, **you need to remove it in the** `didReceiveMemoryWarning` **method.**

Detecting Camera Types

The iPhone shipped with a single camera originally and added a front-facing camera later in iPhone 4. The iPod touch had no camera until the fourth generation. Although the iPhone 4 has a front-facing camera, the iPad 1 (its bigger brother) doesn't have one, whereas the iPad 2 has both a front-facing and a back-facing camera. All this means that you should not write code with device-based assumptions. It's actually far easier to use the API.

The `UIImagePickerController` class has class methods to detect source type availability.

Checking for Camera Presence

```
- (BOOL) cameraAvailable {
  return [UIImagePickerController isSourceTypeAvailable:
UIImagePickerControllerSourceTypeCamera];
}
```

Checking for a Front-Facing Camera

```
- (BOOL) frontCameraAvailable
{
#ifdef __IPHONE_4_0
  return [UIImagePickerController isCameraDeviceAvailable:
UIImagePickerControllerCameraDeviceFront];
#else
  return NO;
#endif
}
```

To detect a front-facing camera, you need to be running on iOS 4 and above. The enumeration `UIImagePickerControllerCameraDeviceFront` is available only on iOS 4 and above because any device that has a front-facing camera (iPhone 4 and iPad 2) always runs iOS 4 and above. So you use a macro and return `NO` if the device runs iOS 3 or below.

Similarly, you can check whether the camera attached has video-recording capabilities. Cameras on iPhone 3GS and above can record videos. You can check that using the following code:

Checking for a Video-Recording-Capable Camera

```
- (BOOL) videoCameraAvailable {
  UIImagePickerController *picker =
[[UIImagePickerController alloc] init];
// First call our previous method to check for camera presence.
if(![self cameraAvailable]) return NO;
NSArray *sourceTypes =
[UIImagePickerController availableMediaTypesForSourceType:
UIImagePickerControllerSourceTypeCamera];

  if (![sourceTypes containsObject:(NSString *)kUTTypeMovie]){

    return NO;
  }
  return YES;
}
```

This code enumerates the available media types for a given camera and determines whether it contains `kUTTypeMovie`.

Detecting Whether a Photo Library Is Empty

If you're using a camera, you almost always use the user's photo library. Before calling `UIImagePicker` to show the user's photo album, ensure that it has photos in it. You can check this the same way you check for camera presence. Just pass `UIImagePickerControllerSourceTypePhotoLibrary` or `UIImagePickerControllerSourceTypeSavedPhotosAlbum` for the source type.

Detecting the Presence of a Camera Flash

You can easily check for camera flash presence using the class method for `UIImagePickerController`.

Checking for a Camera Flash

```
- (BOOL) cameraFlashAvailable {
return [UIImagePickerController isFlashAvailableForCameraDevice:
UIImagePickerControllerCameraDeviceRear];
}
```

Detecting a Gyroscope

The gyroscope is an interesting addition to the iPhone 4. Devices introduced after iPhone 4, including the new iPad and iPhone 5, also have a gyroscope. This device allows developers to measure relative changes to the physical position of the device. By comparison, an accelerometer can measure only force. Twisting movements cannot be measured by an accelerometer. Using a gyroscope, game developers can implement six-axis control like that found in Sony's PlayStation 3 controller or Nintendo's Wii controller. You can detect the presence of a gyroscope using an API provided in the CoreMotion.framework, as shown here:

Detecting the Presence of a Gyroscope

```
- (BOOL) gyroscopeAvailable {
CMMotionManager *motionManager = [[CMMotionManager alloc] init];
  BOOL gyroAvailable = motionManager.gyroAvailable;
  return gyroAvailable;
}
```

> If a gyroscope is a core feature of your app but your target device doesn't have a gyroscope, you have to design your app with alternative input methods, or you can specify them in the UIRequiredDeviceCapablities **key in your app's** info.plist, **preventing devices without a gyroscope from installing the app. You learn more about this key later in the chapter.**

Detecting a Compass or Magnetometer

Compass availability can be checked using the CoreLocation.framework class CLLocationManager. Call the method headingAvailable in CLLocationManager, and if it returns true, you can use a compass in your app. A compass is more useful in a location-based application and augmented reality-based applications.

Detecting a Retina Display

As an iOS developer, you already know that catering to a retina display is as easy as adding an @2x image file for every resource you use in the app. But when you download the image from a remote server, you need to download images at twice the resolution on devices with retina display.

A good example of this is a photo browser app such as a Flickr viewer or Instagram. When your user launches the app in iPhone 4 or the new iPad or iPhone 5, you should be downloading images of double the resolution you do for non-retina-display devices. Some developers choose to ignore this and download higher-resolution images for all devices, but that is a waste of bandwidth and might even be slower to download over EDGE. Instead, download higher-resolution files after determining that the device has a retina display. Checking for this is easy using the following code:

Retina-Display Capable

```
- (BOOL) retinaDisplayCapable  {
  int scale = 1.0;
  UIScreen *screen = [UIScreen mainScreen];
  if([screen respondsToSelector:@selector(scale)])
   scale = screen.scale;
  if(scale == 2.0f) return YES;
  else return NO;
}
```

With this code, you look for the mainScreen of the device and check whether the device is capable of showing high-resolution retina-display-capable graphics. This way, if Apple introduces an external retina display (maybe the newer Apple Cinema Displays) and allows the current generation iPads to project to it in retina mode, your app will still work without changes.

Detecting Alert Vibration Capability

As of this writing, only iPhones are capable of vibrating to alert the user. Unfortunately, there is no public API for checking whether a device is vibration-capable. However, the AudioToolbox.framework has two methods to selectively vibrate only iPhones:

```
AudioServicesPlayAlertSound(kSystemSoundID_Vibrate);

AudioServicesPlaySystemSound(kSystemSoundID_Vibrate);
```

The first method vibrates the iPhone and plays a beep sound on the iPod touch/iPad. The second method just vibrates the iPhone. On devices not capable of vibrating, it doesn't do anything. If you're developing a game that vibrates the device to signify danger or a Labyrinth game where you want to vibrate whenever the player hits the wall, use the second method. The first method is for alerting the user, which includes vibration plus beeps, whereas the second is just for vibrations.

Detecting Remote Control Capability

iOS apps can handle remote control events generated by buttons pressed on the external headset. To handle these events, use the following method to start receiving notifications:

```
[[UIApplication sharedApplication] beginReceivingRemoteControlEvents];
```

Implement the following method in your firstResponder:

```
remoteControlReceivedWithEvent:
```

Be sure to turn this off when you no longer need the events by calling

```
[[UIApplication sharedApplication] endReceivingRemoteControlEvents];
```

Detecting Phone Call Capability

You can check whether a device can make phone calls by checking if it can open URLs of type `tel:`. The `canOpenURL:` method of the `UIApplication` class is handy for checking whether a device has an app that can handle URLs of a specific type. The phone app on iPhone handles `tel:` URLs. You also can use the same method to check whether a specific app that can handle a given URL is installed on a device.

Phone Call Capabilities

```
- (BOOL) canMakePhoneCalls {
    return [[UIApplication sharedApplication]
canOpenURL:[NSURL URLWithString:@"tel://"]];
}
```

> A few words about usability: Developers should completely hide features specific to phones on iPod touch devices. For example, if you're developing a Yellow Pages app that lists phone numbers from the Internet, show the button to place a call only on devices that are capable of making phone calls. Do not simply disable it (because nothing can be done by the user to enable it) or show an error alert. In some cases, showing a "Not an iPhone" error on an iPod touch has led the app review team to reject the app.

In App Email and SMS

Although In App email and In App SMS are technically not sensors or hardware, not every device can send e-mails or SMSs. This includes iPhones as well—even those that run iOS 4 and above. Although `MFMessageViewController` and `MFMailComposeViewController` are available from iOS 4, and even if your app's minimum deployment target is set to iOS 4, you still need to know and understand the common pitfalls when using these classes.

A common case is an iOS device that has no configured e-mail accounts and therefore cannot send e-mail, even when it's technically capable of sending one. The same applies to SMS/MMS. An iPhone that doesn't have a SIM card cannot send text messages. You need to be aware of this and always check for capabilities before attempting to use this feature.

Checking for this capability is easy. Both `MFMessageComposeViewController` (for In app SMS) and `MFMailComposeViewController` (for In App e-mail) have class methods `canSendText` and `canSendMail`, respectively, that you can use.

Obtaining the UIDevice+Additions Category

The code fragments you've seen so far in this chapter are available as a UIDevice category addition. You can download them from the book's website.

This addition has just two files: `UIDevice+Additions.h` and `UIDevice+Additions.m`. You have to link necessary frameworks to avoid those pesky linker errors because this class links to various Apple library frameworks. But don't worry; they are dynamically loaded, so they don't bloat your app.

Supporting the New Class of 4-inch Devices (iPhone 5 and 5s)

iPhone 5 was announced in September 2012, and it poses a new challenge to developers: a bigger screen. iOS developers have never been required to support multiple device resolutions in the past. Fret not, Apple has made things easy for you. The first step is to add a launch image (Default-568h@2x.png). As shown in Figure 15-3, when you build your project with Xcode 4.5 and above, you see a warning: "Missing Retina 4 launch image." Click Add to add a default launch image to your project.

Figure 15-3 Xcode 4.5 prompting for addition of a launch image for iPhone 5

The app then launches in full screen without letter boxing. However, most of your nib files still do not scale properly. If you have not migrated to Auto Layout yet, the next step is to check the autoresizing mask of every nib file and ensure that the view inside the nib file automatically sizes based on the super view's height. Figure 15-4 illustrates this.

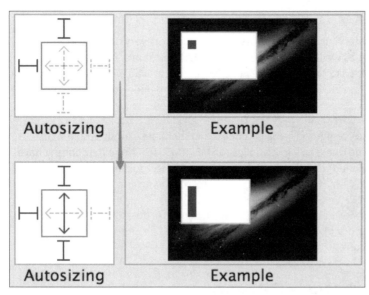

Figure 15-4 Changing the autoresizing mask property using Interface Builder

The properties to use are `UIViewAutoresizingFlexibleTopMargin`, `UIViewAutoresizing FlexibleBottomMargin`, and `UIViewAutoresizingFlexibleHeight`.

You use the `UIViewAutoresizingFlexibleHeight` for the top-most view so that it autosizes with the main window. You use the `UIViewAutoresizingFlexibleTopMargin` and/or

UIViewAutoresizingFlexibleBottomMargin for subviews. Use UIViewAutoresizing FlexibleTopMargin property when you want the subview to be "pinned" to the bottom (the top margin is flexible). Use the UIViewAutoresizingFlexibleBottomMargin when you want the subview to be "pinned" to the top (the bottom margin is flexible).

> Remember, a flexible bottom margin pins the UI element to the top, and a flexible top margin pins the UI element to the bottom. Note too that when you use Cocoa Auto Layout, you don't have to deal with autoresize masks.

Cocoa Auto Layout

Auto Layout is a constraint-based layout engine and is a much better option for handling layout and size adjustments based on screen size or device orientation. Xcode 5 makes working with Auto Layout easier than using its predecessor. For more in-depth understanding of Auto Layout, turn to Chapter 6.

Hard-Coded Screen Size

If you have hard-coded the height of a view to 460 or 480, you might have to change this using bounds. For example, you use

```
self.window = [[UIWindow alloc] initWithFrame:
[[UIScreen mainScreen] bounds]];
```

instead of

```
self.window = [[UIWindow alloc] initWithFrame:CGRectMake(0, 0, 320, 480)];
```

Finally, any CALayer that you added to the view will have to be manually resized. The following code shows how to do this. This code assumes you have a "patternLayer" for all your view controllers.

Resizing a CALayer for iPhone 5 Compatibility

```
- (void)viewWillLayoutSubviews {
self.patternLayer.frame = self.view.bounds;
[super viewWillLayoutSubviews];
}
```

iPhone 5 supports a new instruction set: the armv7s and iPhone 5s supports the new 64-bit instruction set, the arm64. Xcode 4.5 supports generating an armv7s instruction set and only the latest Xcode 5 supports generating arm64 instruction set. In the next section, I show you how to support the 64-bit device.

iPhone 5s and the New 64-bit Instruction Set

iPhone 5s supports a new instruction set: the arm64. The arm64 instruction set is a 64-bit instruction set, and that means your application is now able to address more than 4GB of physical memory. Although that will not improve performance right away (the latest iPhone 5s still have only about 1GB RAM), the support ensures that the hardware shipped is future proof. To generate a 64-bit instruction set, your target's minimum deployment should be at least iOS 7 and above. Xcode 5 doesn't allow you to create a fat binary that runs in 32-bit mode in iOS 6 and 64-bit mode in iOS 7, although Apple promises that a future Xcode update might allow this. You learn about transitioning to iOS 7 and transitioning to the 64-bit architecture in the next couple of sections.

iPhones, Instruction Sets, and Fat Binaries

The original iPhone supported an instruction set called armv6. iPhone 3G supported the same. As the hardware evolved, Apple implemented support for armv7 until iPhone 4S and armv7s in iPhone 5 and now arm64 in iPhone 5s. Now what does an instruction set change mean to you? An instruction set can bring in improved performance. It's entirely optional for you to generate an instruction set. But if you do, the new hardware will use the new instruction set, and the old hardware can still run the same app using the old instruction set.

A binary file that has more than one instruction set is called a *fat binary*. From iPhone 3GS, almost all the binary files generated by Xcode's GCC or LLVM compiler were all fat binaries supporting both armv6 and armv7. With the launch of iPhone 5, Xcode defaulted LLVM code generation to armv7/armv7s. Xcode 5 defaults code generation to armv7/armv7s/arm64 for new projects.

You can easily change the architecture that Xcode generates by changing the Architectures setting from the Targets Build Settings editor. This is shown in Figure 15-5.

Figure 15-5 Changing Architectures from the Build Settings Editor

Transitioning to iOS 7

Migrating your apps to iOS 7 isn't as straightforward as previous releases. The UI changes in the new operating system are profound. Migrating an existing app and supporting just iOS 7 aren't easy either. I show you some of the important changes that iOS 7 has brought forth and how to support them in your app.

Auto Layout

If you have used Auto Layout, things are going to be easy. If you haven't, this is the right time. Although the screen sizes remain the same 3.5" (320×480 points) and 4" (320×568 points), iOS 7 uses the complete screen for layout, whereas iOS 6 excludes the status bar, navigation bar, and tab bar (if you have any of them). That means, in navigation-based apps, your usable screen area in a device with a 3.5" screen was 320×416, and in a tab-bar-based application, it was 320×368. iOS 7 extends the edges of your view controller beneath the status bar, navigation bar, and tab bar. As a result, regardless of the container view, your view controller will always be able to use 320×480 points. You now have to write layout code that adapts well with the 320×416 points (or 320×368 points) layout and 320×480 points and 320×568 points layout.

Auto Layout helps you to do layout without writing a single line of layout code. You express your constraints, and the layout engine takes care of laying out the user interface elements based on the constraints. Auto Layout is covered in depth in Chapter 6.

Many default UIKit elements are now smaller than before. A `UISwitch` in iOS 6 is much bigger in size than the same switch in iOS 7. The same goes with `UISegmentedControl`. Auto Layout also can help you with the layout of your user interface in this context.

Supporting iOS 6

As of this writing, iOS 7 is in use in more than 60% of devices (four days after iOS 7's general availability). That means you can safely forget about iOS 6. But if business requirements demand iOS 6 support, it's best to start by adopting the iOS 7 visual styles and back porting the design changes to the iOS 6 version of the app.

When you use an iOS 7-specific feature, as far as possible, use one of the methods (framework availability, class availability, or method availability) explained earlier in this chapter. If you want to show a completely different view controller/user interface, you can use the following snippet:

Switching between iOS 6 and iOS 7

```
if (floor(NSFoundationVersionNumber) <= NSFoundationVersionNumber_iOS_6_1) {

// iOS 6 specific code
} else {

// iOS 7 specific code
}
```

If you use Interface Builder, Xcode 5 also has the capability to preview the user interface on iOS 6 and iOS 7. Open the Assistant Editor and choose Preview.

Application Icon

iOS 7 icons are slightly bigger. All applications should now have a 120×120 pixel icon for iPhone-only apps and 76×76 (for iPad mini) and a 152×152 pixel icons for iPad apps (and all these three sizes for a universal app) in the Application bundle. The rounded-corner algorithm is also slightly different from iOS 6. My recommendation is not to round the corners at all and let iOS handle this task for you. You just have to ensure that the icon looks as you intended on both iOS 6 and iOS 7.

Borderless Buttons

iOS 7 buttons don't have borders. Embrace borderless buttons. If you use your own custom buttons that have borders, make the borders less obvious.

Tint Color

In iOS 7, view controllers and your application's top-level window now have a property called `tintColor`. The `tintColor` gets inherited by subviews in the view controller hierarchy. For example, a `UISwitch` in your application uses the `tintColor` to show the highlighted state instead of the hard-coded "blue."

A `UITextField` uses the `tintColor` to paint the blinking caret instead of the hard-coded "blue." Many `UIKit` elements such as `UISegmentedControl` and `UIProgressView` make heavy use of `tintColor`. If you are writing your own custom control, you should consider using this `tintColor` to "highlight" stuff.

Artwork Updates

The artwork that Apple uses for iOS 7 is completely different from that used in iOS 6. If you are using custom artwork, ensure that you update it to the new iOS 7 style. In short, try to make your iOS 6 app look as close as possible to the iOS 7 equivalent (excluding the translucency). You can start off by setting the `wantsFullScreenLayout` property of all your view controllers to `YES`. Note that this property is deprecated in iOS 7, and it is set to `YES` by default. Changing it to `NO` has no effect in iOS 7. Leaving it untouched means your view controllers will behave very differently in iOS 6 compared to iOS 7, necessitating you to write layout code.

In fact, using Auto Layout to make your view controllers use the full screen in iOS 6 should complete much of the complexities in porting from iOS 6 to iOS 7.

With iPhone 5s, you should also ensure that your code behaves as expected when it runs on a 64-bit architecture.

Transitioning to 64-bit Architecture

As of this writing, Xcode 5.0 doesn't support 64-bit architectures on iOS 6. So your first step should be to convert your application to run as expected on iOS 7 before attempting to support 64-bit architectures.

The first step in transitioning to 64-bit architecture is to start using `NSInteger`, `NSUInteger`, and `CGFloats` instead of `int` and `float`. The Objective-C runtime defines them as follows:

Code in ObjRuntime.h and CGBase.h

```
#if __LP64__ || (TARGET_OS_EMBEDDED && !TARGET_OS_IPHONE) ||
TARGET_OS_WIN32 || NS_BUILD_32_LIKE_64
    typedef long NSInteger;
    typedef unsigned long NSUInteger;
#else
    typedef int NSInteger;
    typedef unsigned int NSUInteger;
#endif
#if defined(__LP64__) && __LP64__
# define CGFLOAT_TYPE double
# define CGFLOAT_IS_DOUBLE 1
# define CGFLOAT_MIN DBL_MIN
# define CGFLOAT_MAX DBL_MAX
#else
# define CGFLOAT_TYPE float
# define CGFLOAT_IS_DOUBLE 0
# define CGFLOAT_MIN FLT_MIN
# define CGFLOAT_MAX FLT_MAX
#endif
```

When you compile an application, the preprocessor takes care of using 64-bit wide integers (`long`) instead of 32-bit wide integers (`int`) if you use `NSInteger`. The same is true with `CGFloat`.

You should never assign a 64-bit pointer/data to a 32-bit data type. That is, you should not assume that typecasting a `Long` data type to `Int` will make it 32-bit. In fact, the LLVM compiler usually warns you of these lossy typecasts. If you really want to convert 64-bit data to 32-bit data, use a `&` mask instead. This use is illustrated here:

```
NSUInteger lowerOrderData = fullData & 0x00000000FFFFFFFF;
```

Data Overflows

Don't assume data overflows in your algorithmic code: 4294967295 + 1 is 4294967296 in 64-bit architectures and 0 in 32-bit architectures. Also, don't hard-code your upper limits to `INT32_MAX`. If you do so, you might have to rewrite your logic when you port your code over to 64-bit.

Serialized Data

Data that the user creates might get serialized and synced over iCloud. If you encode data in 64-bit format and send it to iCloud and read it back on a 32-bit device, the results might be unexpected. You should use an explicit data type in this case instead of `NSInteger` or `CGFloat`. That doesn't mean you should migrate your existing data to 64-bit. A 64-bit processor can crunch 64-bit data as fast as a 32-bit processor crunches 32-bit data. But a 32-bit processor takes much longer to process 64-bit data. Analyze your data again and determine which fields really need the wider data width.

Conditional Compilation for 64-bit Architectures

If you need to selectively change the code, use the `__LP64__` macro:

```
#if __LP64__
// 64 bit code
#else
// 32 bit code
#endif
```

Optimize your code. Instruments is your friend. After you migrate, try running the 32-bit equivalent (on real 32-bit hardware) and look for memory issues or slowness in performance. If you see an unexpected increase in memory usage, this is the right place to start. Similarly, try running the 64-bit version on real 64-bit hardware. Watch out for an unexpected increase in memory and/or CPU usage and optimize the code. In most cases, these optimizations might be to change to an explicit data type such as `int` or `float` instead of `NSInteger/CGFloat` and vice versa. Pay careful attention to serialized data if you are not using explicit data types.

Although not immediately necessary, you should plan your 64-bit migration as soon as possible. Almost every device that will be released in the future will use the new 64-bit architecture. In iOS 7, all of Apple's own apps are migrated to 64-bit. iOS 7 still maintains the 32-bit runtime and loads it if 32-bit applications are running on the device. If all apps support 64-bit, iOS does not even load the 32-bit runtime, and that means better performance and more memory availability (fewer memory warnings) for all apps. To be a nice citizen on the platform, plan your migration and get it ready by 2014.

UIRequiredDeviceCapabilities

So far, you've learned how to conditionally check a device for specific capabilities and use them if they are present. In some cases, your app depends solely on the presence of particular hardware, and without that hardware, your app will be unusable. Examples include a camera app such as Instagram or Camera+. The core functionality of the app doesn't work without a camera. In this case, you need something more than just checking for device capabilities and hiding specific parts of your app. You normally don't need devices without a camera to use or download your app.

Apple provides a way to ensure this using the `UIRequiredDeviceCapabilities` key in the Info plist file. The following values are supported for this key (this list is not comprehensive):

telephony	wifi	sms, still-camera
auto-focus-camera	front-facing-camera	camera-flash
video-camera	accelerometer	gyroscope
location-services	gps	magnetometer
gamekit	opengles-1	opengles-2
armv6	armv7	peer-peer
accelerometer	Bluetooth-le	microphone
Fetch	Bluetooth-central	Bluetooth-peripheral

You can explicitly require particular device capabilities or prohibit installation of your app on devices without a specific capability. For example, you can prevent your apps from running on devices with `video-camera` by setting the `video-camera` key to `NO`. Alternatively, you can mandate the presence of `video-camera` by setting the `video-camera` key to `YES`.

> Apple doesn't allow you to submit an update to an existing app and prevent it from running on a specific device that was supported before the update. For example, if your app supported both iPhone and iPod touch in version 1.0, you cannot submit an update that prevents it from running on either device. Put another way, you cannot introduce a mandate for the presence of particular hardware later in your app's product life cycle. The submission process on iTunes Connect will fail and show you an error. The opposite is allowed, however. That is, if you were excluding a device previously, you can allow installations on it in a subsequent version. In other words, if version 1 of your app supported only iPhones, you can submit a version 2 to support all devices.

Adding values to the `UIRequiredDeviceCapablities` key prohibits your app from being installed on devices without the capabilities you requested. If you specify that telephony is needed, users cannot even download the app on their iPod touch or iPad. You must be certain that this is your expected behavior before using this key.

Summary

This chapter discussed various techniques and tricks to help run your app on multiple platforms. It also looked at the various hardware and sensors available for iOS developers and how to detect their presence the right way. You incrementally wrote a category extension on `UIDevice` that could be used for detecting most device capabilities. You also learned about supporting the new iPhone 5's taller screen size and supporting the iPhone 5's new 64-bit arm64 architecture; plus, you learned about transitioning your UI to iOS 7. Finally, you learned about the `UIRequiredDeviceCapablities` key and how to completely exclude devices without a required capability. My recommendation is to depend on the methods explained in this chapter and use the `UIRequiredDeviceCapablities` key sparingly.

Further Reading

Apple Documentation

The following documents are available in the iOS Developer Library at `developer.apple.com` or through the Xcode Documentation and API Reference.

iOS 7 Transitioning Guide

64-Bit Transitioning Guide for Cocoa Touch Applications

Understanding the UIRequiredDeviceCapablities Key

iOS Build Time Configuration Details

Other Resources

MK blog. (Mugunth Kumar) "iPhone Tutorial: Better way to check capabilities of iOS devices"

`http://blog.mugunthkumar.com/coding/iphone-tutorial-better-way-to-check-capabilities-of-ios-devices/`

Github. "MugunthKumar/DeviceHelper"

`https://github.com/MugunthKumar/DeviceHelper`

Chapter 16

Reach the World: Internationalization and Localization

Localization is a key concern for any application with a global market. Users want to interact in their own languages, with their familiar formatting. Supporting global languages and cultures this way in your application is called *internationalization* (sometimes abbreviated *i18n* for the 18 characters between the *i* and the *n*) and *localization* (*L10n*). The differences between i18n and L10n aren't really important or consistently agreed upon. Apple says, "Internationalization is the process of designing and building an application to facilitate localization. Localization, in turn, is the cultural and linguistic adaptation of an internationalized application to two or more culturally-distinct markets." (See "Internationalization Programming Topics" at `developer.apple.com`.) This chapter uses the terms interchangeably.

After reading this chapter, you will have a solid understanding of what localization is and how to approach it. Even if you're not ready to localize your application yet, this chapter provides easy steps to dramatically simplify localization later. You find out how to localize strings, numbers, dates, and nib files and how to regularly audit your project to make sure it stays localizable.

What Is Localization?

Localization is more than just translating strings. Localization means making your application culturally appropriate for your target audience. That includes translating language in strings, images, and audio, and formatting numbers and dates.

Here are some general things you can do to improve your iOS localizations:

- **Use Base Internationalization.** See the section "Nib Files and Base Internationalization" later in this chapter. If you're localizing each nib file or storyboard individually, you're doing it the hard way.

- **Internationalize from (nearly) the beginning.** After you have the basics of your program working, go ahead and set it up for internationalization. I don't recommend internationalizing from the beginning because this tends to complicate things when they're in the most flux. After the basic structure of your program settles down, that's when you should turn on Base Internationalization and make sure everything keeps working correctly.

- **Use Auto Layout.** Apple has bet heavily on Auto Layout. Particularly with the iOS 7 focus on text-based UI, Auto Layout has become a crucial part of internationalization.

- **Remember right-to-left languages.** Directionality is one of the hardest things to fix later, especially if you have custom text views.

■ **Don't assume that a comma is the thousands separator or a dot is the decimal point.** These indicators are different in different languages, so build your regular expressions using NSLocale.

■ **Glyphs (drawn symbols) and characters do not always map one-to-one.** If you're doing custom text layout, the way letters map to the shapes that represent them can be particularly surprising. Apple's frameworks generally handle this automatically, but don't try to circumvent systems like Core Text when they force you to calculate the number of glyphs in a string rather than using length. This issue is particularly common in Thai, but exists in many languages (even occasionally in English, as we discuss in Chapter 21).

In my experience, although it is best to set up internationalization near the beginning of the project, it is better to do the actual translation close to release time. Retranslating things every time you tweak the UI is expensive.

Although translation is best done near the time of release, you should line up your localization provider fairly early in the development cycle and prepare for localization throughout the process. A good localization provider does more than just translate a bunch of strings. Ideally, your localization provider will provide testing services to make sure your application "makes sense" in the target culture. Getting the provider involved early in the process can save expensive rework later if your interface is hard to localize.

An example of a "hard-to-localize" application is one that includes large blocks of text. Translating large blocks of text can play havoc with layout, even when using Auto Layout. Remembering that you will often pay by the word for translation may help you focus on reducing the number of words you use.

Another frequent localization problem is icons that assume a cultural background, such as a decorated tree to indicate "winter." Check marks can also cause problems because they are not used in all cultures (French, for instance), and in some cultures a check mark means "incorrect" (Finnish, for instance). Involving a good localization provider before producing your final artwork can save you a lot of money re-creating your assets.

Localizing Strings

The most common tool for localizing strings is NSLocalizedString. This function looks up the given key in Localizeable.strings and returns the value found there, or the key itself if no value is found. Localizeable.strings is a localized file, so a different version is available for each language, and NSLocalizedString automatically selects the correct one based on the current locale. A command-line tool called genstrings automatically searches your files for calls to NSLocalizedString and writes your initial Localizeable.strings file for you.

The easiest approach is to use the string as its own key (the second parameter is a comment to the localizer):

```
NSString *string =
    NSLocalizedString(@"Welcome to the show.",
                      @"Welcome message");
```

To run genstrings, you open a terminal, change to your source code directory, and run it as shown here (assuming an English localization):

```
genstrings -o en.lproj *.m
```

This code creates a file called `en.lproj/Localizeable.string` that contains the following:

```
/* Welcome message */
"Welcome to the show." = "Welcome to the show.";
```

Even if you don't run `genstrings`, this code works in the developer's language because it automatically returns the key as the localized string.

In most cases, I recommend using the string as its own key and automatically generating the `Localizeable.strings` file when you're ready to hand off the project to localizers. This approach simplifies development and helps keep the `Localizeable.strings` file from accumulating keys that are no longer used.

Auditing for Nonlocalized Strings

During development, be sure to periodically audit your program to make sure that you're using `NSLocalizedString` as you should. I recommend a script like this:

find_nonlocalized

```perl
#!/usr/bin/perl -w
# Usage:
#     find_nonlocalized [<directory> ...]
#
# Scans .m and .mm files for potentially nonlocalized
#   strings that should be.
# Lines marked with DNL (Do Not Localize) are ignored.
# String constant assignments of this form are ignored if
#   they have no spaces in the value:
#   NSString * const <...> = @"...";
# Strings on the same line as NSLocalizedString are
#   ignored.
# Certain common methods that take nonlocalized strings are
#   ignored
# URLs are ignored
#
# Exits with 1 if there were strings found

use File::Basename;
use File::Find;
use strict;

# Include the basenames of any files to ignore
my @EXCLUDE_FILENAMES = qw();

# Regular expressions to ignore
my @EXCLUDE_REGEXES = (
    qr/\bDNL\b/,
    qr/NSLocalizedString/,
```

(continued)

```
        qr/NSString\s*\*\s*const\s[^@]*@"[^ ]*";/,
        qr/NSLog\(/,
        qr/@"http/, qr/@"mailto/, qr/@"ldap/,
        qr/predicateWithFormat:@"/,
        qr/Key(?:[pP]ath)?:@"/,
        qr/setDateFormat:@"/,
        qr/NSAssert/,
        qr/imageNamed:@"/,
        qr/NibNamed?:@"/,
        qr/pathForResource:@"/,
        qr/fileURLWithPath:@"/,
        qr/fontWithName:@"/,
        qr/stringByAppendingPathComponent:@"/,
);

my $FoundNonLocalized = 0;

sub find_nonlocalized {
    return unless $File::Find::name =~ /\.mm?$/;
    return if grep($_, @EXCLUDE_FILENAMES);

    open(FILE, $_);

    LINE:
    while (<FILE>) {
        if (/@"[^"]*[a-z]{3,}/) {
            foreach my $regex (@EXCLUDE_REGEXES) {
                next LINE if $_ =~ $regex;
            }
            print "$File::Find::name:$.:$_";
            $FoundNonLocalized = 1;
        }
    }
    close(FILE);
}

my @dirs = scalar @ARGV ? @ARGV : (".");
find(\&find_nonlocalized, @dirs);
exit $FoundNonLocalized ? 1 : 0;
```

Periodically run this script over your source to make sure that it doesn't have any nonlocalized strings. If you use Jenkins (jenkins-ci.org) or another continuous-integration tool, you can make this script part of the build process, or you can add it as a script step in your Xcode build. Whenever the script returns a new string, you can decide whether to fix it, update the regular expressions to ignore it, or mark the line with DNL (Do Not Localize).

Formatting Numbers and Dates

Numbers and dates are displayed differently in different locales. Formatting them is generally straightforward using NSDateFormatter and NSNumberFormatter, which you are likely already familiar with.

For an introduction to `NSDateFormatter` **and** `NSNumberFormatter`, **see the "Data Formatting Guide" in Apple's documentation at** `developer.apple.com`.

You need to keep in mind a few things, however. First, formatters are needed for input as well as output. Most developers remember to use a formatter for date input but may forget to use one for numeric input. The decimal point is not universally used to separate whole from fractional digits on input. Some countries use a comma or an apostrophe. It's best to validate number input using an `NSNumberFormatter` rather than custom logic.

Digit groupings have a bewildering variety. Some countries split thousands groups with a space, comma, or apostrophe. China sometimes groups ten thousands (four digits). Don't guess. Use a formatter. Remember that the way digits are grouped can impact the length of your string. If you leave room for only seven characters for one hundred thousand ("100,000") you may overflow in India, which uses eight ("1,00,000" or one lakh).

Percentages are another place where you need to be careful because different cultures place the percent sign at the beginning or end of the number, and some use a slightly different symbol. Using `NSNumberFormatterPercentStyle` will behave correctly.

Be especially careful with currency. Don't store currency as a float because that can lead to rounding errors as you convert between binary and decimal. Always store currency as an `NSDecimalNumber`, which does its math in decimal. Keep track of the currency you're working in. If your user switches locale from the U.S. to France, don't switch his $1 purchase to €1. Generally, you need to persist in using the currency in which a given value is expressed. The `RNMoney` class is an example of how to do this. First, the following code demonstrates how to use the class to store rubles and euros:

main.m (Money)

```
NSLocale *russiaLocale = [[NSLocale alloc]
                    initWithLocaleIdentifier:@"ru_RU"];

RNMoney *money = [[RNMoney alloc]
                initWithIntegerAmount:100];
NSLog(@"Local display of local currency: %@", money);
NSLog(@"Russian display of local currency: %@",
    [money localizedStringForLocale:russiaLocale]);


RNMoney *euro =[[RNMoney alloc] initWithIntegerAmount:200
                            currencyCode:@"EUR"];
NSLog(@"Local display of Euro: %@", euro);
NSLog(@"Russian display of Euro: %@",
    [euro localizedStringForLocale:russiaLocale]);
```

`RNMoney` is an immutable object that stores an amount and a currency code. If you do not provide a currency code, it defaults to the current locale's currency. This simple data class is designed to be easy to initialize, serialize, and format. Here is the code:

RNMoney.h (Money)

```
#import <Foundation/Foundation.h>

@interface RNMoney : NSObject <NSCoding>
@property (nonatomic, readonly, strong)
                                    NSDecimalNumber *amount;
@property (nonatomic, readonly, strong)
                                    NSString *currencyCode;

- (RNMoney *)initWithAmount:(NSDecimalNumber *)anAmount
        currencyCode:(NSString *)aCode;
- (RNMoney *)initWithAmount:(NSDecimalNumber *)anAmount;

- (RNMoney *)initWithIntegerAmount:(NSInteger)anAmount
                    currencyCode:(NSString *)aCode;
- (RNMoney *)initWithIntegerAmount:(NSInteger)anAmount;

- (NSString *)localizedStringForLocale:(NSLocale *)aLocale;
- (NSString *)localizedString;

@end
```

RNMoney.m (Money)

```
#import "RNMoney.h"

@implementation RNMoney

static NSString * const kRNMoneyAmountKey = @"amount";
static NSString * const kRNMoneyCurrencyCodeKey =
                                    @"currencyCode";

- (RNMoney *)initWithAmount:(NSDecimalNumber *)anAmount
                currencyCode:(NSString *)aCode {
  if ((self = [super init])) {
    _amount = anAmount;
    if (aCode == nil) {
      NSNumberFormatter *formatter = [[NSNumberFormatter alloc] init];
      _currencyCode = [formatter currencyCode];
    }
    else {
      _currencyCode = aCode;
    }
  }
  return self;
}

- (RNMoney *)initWithAmount:(NSDecimalNumber *)anAmount {
  return [self initWithAmount:anAmount
                currencyCode:nil];
}
```

```objc
- (RNMoney *)initWithIntegerAmount:(NSInteger)anAmount
                      currencyCode:(NSString *)aCode {
    return [self initWithAmount:
            [NSDecimalNumber decimalNumberWithDecimal:
             [@(anAmount) decimalValue]]
                  currencyCode:aCode];
}

- (RNMoney *)initWithIntegerAmount:(NSInteger)anAmount {
    return [self initWithIntegerAmount:anAmount
                         currencyCode:nil];
}

- (id)init {
    return [self initWithAmount:[NSDecimalNumber zero]];
}

- (id)initWithCoder:(NSCoder *)coder {

    NSDecimalNumber *amount = [coder decodeObjectForKey:
                               kRNMoneyAmountKey];
    NSString *currencyCode = [coder decodeObjectForKey:
                              kRNMoneyCurrencyCodeKey];
    return [self initWithAmount:amount
                   currencyCode:currencyCode];
}

- (void)encodeWithCoder:(NSCoder *)aCoder {
    [aCoder encodeObject:_amount forKey:kRNMoneyAmountKey];
    [aCoder encodeObject:_currencyCode
                  forKey:kRNMoneyCurrencyCodeKey];
}

- (NSString *)localizedStringForLocale:(NSLocale *)aLocale
{
    NSNumberFormatter *formatter = [[NSNumberFormatter alloc]
                                    init];
    [formatter setLocale:aLocale];
    [formatter setCurrencyCode:self.currencyCode];
    [formatter setNumberStyle:NSNumberFormatterCurrencyStyle];
    return [formatter stringFromNumber:self.amount];
}

- (NSString *)localizedString {
    return [self localizedStringForLocale:
            [NSLocale currentLocale]];
}

- (NSString *)description {
    return [self localizedString];
}

@end
```

Nib Files and Base Internationalization

iOS 6 added a new feature called "Base Internationalization." In the Project Info panel, you can select Use Base Internationalization, and Xcode will convert your project to the new system. Prior to Base Internationalization, you needed a copy of all your nib files for every locale. With Base Internationalization, there is an unlocalized version of your nib and storyboard files, and there is a strings file for every localization. iOS takes care of inserting all the strings into the nib files for you at runtime, greatly simplifying nib localization. You may still want to create individually localized nib files in some cases.

Some languages require radically different layout. For example, visually "large" languages like Russian and dense languages like Chinese may not fit well in the layout used for English or French. Right-to-left languages also may need some special handling. Luckily, you can still create per-language nib files while using Base Internationalization.

Localizing Complex Strings

Sentence structure is radically different among languages. This means that you can almost never safely compose a string from parts like this:

```
NSString *intro = @"There was an error deleting";
NSString *num = [NSString stringWithFormat:@"%d", 5];
NSString *tail = @"objects.";
NSString *str = [NSString stringWithFormat:@"%@ %@ %@",
                    intro, num, tail]; // Wrong
```

The problem with this code is that when you translate "There was an error deleting" and "objects" into other languages, you may not be able to glue them together in the same order. Instead, you need to localize the entire string together like this:

```
NSString *format = NSLocalizedString(
                @"There was an error deleting %d objects",
                @"Error when deleting objects.");
NSString *str = [NSString stringWithFormat:format, 5];
```

Some languages have more complex plurals than English. For instance, there may be special word forms for describing two of something versus more than two. Don't assume you can check for greater-than-one and easily determine linguistic plurals. Solving this issue well can be very difficult, so try to avoid it instead. Don't have special code that tries to add an s to the end of plurals because this is almost impossible to translate. A good translator will help you word your messages in ways that translate better in your target languages.

Talk with your localization provider early on to understand its process and how to adjust your development practice to facilitate working with it. Figure 16-1 demonstrates a good approach.

1. **Pseudo-Localize**—During development, it's a good idea to start doing experimental localization to work out any localization bugs early in the process. Pseudo-localization is the process of localizing into a nonsense language. A common nonsense language is one that substitutes all vowels with the letter *x*. For example, "Press here to continue" would become "Prxss hxrx tx cxntxnxx." This kind of "translation" can be done by developers, generally with a simple script, and will make it more obvious where you have used nonlocalized strings. This won't find every problem. In particular, it is not good at discovering strings that are pieced together from other strings, but it can find many simple problems before you pay for real

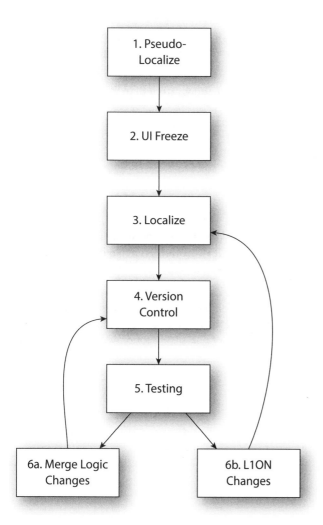

Figure 16-1 Localization workflow

translation services. You will need a language code for this localization. Pick a language that you do not plan to localize your application for. If you're an American English speaker and don't plan to localize for British English, it is particularly useful to use the British English slot for this purpose because you'll still be able to easily read the rest of the iPhone's interface.

2. **UI Freeze**—There should be a clear point in the development cycle at which you freeze the UI. After that point, firmly avoid any changes that affect localizable resources. Many teams ship a monolingual version of their product at this point and then ship a localization update. That's the easiest approach if your market is tolerant of the delay.

3. **Localize**—You will send your resource files to your localizers, and they will send you back localized files.

4. **Version Control**—As you make changes to your nib files, you will need to keep track of the original files your localizer sent to you. Put them into a version control system or save them in a separate directory.

5. **Testing**—You'll need to do extensive testing to make sure that everything is correct. Ideally, you will have native speakers of each of your localized languages test all your UI elements to ensure that they make sense and that there aren't any leftover nonlocalized strings. A good localizer can assist in this task.

6a. **Merge Logic Changes**—Certain nib file changes do not affect localization. Changes to connections or class names don't change the layout or the resources. These are logic changes rather than localization (L10n) changes. You can merge the localized nib files like this:

```
ibtool --previous-file ${OLD}/en.lproj/MyNib.nib
       --incremental-file ${OLD}/fr.lproj/MyNib.nib
       --strings-file ${NEW}/fr.lproj/Localizeable.strings
       --localize-incremental
       --write ${NEW}/fr.lproj/MyNib.nib
       ${NEW}/en.lproj/MyNib.nib
```

This code computes the nonlocalization changes between the old and new English `MyNib.nib`. It then applies these changes to the old French `MyNib.nib` and writes it as the new French nib file. As long as you keep track of the original files you were sent by the localizer, this approach works quite well for nonlayout changes and can be scripted fairly easily.

6b. **L10n Changes**—If you make localization changes such as changing the layout of a localized nib file or changing a string, you'll need to start the process over and send the changes to the localizer. You can reuse the previous string translations, which makes things more efficient but is still a lot of work, so avoid making these changes late in the development cycle.

Summary

Localization is never an easy subject, but if you work with a good localization partner early and follow the best practices detailed here, you can greatly expand the market for your applications.

Further Reading

Apple Documentation

The following documents are available in the iOS Developer Library at `developer.apple.com` or through the Xcode Documentation and API Reference.

Data Formatting Guide

Internationalization Programming Topics

Locales Programming Guide

WWDC Sessions

The following session videos are available at `developer.apple.com`.

WWDC 2012, "Session 244: Internationalization Tips and Tricks"

WWDC 2013, "Session 219: Making Your App World-Ready"

Chapter 17

Those Pesky Bugs: Debugging

The most difficult part of writing software is debugging. Debugging is hard. It's even harder when you're writing software in a low-level language (compared to high-level languages like Java or C#). Instead of a stack trace, you often hear iOS developers using buzzwords like *dSYM files, symbolication,* and *crash dumps.*

This chapter introduces you to LLDB, Apple's Lower Level DeBugger. I explain some commonly used terms such as *dSYM, symbolication,* and others that are traditionally different from other programming languages. A lot changed when Apple replaced GDB with LLDB, and nearly everything that I describe in this chapter is specific to LLDB. If you're still using GDB, it's high time to change to the newer LLDB. I show you some of Xcode's features that will help you with debugging and that will unleash the power of the LLDB console. Later in this chapter, you'll find different techniques for collecting crash reports, including a couple of third-party services.

LLDB

LLDB is a next-generation high-performance debugger built using reusable components from LLVM, including the complete LLVM compiler, that includes LLVM's Clang expression parser and the disassembler. What this means to you, the end user/developer, is that LLDB understands the same syntax that your compiler understands, including Objective-C literals and Objective-C's dot notation for properties. A debugger with compiler-level accuracy means that any new feature added to LLVM will automatically be available to LLDB.

Xcode's previous debugger, GDB, didn't really "understand" Objective-C. As such, something as simple as `po self.view.frame` was more complicated to type. You need to type `po [[self view] frame]`. When the compiler was replaced, there was a need to improve the debugger. Because GDB was monolithic, there was no workaround; it had to be rewritten from scratch. LLDB is modular, and providing the debugger with API support and a scripting interface was one of the design goals. In fact, the LLDB command-line debugger links to the LLDB library through this API. In the "Scripting Your Debugger" section later in this chapter, I show you how to script LLDB to make your debugging session easier.

Debugging with LLDB

Debugging with LLDB offers little advantage over GDB for most developers. In most cases, you won't even notice the difference, except for a few obvious changes such as support for Objective-C properties and literals. However, knowing how LLDB works internally and the subtle differences it brings along will make you a better developer and one who can indeed push the limits. After all, you picked up this book to learn about the advanced concepts that would help you push the limits, right? The next few sections aim at explaining these concepts in detail. With that, it's time to get started.

Debug Information File (dSYM file)

The debug information file (dSYM) stores debugging information about the target. What does it contain? Why do you need a debug information file in the first place? Every programming language you've ever written code for has a compiler that converts your code either to some kind of intermediate language that the runtime can understand or to machine code that gets executed natively on the machine's architecture.

A debugger is commonly integrated within your development environment. The development environment often allows you to place breakpoints that stop the app from running and allows you to inspect the values of variables in your code. That is, a debugger effectively freezes the app in real time and allows you to inspect variables and registers. There are at least two important types of debuggers: symbolic debuggers and machine language debuggers. A machine language debugger shows the disassembly when a breakpoint is hit and lets you watch the values of registers and memory addresses. Assembly code programmers generally use this kind of debugger. A symbolic debugger shows the symbol/variable you use in your application when you're debugging through the code. Unlike a machine language debugger, a symbolic debugger allows you to watch symbols in your code instead of registers and memory addresses.

For a symbolic debugger to work, there should be a link or a mapping between the compiled code and the source code you wrote. This is precisely what a debug information file contains. Some languages, such as Java, for example, inject debug information within the byte code. Microsoft Visual Studio, on the other hand, has multiple formats including a standalone PDB file.

> Languages such as PHP, HTML, and Python are different. They usually don't have a compiler, and to some extent, they're not classified as programming languages. PHP and Python are technically scripting languages, whereas HTML is a markup language.

Debuggers use this debug information file to map the compiled code, either intermediate code or the machine code, back to your source code. Think of a debug information file as a map you would refer to if you were to visit an unknown city as a tourist. The debugger is able to stop at the correct location according to the breakpoints you place in your source code by referring to the debug information file.

Xcode's debug information file is called a dSYM file (because the file's extension is `.dSYM`).

> A dSYM file is technically a package containing a file with your target name in DWARF scheme.

When you create a new project, the default setting is to create a debug file automatically. The Build Options, which is under the Build Settings tab, of your project file (see Figure 17-1) shows this setting.

The dSYM file is created automatically every time you build the project. You can also create a dSYM file using the command-line utility `dsymutil`.

Figure 17-1 The Debug Information Format setting in your target settings

Symbolication

Compilers, including the LLVM compiler, convert your source code to assembly code. All assembly codes have a base address, and the variables you define, the stack you use, and the heap you use are all dependent on this base address. This base address could change every time the application runs, especially on iOS 4.3 and above—operating systems that introduced Address Space Layout Randomization. Symbolication is the process of replacing this base address with method names and variable names (collectively known as *symbols*). The base address is the entry point to your application, normally your "main" method unless you are writing a static library. You symbolicate other symbols by calculating their offset from the base address and mapping them to the dSYM file. Don't worry; this symbolication happens (almost transparently) when you're debugging your app in Xcode or when you profile it using Instruments.

> Instruments also needs the debug information file to symbolicate the running target and locates it using Spotlight. If you added the `.dSYM` extension to the exclusion filter in Spotlight, you do not see the variable names (symbol names) in Instruments. More often, this happens when your disk permissions, especially permissions to the Derived Data directory, are corrupted. A quick run by a disk utility should fix that issue and get you up and running.

Xcode's Symbolication

Sometimes you want to symbolicate a binary or a crash report (more often the latter) manually. Later, I discuss the various kinds of crash reports that you can use to analyze and fix issues in your app, including a homegrown crash reports collection tool and a couple of third-party crash-reporting services. When you're using the Apple-provided iTunes Connect crash reports, you see only addresses and hex codes scattered throughout. Without proper symbolication, you cannot understand what is going on. Fortunately, Xcode symbolicates crash reports when you just drag a crash report to it. Now, how does Xcode know the corresponding dSYM file? For this "automatic" symbolication to work, you use Xcode's Build and Archive option to build your app for submission to the App Store.

Inside the xcarchive

A quick look at the `xcarchive` bundle created by Xcode reveals the following directories: dSYMs, Products, and an `Info.plist` file. The dSYMs directory contains the list of dSYM files for all the target/static libraries you included in your project. The Products directory contains the list of all executable binaries. The `Info.plist` file is the same `plist` file in your project. The `Info.plist` file plays a very important role in identifying the version of the `target/`

dsym in the xcarchive bundle. In fact, when you drag a .crash file from iTunes Connect to your Xcode, Xcode internally looks up the archives for an Info.plist file that matches the crash report and picks up the .dSYM file from the dSYMs directory inside that archive. This is the reason you should *never* delete a submitted archive. When you delete your archives after submission, you'll end up in hot water when you try to symbolicate a crash report.

Committing Your dSYMs to Your Version Control

The other way to store your dSYMs is to commit them to your version control. When you get a crash report from iTunes Connect, check out the dSYM that corresponds to the submitted version and symbolicate your crash reports by matching them with the dSYMs. This way, all developers on a team have access to the dSYM files, and anyone on the team can debug crashes without e-mailing the dSYMs around.

> Instead of committing dSYMs for every commit to your develop or feature branch, consider committing them only when you make a release. If you're using Git as explained in Chapter 2, you should be committing dSYMs to the master branch for every release. You do have some other alternatives, including some third-party services that do server-side symbolication. I discuss them in the "Collecting Crash Reports" section later in this chapter.

Breakpoints

Breakpoints are a way to pause your debugger and inspect your symbols and objects in real time. Some debuggers, including LLDB, allow you to move the instruction pointer and continue debugging from a different location. You can set LLDB breakpoints in your app directly from Xcode. You just scroll to the line where you want the breakpoint to be placed and click Xcode's Product⇨Debug⇨Add Breakpoint at Current Line menu, or press Cmd-\.

The Breakpoint Navigator

Breakpoints that you added to your project are listed automatically in your Breakpoint navigator. You can access the breakpoint navigator using the keyboard shortcut Cmd-6.

The breakpoint navigator also allows you to add breakpoints for exceptions and symbols.

Exception Breakpoints

An exception breakpoint stops program execution when your code causes an exception to be thrown. Some of the methods in the NSArray, NSDictionary, or UIKit classes (such as the UITableView method) of Foundation.framework throw exceptions when certain conditions cannot be met. These conditions include trying to mutate an NSArray or trying to access an array element that is beyond the array's bounds. A UITableView throws an exception when you declare the number of rows to be "n" and don't provide a cell for every row. In theory, debugging exceptions might look easy, but understanding the source of your exception could be fairly complicated. Your app will just crash with a log that prints out the exception that caused the crash. These Foundation.framework methods are used throughout your project, and you won't really understand what happened by looking at the log if you don't have an exception breakpoint set. When you set an exception breakpoint, the debugger pauses the execution of the program just after the exception is thrown, but before it is caught, and you get to see the stack trace of the crashed thread in your breakpoint navigator.

To make things clear, compare debugging an app with and without the exception breakpoint enabled.

Create an empty application in Xcode (any template should work). In your app delegate, add the following line:

```
NSLog(@"%@", [@[] objectAtIndex:100]);
```

This line creates an empty array, accesses the 100th element, and logs it. Because accessing the 100th element in an empty array is not legally allowed, Xcode crashes with the following log in Console and takes you to the infamous `main.m`:

```
2012-08-27 15:25:23.040 Test[31224:c07] (null)
libc++abi.dylib: terminate called throwing an exception
(lldb)
```

No one can understand what is happening behind the scenes from that cryptic log message. To debug exceptions like these, you set an exception breakpoint.

You set an exception breakpoint from the breakpoint navigator. Open the breakpoint navigator and click the + button in the lower-left corner, choose Add Exception Breakpoint, and accept the default settings to set a new exception breakpoint, as shown in Figure 17-2.

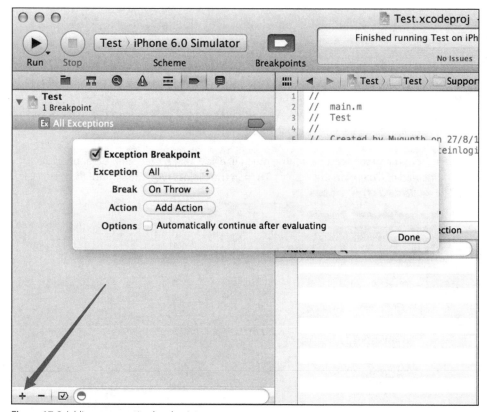

Figure 17-2 Adding an exception breakpoint

Run the same project again. This time, you should see the debugger pausing your application and stopping exactly on the line that raised the exception, as illustrated in Figure 17-3.

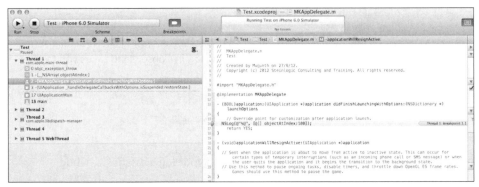

Figure 17-3 Xcode breaking at a breakpoint you just set

Exception breakpoints can help you understand the root cause of the exception. The first thing I do when I start a new project is to set an exception breakpoint. I highly recommend your doing so.

> When you want to run your app quickly without hitting any breakpoints, disable all of them by using the keyboard shortcut Cmd-Y.

Symbolic Breakpoints

Symbolic breakpoints pause your program's execution when a particular symbol gets executed. A symbol can be a method name or a method in a class or any C method (`objc_msgSend`).

You set a symbolic breakpoint from the breakpoint navigator, much like you set an exception breakpoint. Just choose Symbolic Breakpoint instead of Exception Breakpoint. Then in the dialog, type the symbol that you are interested in, as shown in Figure 17-4.

Figure 17-4 Adding a symbolic breakpoint

Now type `application:didFinishLaunchingWithOptions:` and press Enter. Then build and run your application. You should see the debugger stopping your application as it starts and showing the stack trace.

The symbol you watched didn't provide any advantage besides placing a normal breakpoint in your `applica tion:didFinishLaunchingWithOptions:`. Symbolic breakpoints are often used to watch interesting methods such as

```
-[NSException raise]
malloc_error_break
-[NSObject doesNotRecognizeSelector:]
```

In fact, the first exception breakpoint you created in the previous section is synonymous with a symbolic breakpoint on `[NSException raise]`.

The symbols `malloc_error_break` and `[NSObject doesNotRecognizeSelector:]` are useful for debugging memory-related crashes. If your app crashes with an `EXC_BAD_ACCESS`, setting a symbolic breakpoint in one or both of these symbols helps you localize the issue.

Editing Breakpoints

Every breakpoint you create can be edited from the breakpoint navigator. You edit a breakpoint by Ctrl-clicking the breakpoint and choosing Edit Breakpoint from the menu. You then see a breakpoint-editing sheet, as shown in Figure 17-5.

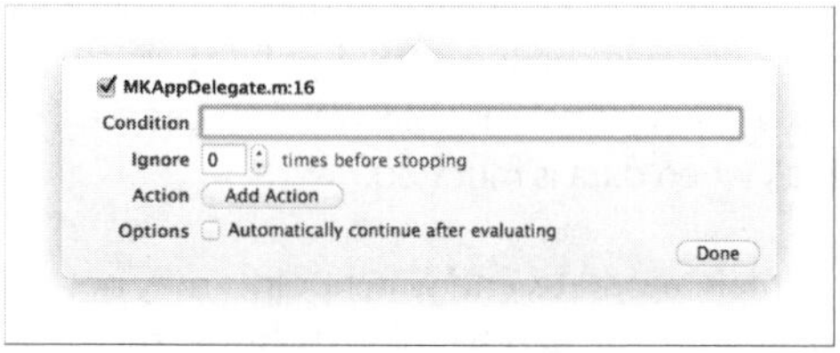

Figure 17-5 Editing breakpoints

Normally, a breakpoint stops program execution every time the line is executed. You can edit a breakpoint to set a condition and create a conditional breakpoint that is executed only when the set condition is reached. Why would this capability be useful? Imagine you are looping through a large array (n>10000). You're sure that the objects after 5500 are malformed and want to find out what is going wrong. The traditional approach is to write additional code (in your application's code) to check for index values greater than 5500 and remove this code after your debugging session is over.

For example, you could write something like this:

```
for(int i = 0 ; i < 10000; i ++) {

  if(i>5500) {
    NSLog(@"%@", [self.dataArray objectAtIndex:i]);
  }
}
```

Then you could place a breakpoint in the NSLog. A cleaner way is to add this condition to the breakpoint itself. In Figure 17-5, the sheet has a text field for adding a condition. Set this condition to i>5500 and run your app. Now, instead of breaking for every iteration, your breakpoints stop only when your condition is met.

You can customize your breakpoints to print out a value or play an audio file or execute an action script by adding an action. For example, if the objects that you're iterating are a list of users, and you want to know if a said user is in your list, you can edit your breakpoint to break when the object that you're interested in is reached. Additionally, in the action, you can select from a list of audio clips to play, execute an Apple script, or perform a variety of other functions. Click the Action button (refer to Figure 17-5) and choose the custom action named Sound. Now instead of stopping at your breakpoint, Xcode plays the audio clip you selected. If you're a game programmer, you might be interested in capturing an Open GL ES frame when a particular condition occurs, and this option is also available through the Action button.

Sharing Breakpoints

Your breakpoints now have code (or rather code fragments) associated with them that you want to save to your version control. Xcode 4 (and above) allows you to share your breakpoints with your coworkers by committing them to your version control. All you need to do is Ctrl-click a breakpoint and click Share. Your breakpoints are now saved to the xcshareddata directory inside the project file bundle. Commit this directory to version control to share your breakpoints with all other programmers in your team.

Watchpoints

Breakpoints provide you with the ability to pause the program execution when a given line is executed. Watchpoints provide a way to pause the program execution when the value stored at a variable changes. Watchpoints help to solve issues related to global variables and track which method updates a given global variable. Watchpoints are like breakpoints, but instead of breaking when a code is executed, they break when data is mutated.

In an object-oriented setting, you don't normally use global variables to maintain state, and watchpoints may not be used often. However, you may find it useful to track state changes on a singleton or other global objects such as your Core Data persistent store coordinator or API engines like the one you created in Chapter 12. You can set a watchpoint on accessToken in the RESTfulEngine class to know the engine's authentication state changes.

> You cannot add a watchpoint without running your application first. Start your app and open the watch window. The watch window, by default, lists the variables in your local scope. Ctrl-click a variable in the watch window. Now click on the watch <var> menu item to add a watchpoint on that variable. Your watchpoints are then listed in the breakpoint navigator.

The LLDB Console

Xcode's debugging console window is a full-featured LLDB debugging console. When your app is paused (at a breakpoint), the debugging console shows the LLDB command prompt. You can type any LLDB debugger command into the console to help you with debugging, including loading external python scripts.

The most frequently used command is po, which stands for print object. When your application is paused in debugger, you can print any variable that is in the current scope. This includes any stack variables, class variables, properties, ivars, and global variables. In short, any variable that your application can access at the breakpoint can be accessed via the debugging console.

Printing Scalar Variables

When you're dealing with scalars such as integers or structs (CGRect, CGPoint, and so on), instead of using po, you use p, followed by the type of struct.

Examples include

```
p (int) self.myAge
p (CGPoint) self.view.center
```

Printing Registers

Why do you need to print values on registers? You don't store variables in CPU registers, right? Yes, but the registers hold a wealth of information about the program state. This information is dependent on the subroutine calling convention on a given processor architecture. Knowledge of this information reduces your debug cycle time tremendously and makes you a programmer who can push the limits.

Registers in your CPU are used for storing variables that have to be accessed frequently. Compilers optimize frequently used variables such as the loop variable, method arguments, and return variables in the registers. When your app crashes for no apparent reason (apps always crash for no apparent reason until you find the problem, right?), probing the register for the method name or the selector name that crashed your app is very useful.

> The C99 language standard defines the keyword register that you can use to instruct the compiler to store a variable in the CPU registers. For example, declaring a for loop like for (register int i = 0 ; i < n ; i ++) stores the variable i in the CPU registers. Note that the declaration isn't a guarantee, and the compiler is free to store your variable in memory if no free registers are available.

You can print the registers from the LLDB console using the command register read. Now, create an app and add a code snippet that causes a crash:

```
int *a = nil;
NSLog(@"%d", *a);
```

You create a nil pointer and try accessing the value at the address. Obviously, this is going to throw an EXC_BAD_ACCESS. Write the preceding code in your application:didFinishLaunchingWithOptions: method (or any method you like) and run the app in your *simulator*. Yes, I repeat, in your *simulator*. When the app crashes, go to the LLDB console and type the command to print the register values:

```
register read
```

Your console should show something like the following:

Register Contents (Simulator)

```
(lldb) register read
General Purpose Registers:
       eax = 0x00000000
       ebx = 0x07408520
       ecx = 0x00001f7e Test`-[MKAppDelegate
       application:didFinishLaunchingWithOptions:] + 14 at
       MKAppDelegate.m:13
       edx = 0x00003604  @"%d"
       edi = 0x07122070
       esi = 0x0058298d  "application:didFinishLaunchingWithOptions:"
       ebp = 0xbfffde68
       esp = 0xbfffde30
        ss = 0x00000023
    eflags = 0x00010286 UIKit`-[UIApplication _
addAfterCACommitBlockForViewController:] + 23
       eip = 0x00001fca Test`-[MKAppDelegate
       application:didFinishLaunchingWithOptions:] + 90 at
       MKAppDelegate.m:19
        cs = 0x0000001b
        ds = 0x00000023
        es = 0x00000023
        fs = 0x00000000
        gs = 0x0000000f
(lldb)
```

The equivalent on a device (ARM processor) looks like the following:

Register Contents (Device)

```
(lldb) register read
General Purpose Registers:
        r0 = 0x00000000
        r1 = 0x00000000
        r2 = 0x2fdc676c
        r3 = 0x00000040
        r4 = 0x39958f43  "application:didFinishLaunchingWithOptions:"
        r5 = 0x1ed7f390
        r6 = 0x00000001
        r7 = 0x2fdc67b0
        r8 = 0x3c8de07d
        r9 = 0x0000007f
       r10 = 0x00000058
       r11 = 0x00000004
       r12 = 0x3cdf87f4 (void *)0x33d3eb09: OSSpinLockUnlock$VARIANT$mp + 1
        sp = 0x2fdc6794
        lr = 0x0003a2f3 Test`-[MKAppDelegate
       application:didFinishLaunchingWithOptions:] + 27 at
       MKAppDelegate.m:13
        pc = 0x0003a2fe Test`-[MKAppDelegate
```

```
        application:didFinishLaunchingWithOptions:] + 38 at
        MKAppDelegate.m:18
    cpsr = 0x40000030
```

(lldb)

Your output may vary, but pay close attention to the eax, ecx, and esi on the simulator or r0-r4 registers when running on a device. These registers store some of the values that you're interested in. In the simulator (running on your Mac's Intel processor), the ecx register holds the name of the selector that is called when your app crashed. You print an individual register to the console by specifying the register name as shown here:

```
register read ecx
```

You can also specify multiple registers like this:

```
register read eax ecx
```

The ecx register on the Intel architecture and the r15 register on the ARM architecture hold the program counter. Printing the address of the program counter shows the last executed instruction. Similarly, eax (r0 on ARM) holds the receiver address, ecx (r4 on ARM), and holds the selector that was called last (in this case, it's the application:didFinishLaunchingWithOptions: method). The arguments to the methods are stored in registers r1-r3. If your selector has more than three arguments, they are stored on stack, accessible via the stack pointer (r13). sp, lr, and pc are actually aliases to the r13, r14, and r15 registers, respectively. Hence, register read r13 is equivalent to register read sp.

So *sp, *sp+4, and so on, contain the address of your fourth and fifth arguments, and so on. On the Intel architecture, the arguments start at the address stored in ebp register.

When you download a crash report from iTunes Connect, it normally has the register state, and knowing the register layout on ARM architecture can help you better analyze the crash report. The following is a register state from a crash report:

Register State in a Crash Report

```
Thread 0 crashed with ARM Thread State:
    r0: 0x00000000    r1: 0x00000000        r2: 0x00000001       r3:
0x00000000
    r4: 0x00000006    r5: 0x3f871ce8        r6: 0x00000002       r7:
0x2fdffa68
    r8: 0x0029c740    r9: 0x31d44a4a        r10: 0x3fe339b4      r11:
0x00000000
    ip: 0x00000148    sp: 0x2fdffa5c        lr: 0x36881f5b       pc:
0x3238b32c
  cpsr: 0x00070010
```

Using otool, you can print the methods used in your app. Match the address in your program counter using grep to see which exact method was being executed when the app crashed.

```
otool -v -arch armv7 -s __TEXT __cstring <your image> | grep 3238b32c
```

Replace <your image> with the image of the app that crashed. (You have either committed this to your repository or stored it in the application archives in Xcode.)

Note that what you learned in this section is highly processor-specific and may change in the future if Apple changes the processor specification (from ARM to something else). However, when you understand the basics, you should be able to apply this knowledge to any new processor that comes along.

Scripting Your Debugger

The LLDB debugger was built from the ground up to support APIs and pluggable interfaces. Python scripting is one benefit of these pluggable interfaces. If you're a Python programmer, you'll be pleasantly surprised to learn that LLDB supports importing Python scripts to help you with debugging, which means that you can write a script in Python, import it into your LLDB, and inspect variables in LLDB using your script. If you're not a Python programmer, you can skip this section and probably won't lose anything.

Assume that you want to search for an element in a large array containing, say, 10,000 objects. A simple po on the array is going to list all 10,000, which is tedious to manually go through. If you have a script that takes this array as a parameter and searches for the presence of your object, you can import it into LLDB and use it for debugging.

You start the Python shell from the LLDB prompt by typing `script`. The prompt changes from `(lldb)` to `>>>`. Within the script editor, you can access the LLDB frame using the `lldb.frame` Python variable. So `lldb.frame.FindVariable("a")` will get the value of the variable a from the current LLDB frame. If you're iterating an array to look for a specific value, you can assign the `lldb.frame.FindVariable("myArray")` to a variable and pass it to your Python script.

The following code illustrates this:

Invoking the Python Script to Search for Occurrence of an Object

```
>>> import mypython_script
>>> array = lldb.frame.FindVariable ("myArray")
>>> yesOrNo = mypython_script.SearchObject (array, "<search element>")
>>> print yesOrNo
```

The preceding code assumes that you wrote a `SearchObject` function in a file `mypython_script`. Explaining the implementation details of the Python script is outside the scope of this book.

NSZombieEnabled Flag

A debugging chapter is not complete without mentioning `NSZombieEnabled`. This environment variable is used to debug memory-related bugs and to track over release of objects. `NSZombieEnabled`, when set, swizzles out the default `dealloc` implementation with a zombie implementation that converts an object to a zombie object when the retain count reaches zero. The functionality of the zombie object is to display a log and break into the debugger whenever you send a message to it.

So when you enable NSZombies in your app, instead of crashing, a bad memory access is just an unrecognized message sent to a zombie. A zombie displays the received message and breaks into the debugger allowing you to debug what went wrong.

You enable the `NSZombiesEnabled` environment variable from Xcode's schemes sheet. Click on Product⇨Edit Scheme to open the sheet and set the Enable Zombie Objects check box, as shown in Figure 17-6.

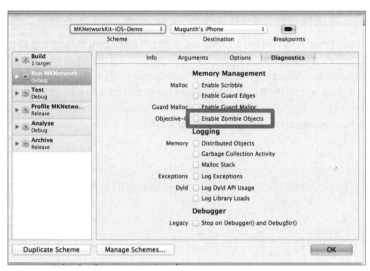

Figure 17-6 Enabling Zombie objects

Zombies were helpful back in the olden days when there was no Automatic Reference Counting (ARC). With ARC, if you're careful with your ownerships, normally you won't have memory-related crashes.

Different Types of Crashes

Software written using a programming language crashes as opposed to web applications written (or scripted) in a scripting or a markup language. Because a web application runs within the context of a browser, there is little possibility that a web app can corrupt memory or behave in a way that could crash the browser. If you're coming from a high-level language background, terms used by Xcode to denote various crashes may be cryptic to you. The following sections attempt to shed light on some of them. Crashes are usually signals sent by the operating system to the running program.

EXC_BAD_ACCESS

An `EXC_BAD_ACCESS` crash occurs whenever you try to access or send a message to a deallocated object. The most common cause of `EXC_BAD_ACCESS` is when you initialize a variable in one of your initializer methods but use the wrong ownership qualifier, which results in deallocation of your object. For example, you create an `NSMutableArray` of elements for your `UITableViewController` in the `viewDidLoad` method but set the ownership qualifier of the list to `unsafe_unretained` or `assign` instead of `strong`. Now in `cellForRowAtIndexPath:`, when you try to access the deallocated object, you crash with `EXC_BAD_ACCESS`. Debugging `EXC_BAD_ACCESS` is made easy with the `NSZombiesEnabled` environment variable described previously.

SIGSEGV

A signal segment fault (SIGSEGV) is a more serious issue that the operating system raises. It occurs when a hardware failure occurs or when you try to access a memory address that cannot be read or when you try to write to a protected address.

The first case, a hardware failure, is uncommon. When you try to read data stored in RAM and the RAM hardware at that location is faulty, you get a SIGSEGV crash. But more often than not, a SIGSEGV occurs for the latter two reasons. By default, code pages are protected from being written into and data pages are protected from being executed. When one of your pointers in your application points to a code page and tries to alter the value pointed to, you get a SIGSEGV. You also get a SIGSEGV when you try to read the value of a pointer that was initialized to a garbage value pointing to an address that is not a valid memory location.

SIGSEGV faults are more tedious to debug, and the most common reason a SIGSEGV happens is an incorrect typecast. Avoid hacking around with pointers or trying to a read a private data structure by advancing the pointer manually. When you do that and don't advance the pointer to take care of the memory alignment or padding, you get a SIGSEGV.

SIGBUS

A signal bus (SIGBUS) error is a kind of bad memory access where the memory you tried to access is an invalid memory address. That is, the address pointed to is not a physical memory address at all (it could be the address of one of the hardware chips). Both SIGSEGV and SIGBUS are subtypes of EXC_BAD_ACCESS.

SIGTRAP

SIGTRAP stands for signal trap. This is not really a crash signal. It's sent when the processor executes a trap instruction. The LLDB debugger usually handles this signal and stops at a specified breakpoint. If you get a SIGTRAP for no apparent reason, a clean build will usually fix it.

EXC_ARITHMETIC

Your application receives an EXC_ARITHMETIC error when you attempt a division by zero. The fix for this problem should be straightforward.

SIGILL

SIGILL stands for SIGNALILLEGALINSTRUCTION. This crash happens when you try to execute an illegal instruction on the processor. You execute an illegal instruction when you're trying to pass a function pointer to another function, but for one reason or other, the function pointer is corrupted and is pointing to a deallocated memory or a data segment. Sometimes you get EXC_BAD_INSTRUCTION instead of SIGILL, and although both are synonymous, EXC_* crashes are machine-independent equivalents of this signal.

SIGABRT

SIGABRT stands for SIGNAL ABORT. This is a more controlled crash in which the operating system has detected a condition that is not safe and asks the process to perform cleanup if any is needed. There is no one silver bullet to help you debug the underlying error for this signal. Frameworks like cocos2d or UIKit often call the C function abort (which in turn sends this signal) when certain preconditions are not met or when something really bad happens. When a SIGABRT occurs, the console usually has a wealth of information about what went wrong. Because it's a controlled crash, you can print the backtrace by typing bt into the LLDB console.

The following shows a SIGABRT crash on the console:

Printing the Backtrace Often Points to the Reason for a SIGABRT

```
(lldb) bt
* thread #1: tid = 0x1c03, 0x97f4ca6a libsystem_kernel.dylib`__pthread_
kill
  + 10, stop reason = signal SIGABRT
      frame #0: 0x97f4ca6a libsystem_kernel.dylib`__pthread_kill + 10
      frame #1: 0x92358acf libsystem_c.dylib`pthread_kill + 101
      frame #2: 0x04a2fa2d libsystem_sim_c.dylib`abort + 140
      frame #3: 0x0000200a Test`-[MKAppDelegate
       application:didFinishLaunchingWithOptions:] + 58 at
MKAppDelegate.m:16
```

Watchdog Timeout

A watchdog timeout is normally distinguishable because of the error code 0x8badf00d. (Programmers, too, have a sense of humor—this is read as *Ate Bad Food*). On iOS, this situation happens mostly when you block the main thread with a synchronous networking call. For that reason, don't ever do a synchronous networking call.

For more in-depth information about networking, read Chapter 12 of this book.

Custom Error Handling for Signals

You can override signal handling by using the C function sigaction and providing a pointer to a signal handler function. The sigaction function takes a sigaction structure that has a pointer to the custom function that needs to be invoked, as shown in the following code:

Custom Code for Processing Signals

```
void SignalHandler(int sig) {

    // your custom signal handler goes here
}

// Add this code at startup
struct sigaction newSignalAction;
memset(&newSignalAction, 0, sizeof(newSignalAction));
newSignalAction.sa_handler = &SignalHandler;
sigaction(SIGABRT, &newSignalAction, NULL);
```

In this method, a custom C function for `SIGABRT` signal is added. You can add more handler methods for other signals as well using code similar to this.

A more sophisticated method is to use an open source class written by Matt Gallagher. Matt, author of the popular cocoa blog *Cocoa with Love,* made the open source class `UncaughtExceptionHandler` that you can use to handle uncaught exceptions. The default handler shows an error alert. You can easily customize it to save the application state and submit a crash report to your server. A link to Matt's post is provided in the "Further Reading" section at the end of this chapter.

Assertions

Assertions are an important defense against programming errors. An assertion requires that something must be true at a certain point in the program. If it is not true, the program is in an undefined state and should not proceed. Consider the following example of `NSAssert`:

```
NSAssert(x == 4, @"x must be four");
```

`NSAssert` tests a condition, and if it returns `NO`, it raises an exception, which is processed by the current exception handler, which by default calls `abort` and crashes the program. If you're familiar with Mac development, you may be used to exceptions terminating only the current run loop, but iOS calls `abort` by default, which terminates the program no matter what thread it runs on.

> **Technically, `abort` sends the process a `SIGABRT`, which can be caught by a signal handler. Generally, I don't recommend catching `SIGABRT` except as part of a crash reporter. See "Catching and Reporting Crashes," later in this chapter, for information about how to handle crashes.**

You can disable `NSAssert` by setting `NS_BLOCK_ASSERTIONS` in the build setting "Preprocessor Macros" (`GCC_PREPROCESSOR_DEFINITIONS`). Opinions differ on whether `NSAssert` should be disabled in release code. It really comes down to this: When your program is in an illegal state, would you rather it stop running, or would you prefer that it run in a possibly random way? Different people come to different conclusions here. My opinion is that it's generally better to disable assertions in release code. I've seen too many cases where the programming error would have caused only a minor problem, but the assertion causes a crash. Xcode 4 templates automatically disable assertions when you build for the Release configuration.

That said, although I like removing assertions in the Release configuration, I don't like ignoring them. They're exactly the kind of "this should never happen" error condition that you would want to find in your logs. Setting `NS_BLOCK_ASSERTIONS` completely eliminates them from the code. My solution is to wrap assertions so that they log in all cases. The following code assumes you have an `RNLogBug` function that logs to your log file. It's mapped to `NSLog` as an example. Generally, I don't like to use `#define`, but it's necessary here because `__FILE__` and `__LINE__` need to be evaluated at the point of the original caller.

This code also defines `RNCAssert` as a wrapper around `NSCAssert` and a helper function called `RNAbstract`. `NSCAssert` is required when using assertions within C functions, rather than Objective-C methods.

RNAssert.h

```
#import <Foundation/Foundation.h>

#define RNLogBug NSLog // Use DDLogError if you're using Lumberjack

// RNAssert and RNCAssert work exactly like NSAssert and NSCAssert
// except they log, even in release mode

#define RNAssert(condition, desc, ...) \
  if (!(condition)) { \
    RNLogBug((desc), ## __VA_ARGS__); \
    NSAssert((condition), (desc), ## __VA_ARGS__); \
  }

#define RNCAssert(condition, desc) \
  if (!(condition)) { \
    RNLogBug((desc), ## __VA_ARGS__); \
    NSCAssert((condition), (desc), ## __VA_ARGS__); \
  }
```

Assertions often precede code that would crash if the assertion were not valid. For example (assuming you're using `RNAssert` to log even in the Release configuration):

```
RNAssert(foo != nil, @"foo must not be nil");
[array addObject:foo];
```

The problem is that this code still crashes, even with assertions turned off. What is the point of turning off assertions if you're going to crash anyway in many cases? That leads to code like this:

```
RNAssert(foo != nil, @"foo must not be nil");
if (foo != nil) {
  [array addObject:foo];
}
```

This approach is a little better, using `RNAssert` so that you log, but you have duplicated code. This raises more opportunities for bugs if the assertion and conditional don't match. Instead, I recommend this pattern when you want an assertion:

```
if (foo != nil) {
  [array addObject:foo];
}
else {
  RNAssert(NO, @"foo must not be nil");
}
```

This pattern ensures that the assertion always matches the conditional. Sometimes assertions are overkill, but this is a good pattern in cases in which you want one. I almost always recommend an assertion as the default case of a `switch` statement, however.

```
switch (foo) {
  case kFooOptionOne:
    ...
    break;
  case kFooOptionTwo:
    ...
    break;
  default:
    RNAssert(NO, @"Unexpected value for foo: %d", foo):
    break;
}
```

This way, if you add a new enumeration item, it helps you catch any `switch` blocks that you failed to update.

Exceptions

Exceptions are not a normal way of handling errors in Objective-C. From *Exception Programming Topics* (`developer.apple.com`):

> *The Cocoa frameworks are generally not exception-safe. The general pattern is that exceptions are reserved for programmer error only, and the program catching such an exception should quit soon afterwards.*

In short, exceptions are not for handling recoverable errors in Objective-C. Exceptions are for handling those things that should never happen and which should terminate the program. This is similar to `NSAssert`, and in fact, `NSAssert` is implemented as an exception.

Objective-C has language-level support for exceptions using directives such as `@throw` and `@catch`, but you generally should not use them. There is seldom a good reason to catch exceptions except at the top level of your program, which is done for you with the global exception handler. If you want to raise an exception to indicate a programming error, it's best to use `NSAssert` to raise an `NSInternalInconsistencyException`, or create and raise your own `NSException` object. You can build these by hand, but I recommend `+raise:format:` for simplicity:

```
[NSException raise:NSRangeException
          format:@"Index (%d) out of range (%d...%d)",
            index, min, max];
```

There seldom is much reason to use this approach. In almost all cases, using `NSAssert` would be just as clear and useful. Because you generally shouldn't catch exceptions directly, the difference between `NSInternalInconsistencyException` and `NSRangeException` is rarely useful.

Automatic Reference Counting is not exception-safe by default in Objective-C. You should expect significant memory leaks from exceptions. In principle, ARC is exception-safe in Objective-C++, but `@autoreleasepool` blocks are still not released, which can lead to leaks on background threads. Making ARC exception-safe incurs performance penalties, which is one of many reasons to avoid significant use of Objective-C++. The Clang flag `-fobjc-arc-exceptions` controls whether ARC is exception-safe.

Collecting Crash Reports

When you're developing for iOS, you have multiple ways to collect crash dumps. In the following sections, I explain the most common way to collect crash reports, namely, iTunes Connect. I also explain about two other third-party services that also offer server-side symbolication in addition to crash reporting.

iTunes Connect

iTunes Connect allows you to download crash reports for your apps. You can log in to iTunes Connect and go to your app's details page to download crash reports. Crash reports from iTunes are not symbolicated, and you should symbolicate them using the exact dSYM that was generated by Xcode when you built the app for submission. You can do this either automatically using Xcode or manually (using the command-line utility `symbolicatecrash`).

Collecting Crash Reports

When your app crashes on your customer's device, Apple takes care of uploading the crash dumps to its servers. But do remember that this happens only when the user accepts to submit crash reports to Apple. Although most users submit crash reports, some may opt out. As such, iTunes crash reports may not represent your actual crash count.

Symbolicating iTunes Connect Crash Reports Using Xcode

If you're the sole developer, using Xcode's automatic symbolication makes more sense, and it's easier. There is almost no added advantage of using manual symbolication. As a developer, you only have to use Xcode's Build⇨Archive option to create your product's archive and submit this archive to the App Store. This archive encapsulates both the product and the dSYM. Don't ever delete this archive file, even after your app is approved. For every app (including multiple versions) that you submit, you need to have a matching archive in Xcode.

Symbolicating iTunes Connect Crash Reports Manually

When there's more than one developer, every developer on the team needs to be able to symbolicate a given crash report. When you depend on Xcode's built-in Archive command to store your dSYMs, the only developer able to symbolicate a crash is the one who submits apps to the App Store. To allow other developers to symbolicate, you may have to e-mail archive files around, which is probably not the right way to do it. My recommendation is to commit the archives to your version control for every release build.

From the Organizer, go to the Archives tab and Cmd-click the archive that you just submitted. Reveal it in Finder, copy it to your project directory, and commit your release branch. Now when you get a crash report from Apple, check out the archive file from your version control that corresponds to the crash report. That is, if your crash report is for version 1.1, get the archive from your version control that corresponds to version 1.1. Open the archive file's location in Terminal.

You can now symbolicate your crash report manually using the following `symbolicatecrash.sh` shell script:

```
symbolicatecrash MyApp.crash MyApp.dSYM > symbolicated.txt
```

Alternatively, you can use `atos` in interactive mode, as shown here:

```
atos -arch armv7 -o MyApp
```

If your crash report matches your dSYM file, you should see your memory addresses symbolicated in the text file.

Using `atos` in interactive mode is handy in many cases in which you just want to know the address of the thread's stack trace that crashed. `atos` assumes that your dSYM file is located in the same location as your application.

`symbolicatecrash` is a script that's not in your `%PATH%` by default. As such, you might get a Command Not Found error when you type the preceding command on Terminal. As of Xcode 4.3, this script resides in `/Applications/Xcode.app/Contents/Developer/Platforms/iPhoneOS.platform/Developer/Library/PrivateFrameworks/DTDeviceKit.framework/Versions/A/Resources/symbolicatecrash`. Even if this location changes in the near future, you should be able to run a quick search for this file from the command line using the command `find /Applications/Xcode.app -name symbolicatecrash -type f`.

Third-Party Crash Reporting Services

Whew, that was painful! But as the saying goes, necessity is the mother of invention. A painful crash analysis process leads plenty of third-party developers to make alternative services that, apart from crash log collection and analysis, do symbolication on the server side.

The best replacement I've found is QuincyKit (`quincykit.net`), which is integrated with HockeyApp (`hockeyapp.net`). It's easy to integrate into an existing project, and it uploads reports to your own web server or the HockeyApp server after asking user permission. Currently, it doesn't handle uploading logs to go along with the crash report.

QuincyKit is built on top of PLCrashReporter from Plausible Labs. PLCrashReporter handles the complex problem of capturing crash information. QuincyKit provides a friendly front end for uploading that information. If you need more flexibility, you might consider writing your own version of QuincyKit. It's handy and nice, but not all that complicated. You probably should not try to rewrite PLCrashReporter. While a program is in the middle of crashing, it can be in a bizarre and unknown state. Properly handling all the subtle issues that go with that is not simple, and Landon Fuller has been working on PLCrashReporter for years. Even something as simple as allocating or freeing memory can deadlock the system and rapidly drain the battery. That's why QuincyKit uploads the crash files when the program restarts rather than during the crash. You should do as little work as possible during the crash event.

When you get your crash reports, depending on how your image was built, they may have symbols or they may not. Xcode generally does a good job of automatically symbolicating the reports (replacing addresses with method names) in Organizer as long as you keep the `.dSYM` file for every binary you ship. Xcode uses Spotlight to find these files, so make sure they're available in a place that Spotlight can search. You can also upload your symbol files to HockeyApp if you're using it for managing crash reports.

There is one other third party crash reporter, TestFlight. While HockeyApp is a paid service, TestFlight is free. Both have their own advantages and disadvantages, but an in-depth discussion/comparison of them is outside the scope of this chapter. The following section briefly compares them to iTunes Connect.

Advantages of TestFlight or HockeyApp over iTunes Connect

Both TestFlight and HockeyApp provide an SDK that you normally integrate with your app. These SDKs take care of uploading crash reports to their servers. Although Apple uploads crash reports only when the user consents, these SDKs always upload the crash reports. That means you get more accurate statistics on the number of times a particular kind of crash occurs. Because crash reports normally don't have personally identifiable information, uploading them without the user's consent is allowed within the rules of the App Store.

You can upload a dSYM to TestFlight and symbolicate crash reports for the live versions of your app. For ad hoc builds, TestFlight's desktop client automatically uploads dSYMs. Server-side symbolication means that you don't have to know anything about `symbolicatecrash` or `atos` commands or use the Terminal. In fact, both of these services upload your dSYM files to their server. You and their SDK collect crash reports from users' devices.

Summary

In this chapter, you read about LLDB and debugging. You learned about breakpoints, watchpoints, and ways to edit and share those edited breakpoints. You discovered the power behind the LLDB console and how to use a Python script to speed up your debugging. You also read about the various types of errors, crashes, and signals that normally occur in your iOS app, and you found out how to avoid and overcome them. Finally, you learned about some third-party services that allow you to symbolicate crash reports on the server.

Further Reading

Apple Documentation

The following documents are available in the iOS Developer Library at *developer.apple.com* or through the Xcode Documentation and API Reference.

Xcode 4 User Guide: "Debug Your App"

Developer Tools Overview

LLVM Compiler Overview

Read the header documentation in the following header files. You can get these files by searching for them using Spotlight.

```
exception_types.h

signal.h
```

WWDC Session

The following session video is available at developer.apple.com.

WWDC 2012: "Session 415: Debugging with LLDB"

Other Resources

Writing an LLVM Compiler Backend

`http://llvm.org/docs/WritingAnLLVMBackend.html`

Apple's "Lazy" DWARF Scheme

`http://wiki.dwarfstd.org/index.php?title=Apple's_%22Lazy%22_DWARF_Scheme`

Hamster Emporium archive. "[objc explain]: So You Crashed in objc_msgSend()"

`http://www.sealiesoftware.com/blog/archive/2008/09/22/objc_explain_So_you_crashed_in_objc_msgSend.html`

Cocoa with Love. "Handling Unhandled Exceptions and Signals"

`http://cocoawithlove.com/2010/05/handling-unhandled-exceptions-and.html`

furbo.org. "Symbolicatifination"

`http://furbo.org/2008/08/08/symbolicatifination/`

LLDB Python Reference

`http://lldb.llvm.org/python-reference.html`

## Chapter 18

Performance Tuning Until It Flies

Performance is one of those things that separates acceptable apps from extraordinary apps. Of course, some performance bugs make an app unusable, but many more just make an app sluggish. Even apps that have good UI responsiveness may be using much more battery life than they should. Every developer needs to spend time periodically evaluating performance and making sure that nothing is being wasted. In this chapter, you discover how to best prioritize, measure, and improve the performance of your apps, including memory, CPU, drawing, and disk and network performance. Through it all, you learn how to best use one of the most powerful profiling tools available: Instruments. This chapter also provides tips on how to get the most from the powerful math frameworks available in iOS.

The Performance Mindset

Before going down the road of optimizing an application, you need to adopt the right mindset. This mindset can be summarized in a few rules.

Rule 1: The App Exists to Delight the User

I could really stop the rules right here. Everything else is a corollary. As a developer, never forget that your app exists to do more than just "provide value" to the user. It's a competitive market out there. You want an app that the user really loves. Even minor annoyances detract from that goal. Scrolling that stutters, a slow launch, poor response to buttons, all these things accumulate in the user's mind and turn a great app into a "good-enough" app.

Rule 2: The Device Exists for the Benefit of the User

It's very unlikely that your app is why users bought an iDevice. Users likely have many apps and many reasons for using their devices. Never let your app get in the way of that reality. Your users may really enjoy your app, but that doesn't mean it's okay to drain the battery. It's not okay to fill the disk. You live in a shared environment. Behave accordingly.

Rule 3: Go to Extremes

When your app is working for the user, use every resource available. When it's not doing work for the user, use as few resources as possible. This means that it's okay to use a lot of CPU to maintain a high frame rate while the user is active. But when the user is not using your app directly, it shouldn't be using much of the CPU. As a general rule, it's better for the app to use 100% of available resources for a short period and then go completely to sleep than to use 10% of available resources all the time. As many have pointed out, a microwave can use more power illuminating its clock than it does heating food because the clock is on all the time. Look for these low-power, always-on activities and get rid of them.

Rule 4: Perception Is Reality

Much of iOS is devoted to giving the illusion of performance that's unrealistic. For instance, launch images that display the UI give the user the impression that the application is immediately available, even though it hasn't finished launching yet. Similar approaches are acceptable and even encouraged in your apps. If you can display some information while waiting for full data, do so. Keep your application responsive, even if you can't actually let the user do anything yet. Swap images for hard-to-draw views. Don't be afraid to cheat, as long as cheating makes the user experience better.

Rule 5: Focus on the Big Wins

Despite everything I just said, I don't think you must chase every leaked byte or every wasted CPU cycle. Doing so isn't worth the effort. Your time is better spent writing new features, squashing bugs, or even taking a vacation. Cocoa has small leaks, which means that even if your program is perfect, it may still have some leaks you can't remove. Focus on the things that impact the user. A thousand bytes leaked over the run of the program don't matter. A thousand bytes lost every second matter quite a lot.

Welcome to Instruments

Instruments is one of the most powerful performance tools ever developed, and it can be one of the easiest to use. That doesn't mean that using it is *always* easy. Many things in Instruments don't work entirely correctly because of default configurations, but when you know the handful of things to reconfigure, no tool is more adept at finding performance problems quickly.

In this chapter, you evaluate a fairly simple piece of code that has several problems. It's an app that reads a file and displays the contents one character at a time in its main view. You can find the full listing in the ZipText project for this chapter on the book's website. You can test each revision by changing the REVISON value in ZipTextView.h. Here is the code you focus on:

ZipTextView1.m (ZipText)

```
- (id)initWithFrame:(CGRect)frame text:(NSString *)text {
    ... Load long string into self.text ...
    ... Set timer to repeatedly call appendNextCharacter ...
}

- (void)appendNextCharacter {
    for (NSUInteger i = 0; i <= self.index; i++) {
        if (i < self.text.length) {
            UILabel *label = [[UILabel alloc] init];
            label.text = [self.text substringWithRange:NSMakeRange(i,1)];
            label.opaque = NO;
            [label sizeToFit];
            CGRect frame = label.frame;
            frame.origin = [self originAtIndex:i
                                      fontSize:label.font.pointSize];
            label.frame = frame;
            [self addSubview:label];
        }
    }
}
```

```
    self.index++;
  }

- (CGPoint)originAtIndex:(NSUInteger)index
               fontSize:(CGFloat)fontSize {
  CGPoint origin;
  if (index == 0) {
    return CGPointZero;
  }
  else {
    origin = [self originAtIndex:index-1 fontSize:fontSize];
    NSString *
    prevCharacter = [self.text
                     substringWithRange:NSMakeRange(index-1,1)];
    CGSize
    prevCharacterSize = [prevCharacter
                          sizeWithAttributes:@{ NSFontAttributeName:
                                  UIFont systemFontOfSize:fontSize]
                                 }];
    origin.x += prevCharacterSize.width;
    if (origin.x > CGRectGetWidth(self.view.bounds)) {
      origin.x = 0;
      origin.y += prevCharacterSize.height;
    }
    return origin;
  }
}
```

This program starts out well but quickly slows to a crawl and starts displaying memory warnings. To figure out what is going wrong, start by launching Instruments using Product⇨Profile in Xcode. Here are a few things to note before diving into the details:

■ By default, Instruments builds in Release mode. This can display radically different performance than Debug mode, but it's usually what you want. If you want to profile in Debug mode, you can modify the mode in the scheme.

■ Performance in the simulator and on a device can be radically different. Most of the time, it's important to profile only on a device.

■ Instruments often has a difficult time finding the symbols for the current application. You'll often need to re-symbolicate the document using File⇨Re-Symbolicate Document.

The last bug in Instruments is extremely frustrating and difficult to eliminate completely. Apple says that Instruments is supposed to find its symbols using Spotlight, but this almost never works. I recommend duplicating rdar://10158512 at bugreport.apple.com. See http://openradar.appspot. com/10158512 for the text of this bug report.

In my experience, if you add the following directory to Preferences⇨Search Paths, you can at least press the Symbolicate button rather than search for the dSYM by hand:

```
~/Library/Developer/Xcode/DerivedData
```

If you have to search by hand, note that the dSYM is in the following directory:

```
~/Library/Developer/Xcode/DerivedData/<app>/Build/Products/
   Release-iphoneos
```

In the rest of this chapter, assume that I have symbolicated whenever required.

Finding Memory Problems

Many performance problems can be tracked to memory problems. If you're seeing unexpected memory warnings, it's probably best to track those down first. Use the Allocations template in Instruments. Figure 18-1 shows the result.

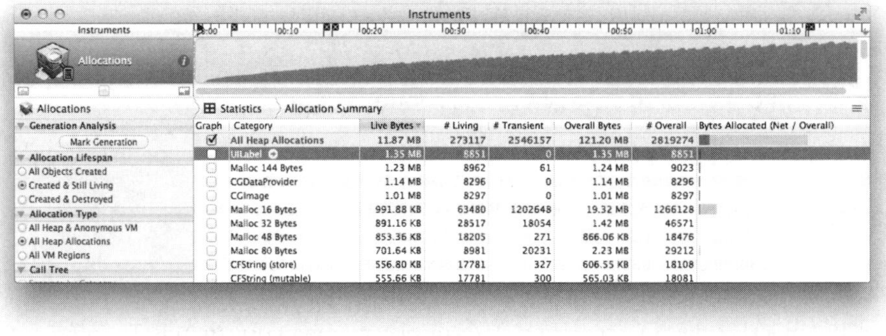

Figure 18-1 Allocations Instrument

Looking at the graph, you can see that memory allocations are clearly out of control in this app. Memory use is constantly growing. By sorting the Live Bytes column, you see that the largest individual memory use is in `UILabel` and that there are more than 8,000 of them currently allocated after just one minute. That's very suspicious. Click the arrow beside `UILabel` to get more information and press Shift-Cmd-E to show Extended Detail (see Figure 18-2).

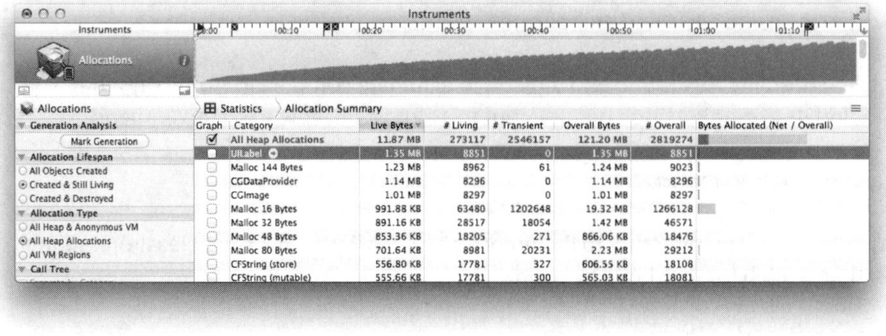

Figure 18-2 Extended Detail for Allocations

Here, you see that the `UILabel` is created in `-[ZipTextView appendNextCharacter]`. If you double-click that stack frame, you see the code along with hotspot coloring indicating how much time is spent in each line of code (see Figure 18-3).

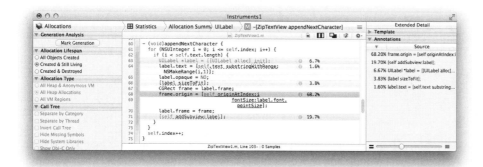

Figure 18-3 Code View

Before investigating further, take a moment to note some new features in Xcode 5. The new Allocation Type selector shows allocations in anonymous virtual memory regions, particularly in Core Animation and Core Data. Selecting All Heap & Anonymous VM adds the information shown in Figure 18-4.

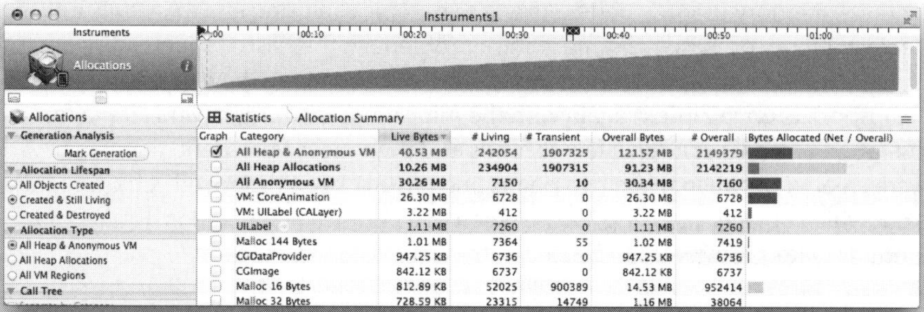

Figure 18-4 All Heap & Anonymous VM

Although the UILabel instances only take a little more than 1MB of memory, the core animation layers add at least three times as much. You can also see a more complete view of the app's memory usage. Rather than just the 10MB of heap allocations, this app uses at least 40MB of total memory.

Instruments now also show you important events that may impact your profiling. By clicking on the flags above the time line, you can see when other applications were launched or terminated. Due to the high memory use of this application, other applications are being terminated, as shown in Figure 18-5.

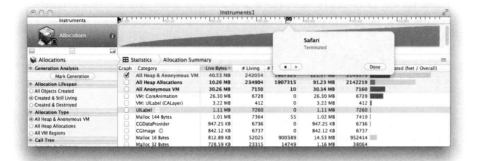

Figure 18-5 Event notifications in Instruments

If you turn back to the investigation at hand, a little digging shows the problem. Every time this method is called, it re-creates all the labels instead of just creating one new one. You just need to remove the loop, as shown here:

ZipTextView2.m (ZipText)

```
for (NSUInteger i = 0; i <= self.index; i++)
```

and replace it with the assignment

```
NSUInteger i = self.index;
```

Now, when you rerun Instruments, memory usage is better. After a minute, you're using less than 2MB, rather than more than 9MB, but steady memory growth over time is still a problem. In this case, it's pretty obvious that the problem is the `UILabel` views, but this is a good opportunity to demonstrate how to use generation analysis (previously called heapshot). Launch ZipText again with the Allocations instrument, and after a few seconds, press the Mark Generation button. Let it run a few seconds and then press Mark Generation again. This shows all the objects that were created between those two points and haven't been destroyed yet. Using this tool is a great way to figure out which objects are being leaked when you perform an operation. Figure 18-6 shows that in about 10 seconds, I created 73 new `UILabel` objects.

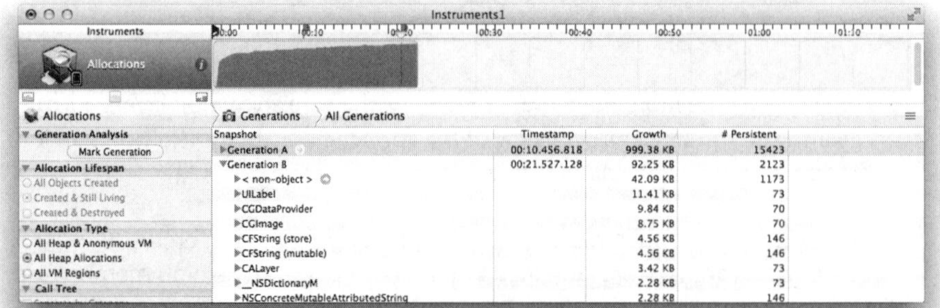

Figure 18-6 Improved memory management

At this point, you should be reevaluating the choice to use a separate `UILabel` for every character. Maybe a single `UILabel`, `UITextView`, or even a custom `drawRect:` would have been a better choice here. `ZipTextView3` demonstrates how to implement this with a custom `drawRect:`.

ZipTextView3.m (ZipText)

```
- (void) appendNextCharacter {
    self.index++;
    [self setNeedsDisplay];
}

- (void) drawRect:(CGRect) rect {
    for (NSUInteger i = 0; i <= self.index; i++) {
        if (i < self.text.length) {
            NSString *character = [self.text substringWithRange:
                                      NSMakeRange(i, 1)];
            [character drawAtPoint:origin
                    withAttributes:@{ NSFontAttributeName:
                                        [UIFont systemFontOfSize:kFontSize] }];
        }
    }
}
```

One more run, and as you see in Figure 18-7, there's almost no memory growth. The 1.21KB allocation is in the text rendering engine in Core Graphics, and is a good example of the small memory growths that are impossible to completely eliminate. But now performance has slowed to a crawl. Before you move on to CPU performance problems, here are a few more useful tips for memory performance investigations:

Figure 18-7 Memory footprint of ZipTextView with custom drawing

▨ Clicking the small *i* beside the Allocations instrument enables you to set various options. One of the most useful options is to change the Track Display from Current Bytes to Allocation Density. This graphs how many allocations happen during a sample period, indicating where you may have excessive memory churn. Allocating memory can be expensive, so you don't want to rapidly create and destroy thousands of objects if you can avoid doing so.

■ The Leaks instrument can be useful occasionally, but don't expect too much from it. It only looks for unreferenced memory. It doesn't detect retain loops, which are more common. Nor does it detect memory you fail to release, such as the `UILabel` views in the previous example. It also can have false positives. For example, it often shows one or two small leaks during program startup. If you see new leaks show up every time you perform an action or regularly at some time interval, it's definitely worth investigating. Generally, generation analysis is a more useful tool in tracking down lost memory problems.

■ Keep track of whether you're looking at Live Bytes. Instruments keeps track of both total and net allocations. Net allocations are allocated memory minus freed memory. If you create and destroy objects regularly, total allocations may be orders of magnitude greater than net allocations. Instruments sometimes refers to net allocations as *live* or *still living*. In the main screen (refer to Figure 18-1), the graph in the final column shows total allocations as a light bar and net allocations as a darker bar.

■ Remember that the Leaks instrument and the Allocations instrument can tell you only where memory was allocated. This has nothing to do with when memory was leaked. The allocation point probably isn't where your bug is.

Finding CPU Problems

Although you fixed the memory problems in the example, it's still running much too slowly. After a few paragraphs, it slows to a crawl. Instruments also can help with this problem. Profile ZipText again using the Time Profiler Instrument. You will see a Time Profiler graph similar to Figure 18-8.

What's important about this graph is that it's staying at 100%. Programs should seldom run at 100% CPU for very long. Using Option-Drag, you can select part of the graph. The Call Tree list will update to show you the methods and functions called during those samples.

A Word on Memory Allocation

All objects in Objective-C are allocated on the heap. Calls to `+alloc` eventually turn into calls to `malloc()`, and calls to `-dealloc` eventually turn into calls to `free()`. The heap is a shared resource across all threads, and modifying the heap requires a lock. This means that one thread allocating and freeing memory can block another thread allocating and releasing memory. Doing this quickly is called *memory churn* and should be avoided, particularly in multithreaded applications (including ones using Grand Central Dispatch). iOS has many tricks to minimize the impact of this churn. Small blocks of memory can often be allocated without modifying the heap. But as a rule, avoid rapidly creating and freeing memory if you can help it. Reusing objects can be better than creating new ones rapidly.

As an example, when working with `NSMutableArray` and `NSMutableDictionary`, consider using `removeAllObjects` rather than throwing away the collection and making a new one.

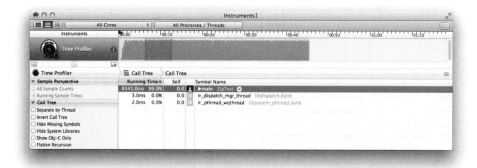

Figure 18-8 Time Profiler analysis of ZipText

In the Call Tree list, select Invert Call Tree. You almost always want an inverted call tree. It is bizarre that it's not the default. The "standard" call tree puts the top-level function at the top. This is always going to be `main()` for the main thread, which is useless information. You usually want to know the *lowest*-level method or function in the stack. Inverting the call tree puts that on top.

Several other Call Tree options are useful at different times:

- **Top Functions**—This option appeared in Instruments at some point, and I've never seen documentation on it, but it's great. It makes a list of the functions (or methods) that are taking the most time. This has become the first place I look when investigating performance problems.

- **Separate by Thread**—There are reasons to turn this option on or off. You definitely want to know what's taking the most time on your main thread, because this can block the UI. You also, however, want to know what's taking the most time across all threads. On iPhone and older iPads, there is only one core, so saying that something is being done "in the background" is a bit misleading. The UI thread still has to context-switch to service background threads. Even with the multicore iPads, "background" threads cost just as much battery life as the UI thread, so you want to keep it under control.

- **Hide System Libraries**—This option shows only "your" code. Many people turn this option on immediately, but I generally find it useful to leave it turned off most of the time. You generally want to know where the time is being spent, whether or not it's in "your" code. Often the biggest time impacts are in system actions like file writes. Hiding system libraries can hide these problems. That said, sometimes it's easier to understand the call tree if you hide system libraries.

- **Show Obj-C Only**—This option is similar to Hide System Libraries in that it tends to focus on higher-level code. I tend to leave it turned off for the same reasons I leave Hide System Libraries turned off. But again, it can be useful sometimes to help you find your place in the code.

For ZipText, select Invert Call Tree and Hide System Libraries. In Figure 18-9, you see that `originAtIndex:fontSize:` is taking most of the time and that it's being called recursively. A little investigation shows that every character recomputes the location of every previous character. No wonder this app runs slowly. You can modify `originAtIndex:fontSize:` to cache previous results as shown in `ZipTextView4.m`.

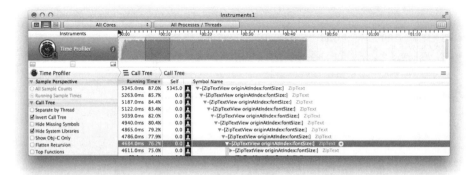

Figure 18-9 ZipText with caching

ZipTextView4.m (ZipText)

```objc
@property (nonatomic) NSMutableArray *locations;
...
- (id)initWithFrame:(CGRect)frame text:(NSString *)text {
    ...
    _locations = [NSMutableArray
                arrayWithObject:[NSValue
                                    valueWithCGPoint:CGPointZero]];
}

- (CGPoint)originAtIndex:(NSUInteger)index
              fontSize:(CGFloat)fontSize {
    if ([self.locations count] &gt; index) {
        return [self.locations[index] CGPointValue];
    }
    else {
        CGPoint origin = [self originAtIndex:index-1 fontSize:fontSize];
        NSString *
        prevCharacter = [self.text
                        substringWithRange:NSMakeRange(index-1,1)];
        CGSize
        prevCharacterSize = [prevCharacter
                            sizeWithAttributes:@{ NSFontAttributeName:
                                                    [UIFont
systemFontOfSize:fontSize]
                                                }];
        origin.x += prevCharacterSize.width;
        if (origin.x > CGRectGetWidth(self.bounds)) {
            origin.x = 0;
            origin.y += prevCharacterSize.height;
        }
        self.locations[index] = [NSValue valueWithCGPoint:origin];
        return origin;
    }
}
```

Now things are looking better, but drawing still seems to slow down when it gets about halfway down the screen. You dig into that in the next section.

Here are some other thoughts on hunting down CPU performance issues:

- Operations can take a long time because they are expensive on the CPU or because they block. Blocking operations are a serious problem on the main thread because the UI can't update. To look for blocking operations, use the Time Profiler instrument, click the *i* button to display the options, and select Record Waiting Threads. In the Sample Perspective, select All Sample Counts. This lets you see what is taking the most time on each thread (particularly the main thread), even if it's blocking.

- By default, the Time Profiler instrument records only User time. It doesn't count Kernel time against you. Most of the time, this is what you want, but if you're making a lot of expensive kernel calls, this can skew the results. Click the *i* button to display the options, and select User & Kernel in Callstacks. You generally also want to set the Track Display to User and System Libraries, which will separate your code from the system code with a small yellow line.

- Function calls in tight loops are the enemy of performance. An ObjC method call is even worse. The compiler will generally inline functions for you when it makes sense, but just marking a function `inline` doesn't guarantee that it will be inlined. For instance, if you ever take a pointer to the function, that function cannot be inlined anymore. Unfortunately, there's no way to know for certain whether the compiler has inlined a function other than compiling in Release mode (with optimizations) and going and looking at the assembly. The only way to be absolutely certain that something will inline is to make it a macro with `#define`. I don't generally like macros, but if you have a piece of code that absolutely must be inlined for performance reasons, you may want to consider it.

- Always do all performance testing in Release mode with optimizations turned on. Performance testing on unoptimized Debug code is meaningless. Optimizations are not linear. One part of your program may be no faster, whereas another may speed up by orders of magnitude. Tight loops are particularly subject to heavy compiler optimization.

- Although Instruments is incredibly powerful, don't neglect an old standby: small test programs that subtract the end time from the start time. For pure computational work, pulling the algorithm into a test program is often the most convenient approach. You can use a standard Single View template and just put the call to your code in `viewDidLoad`.

The Accelerate Framework

The most cited framework for improving performance is Accelerate. In my experience, it's a mixed bag, and you need to approach it as you would any other performance fix—with skepticism and testing. The Accelerate framework is made up of several pieces, including vecLib, vImage, vDSP, BLAS, and LAPACK. Each is distinct, with many similar but slightly different feature sets and data types. This mix makes it difficult to use them together. Accelerate isn't so much a single framework as a collection of mostly preexisting math libraries that have been stuck together.

None of the Accelerate libraries are good at working on small data sets. If you just need to multiply small matrices (particularly 3×3 or 4×4, which are very common), Accelerate will be much slower than trivial C code. Calling an Accelerate function in a loop will often be much slower than a simple C implementation. If the function offers a *stride* option and you don't need a stride, writing your own C code may be faster.

> **A stride lets you skip through a list by a specific size. For example, a stride of 2 uses every other value. This capability is very useful when you have interleaved data (such as coordinates or image color information).**

On multicore laptops, some Accelerate functions can be very fast because they are multithreaded. On mobile devices with one or two cores, this advantage may not outweigh its cost for moderate-sized data sets. The Accelerate libraries are also portable, with a history of focusing on the Altivec (PPC) and SSE3 (Intel) vector processors rather than the NEON processor that's in iOS devices.

For all the Accelerate libraries, you want to gather all your data into one convenient layout so that you can call the Accelerate functions as few times as possible. If at all possible, avoid copying memory and store your data internally in the format that is fastest for Accelerate, rather than rearrange the data when it's time to make the performance-critical call.

You'll generally need to do significant performance testing to determine the fastest approach. Be sure to do all your testing in Release with optimizations turned on. Always start with a simple C implementation as a baseline to make sure you're getting good payback from Accelerate.

The vImage library is a large collection of image-processing functions and, in my experience, is the most useful portion of Accelerate. In particular, vImage is good at applying matrix operations on high-resolution images very quickly. For most common problems, consider Core Image rather than vImage, but if you need to do the raw math yourself, vImage is quite good at applying matrices and at converting images. Like the other parts of Accelerate, vImage is designed for large data sets. If you're working with a small number of moderate-sized images, Core Image is likely a better choice.

To use vImage effectively, you need to be aware of the difference between an interleaved and a planer format. Most image formats are interleaved. The red data for a pixel is followed by the green data for that pixel, which is then followed by the blue data, which is followed by the alpha data for the same pixel. That's the format you're probably accustomed to from `CGBitmapContext` with `kCGImageAlphaLast`. That's RGBA8888. If the alpha comes first, it's ARGB8888. Alpha generally comes first in vImage.

You can also store color information as 16-bit floats rather than 8-bit integers. If you store the alpha first, the format is called ARBGFFFF.

Planer formats separate the color information into *planes*. Each plane holds one kind of information. There's an alpha plane, a red plane, a green plane, and blue plane. If the values are 8-bit integers, this format is called Planar8. If it's 16-bit floats, it's PlanarF. Although there are planar image formats, most vImage functions work on a single plane at a time, and the functions often don't care what kind of data the plane holds. There's no difference between multiplying red information and multiplying blue information.

It's generally much faster to work on planar formats than interleaved formats. If at all possible, get your data into a planar format as soon as you can, and leave it in that format through the entire transformation process. Generally, you convert back to RGB only when you need to display the information.

vecLib and vDSP are vector libraries with some overlap in their functionality. vecLib tends to be a bit easier to use and is focused on applying a single operation to a large list of numbers. vDSP tends to have much more flexible (and therefore slower) functions, and it also has very specialized functions relevant to signal processing. If you need those kinds of functions on large tables of numbers, vDSP is likely much faster than your hand-built solutions. The key word here is *large*. If you're computing only several hundred or a few thousand numbers, simple C solutions can often be faster. The cost of making the function call overwhelms everything else.

BLAS and LAPACK are C versions of well-established FORTRAN libraries. They're primarily focused on linear algebra. As with other parts of Accelerate, they are best applied for very complex problems. They tend to be slow for solving small systems of equations because of the overhead of the function call. Apple does not provide good documentation for BLAS and LAPACK. If you're interested in these functions, go to `netlib.org` and find the FORTRAN documentation. So far, I've seldom found them to be really helpful in solving my common iOS problems.

GLKit

GLKit is great for integrating OpenGL into your applications, but it also has some hidden gems for fast math. In particular, GLKit offers vector-optimized functions for working with 3×3 and 4×4 matrices as well as 2-, 3-, and 4-element vectors. As I mentioned in the previous section on the Accelerate framework, flexibility is often the enemy of performance. Functions that can take matrices of arbitrary sizes can never be as optimized as a hard-coded solution for multiplying a 3×3 matrix. So, for many of these kinds of problems, I recommend GLKit.

Before you get too excited, GLKit's math optimizations are not particularly faster than what you would get by hand-coding a C solution and letting the compiler optimize it. But they're no slower, and they're easy to use and easy to read. Look for `GLKMatrix3`, `GLKMatrix4`, `GLKVector2`, `GLKVector3`, and `GLKVector4`.

Compiler Optimizations

Several very useful compiler optimizations are turned off by default. They are primarily vector and floating-point optimizations, and are collectively called `-Ofast`.

The default optimization level is "Fastest, Smallest [-Os]." This optimization level requests the compiler produce the fastest code it can without increasing code size. Many good optimizations can make programs larger. For example, it is often much faster to unroll small loops than to use conditional branches. However, because this would usually make the program larger, it is forbidden under "Fastest, Smallest." More substantially, many vectorization optimizations introduce a small amount of code overhead, so the compiler can seldom automatically vectorize code in the default configuration. "Fastest, Smallest" has been the default for a very long time, and Apple was not willing to change that default in Xcode 5.

Two available optimization levels are faster than the default: "Fastest [-O3]" and "Fastest, Aggressive Optimizations [-Ofast]." I explain both, but for most nonscientific applications, -Ofast is the best option. For scientific applications or other applications where IEEE floating-point math is required, -O3 is a better choice.

The "Fastest [-O3]" optimization is similar to the default, except that it allows the compiler to increase code size to improve performance. In most cases, this represents very small increases, perhaps a few percent at most. In exchange, you can get improved loop vectorization of integer computations with no code changes. Compared to the size of resources such as images and sounds, binary size is generally a modest component of the total package size, so it's generally worth getting the performance boost.

The "aggressive optimizations" option adds optimizations that loosen some strict requirements on floating-point math. Floating-point math is much more complicated than integer math. For example, consider the following code:

```
float x = a * 0.0;
```

The compiler cannot simply optimize this to x=0. It is possible that a is NaN ("not a number") or infinity, either of which would result in x being NaN. Similarly, consider this simple code:

```
float x = a + b - a;
```

Again, the compiler cannot optimize this to x=b. The expression a+b may overflow or may cause rounding. To maintain correct IEEE math, the compiler has very few options. Similarly, the compiler generally cannot automatically vectorize floating-point math because it may introduce slightly different rounding behavior if the operations are performed in a different order.

For most applications, these kinds of corner cases don't matter. A small rounding error in the fourth or fifth decimal place is usually of no concern. Even in games, where floating-point math is common, such a small error is unlikely to have any impact. But it is a violation of the floating-point specification, so Apple does not turn it on by default. For most iOS programs, "Fastest, Aggressive Optimizations [-Ofast]" is a good choice.

Linker Optimizations

The compiler has a very powerful optimization engine, but it has a significant limitation: it sees only one compile unit (.m file) at a time. The compiler has to assume the worst possible behavior of any function it cannot see. If the compiler knew what was happening elsewhere in the program, it could apply many more optimizations.

LLVM has a new feature called Link-Time Optimization (LTO) that addresses this. It is available in the "Apple LLVM 5.0 – Code Generation" section of the Build Settings. It causes the compiler to include additional information in the intermediate files. The linker can then use that information to perform more aggressive optimizations.

For most moderate-sized programs, LTO is a good choice. For very large programs, however, it can significantly increase the build time and can even cause the linker to run out of memory. I recommend trying it with your projects and turning it off only if it causes a problem.

Drawing Performance

Set REVISION to 4 in ZipTextView.h and select the Core Animation instrument. You'll see results like those in Figure 18-10.

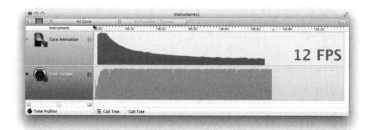

Figure 18-10 Core Animation instrument

The Core Animation instrument displays frames per second, and it's falling over time. That suggests you still have a problem.

The Core Animation instrument has several options that can help debug drawing problems. I discuss them shortly. Unfortunately, not one of them is particularly helpful in the case of ZipText. The fps output does, however, provide a good metric for testing whether changes are effective. You at least want the fps output to stay constant over time.

Looking at `drawRect:`, you can see that it always draws every character, even though most of the characters never change. In fact, only one small part of the view changes during each update. You can improve performance here by drawing only the part that actually needs updating.

In `appendNextCharacter`, calculate the rectangle impacted by the new character:

ZipTextView5.m (ZipText)

```
- (void)appendNextCharacter {
  self.index++;
  if (self.index < self.text.length) {
    CGRect dirtyRect;
    dirtyRect.origin = [self originAtIndex:self.index fontSize:kFontSize];
    dirtyRect.size = CGSizeMake(kFontSize, kFontSize);
    [self setNeedsDisplayInRect:dirtyRect];
  }
}
```

In `drawRect:`, draw only the character that intersects that rectangle:

```
- (void)drawRect:(CGRect)rect {
  ...
  CGPoint origin = [self originAtIndex:i fontSize:kFontSize];
  if (CGRectContainsPoint(rect, origin)) {
    [character drawAtPoint:origin
            withAttributes:@{ NSFontAttributeName:
                                [UIFont systemFontOfSize:kFontSize]}];
  }
  ...
}
```

Instruments now reads a steady frame rate of nearly 60 fps. That's excellent. But don't misread high frame rates. You want high frame rates when you're displaying new data, but it's normal to have low frame rates when nothing is happening. When your UI is stable, it's normal and ideal for your frame rate to drop to zero.

The Core Animation instrument has several Debug options in the left pane that can be very useful. The following are some of my most-used options:

- **Color Blended Layers**—This option applies red to any parts of the screen that require color blending. Parts of the screen that don't require color blending will be green. Color blending generally means nonopaque views. If you have a lot of red on the screen, investigate which views you can make fully opaque.

- **Color Misaligned Images**—As I discuss in Chapter 7, you always want to draw on pixel boundaries when possible. This indicates areas of the screen that aren't aligned with pixel boundaries.

- **Flash Updated Regions**—This is one of my favorite options. It flashes yellow for any layer that is being updated. For UIs that aren't games, very little of the UI should update from moment to moment. If you turn this option on and see a lot of yellow, you're probably updating the screen more often than you mean to. The granularity of this is the layer, however. So, if you use `setNeedsDisplayInRect:` to draw only part of a view, the entire view (which is backed by a layer) flashes yellow.

> Some of the same options are available in the Debug menu of the iOS Simulator. Although most performance testing should almost always be done on the device, you can generally debug these kinds of drawing mistakes in the simulator.

Optimizing Disk and Network Access

Memory and CPU are the most common performance bottlenecks, but they're not the only resources to be optimized. Disk and network access are also critical, particularly in prolonging battery life.

The I/O Activity instrument is particularly useful for checking whether you're hitting the disk more often than you should be. This instrument is most easily accessed through the System Usage template.

Network access is particularly expensive on iOS devices, and it's worth serious consideration. For instance, creating a new network connection is very expensive. The DNS lookup alone can be incredibly, surprisingly slow. Whenever possible, reuse existing connections. `NSURLSession` does this for you automatically if you use HTTP/1.1. This is yet another reason to prefer it versus hand-built network solutions.

Network access is also a serious drain on the battery. It's much better for battery life if you batch all your network activity into short bursts. The cost of network utilization is much more proportional to the length of time than to the amount of data. The device reduces power on its radios when the network is not in use, so providing long periods of quiet with short bursts of activity is the best design.

The Connections instrument can be useful in tracking down network usage, but keep in mind that it tracks all processes on the device, not just yours.

Summary

You need to strongly consider the limited resources of the platform when designing mobile applications. Things that are acceptable on a desktop device will overwhelm an iPhone, bringing it to a crawl and draining its battery. Continual analysis of performance and improvement should be part of every developer's workflow.

Instruments is one of the best tools available for analyzing your application's performance. Despite being slightly buggy and at times infuriating, it is nevertheless very powerful and a critical tool in tracking down memory and CPU bottlenecks. It's also useful for detecting I/O and network overuse. Spend some time with it and experiment. You'll find it a worthwhile investment.

Accelerate can be useful for very large math operations, but simpler code and higher-level frameworks are often faster for more common operations. Core Image is particularly useful for speeding image processing, and GLKit is useful for small-matrix operations.

In any case, always carefully test the changes you make to improve performance. Many times, the simplest code is actually the best. Find your hotspots with Instruments, and test, test, test.

Further Reading

Apple Documentation

The following documents are available in the iOS Developer Library at `developer.apple.com` or through the Xcode Documentation and API Reference.

Accelerate Framework Reference

Core Image Programming Guide

Core Image Filter Reference

File-System Performance Tips

GLKit Framework Reference

Instruments User Guide

Memory Usage Performance Guidelines

Performance Overview

WWDC Sessions

The following session videos are available at `developer.apple.com`.

WWDC 2012, "Session 511: Core Image Techniques"

WWDC 2013, "Session 713: The Accelerate Framework"

WWDC 2013, "Session 408: Optimizing Your Code Using LLVM"

Other Resources

Bumgarner, Bill. *bbum's weblog-o-mat*. "When Is a Leak Not a Leak?" Good introduction to using heapshot analysis (now called generation analysis) to find lost memory.

`www.friday.com/bbum/2010/10/17`

Netlib. "Netlib Repository at UTK and ORNL." Documentation of LAPACK and BLAS, which are offered by Accelerate.

`www.netlib.org`

Part IV

Pushing the Limits

## Chapter 19

Almost Physics: UIKit Dynamics

One of the most interesting additions in iOS 7 is UIKit Dynamics. If you apply some simple "physics-like" rules, views can bounce around the screen, connect to other views, collide with boundaries, and behave more like real objects. Apple takes great pains to call this "physics-like." The tools available are meant to provide dynamic, interactive user interfaces. They're not meant to completely simulate real-world physical behaviors. In this chapter, you learn the parts of UIKit Dynamics, including animators, behaviors, and dynamic items. You learn what behaviors iOS 7 provides and how to create new, custom behaviors. You finally learn how to integrate dynamics with collection views to create interesting and engaging layouts.

UIKit Dynamics is extremely powerful, but it is also a very new technology. There are still a lot of rough edges that can make it challenging to use well. It is easy to leak memory if you're not careful, and maintaining good performance is a constant concern. The process of adding and removing views in the view hierarchy is tedious and error prone.

The most important part of using dynamics well is to build your behaviors one piece at a time, testing carefully as you go. Keep each piece simple. The main example in this chapter contains five custom behaviors and half a dozen standard behaviors. Each one solves a single problem, which keeps them as simple as possible. Complex behaviors are very difficult to understand and debug. As much as you can, keep them simple.

Despite all these caveats, UIKit Dynamics is an exciting new technology and can create very exciting user interactions. Used well, UIKit Dynamics can be a very valuable tool.

You can find the examples in this chapter in the online sample code.

Animators, Behaviors, and Dynamic Items

UIKit Dynamics is primarily composed of three pieces:

- **Dynamic items**—The elements that will be animated. The most common items are views, but an item can be anything that conforms to the `UIDynamicItem` protocol. This protocol requires the `bounds`, `center`, and `transform` properties.

- **Behaviors**—Rules that influence `bounds`, `center`, and `transform` over time for one or more items. Examples include attachment of an item to a point, gravity, and friction.

- **Animators**—Engines that apply behaviors to items over time.

Items are attached to behaviors, and behaviors are attached to animators. A surprising point of this hierarchy is that items do not know their behaviors. If you attach a view to a gravity behavior and then remove the view from its superview, the gravity behavior continues to retain the view and animate it, even though it is not visible. It is up to you to manage removing the view from the behavior or making sure the behavior is deallocated when appropriate. Behaviors do know their animator, and animators do know their behaviors.

The fact that items do not know their behaviors bears repeating. This fact leads to many design headaches when building complicated behaviors. For example, you cannot simply say that a view and all its subviews are subject to gravity. You must keep track of the gravity behavior and attach each view to it. Creating multiple gravity behaviors in a single animator is undefined, so you can't just create a view subclass that has its own gravity behavior. You really must keep track of a single gravity behavior for the view hierarchy and synchronize its list of dynamic items with the view hierarchy's list of views.

This problem multiplies as you have more behaviors. If every item is subject to gravity and collisions, you must attach every view to two behaviors. There are some ways to improve this situation with custom behaviors, as you see in the "Behavior Hierarchies" section, but in general adding and removing views to the view hierarchy can be tedious and error prone. It is difficult to design a general solution to this problem. You need to design a solution in light of your particular UIKit Dynamics problem.

A dynamic animator can animate views, collection views, or custom dynamic items.

To animate views, create a dynamic animator using `initWithReferenceView:`. This is probably the most common approach. All the animated views must descend from the reference view. The reference view provides the coordinate system. For example, if your reference view has a flip transform, gravity pulls up rather than down.

Even though the animator knows the bounds of the reference view, it still has to animate items that are outside the reference view. It is possible that those items affect other items, or that those items will reenter the reference view. For example, if you apply gravity to a view, it will fall forever, or at least until its vertical coordinate reaches `FLT_MAX`. There is no "ground" in UIKit.

You can create a "ground" fairly easily, though. If there is a reference view, you can use it to automatically create a boundary for any objects that are part of a collision behavior. By using the dynamic animator's `setTranslatesReferenceBoundsIntoBoundaryWithInsets:` method, you can fairly easily keep objects inside their reference view.

To animate collection views, create a dynamic animator using `initWithCollectionViewLayout:`. This kind of animator animates `UICollectionViewLayoutAttributes` objects. See the "Interacting with UICollectionView" section for an example.

If you create a dynamic animator using `init`, it doesn't have a reference view and can animate any object that conforms to `<UIDynamicItem>`. Because `UIView` conforms to `<UIDynamicItem>`, you can animate views this way if you don't need a reference view.

UIKit "Physics"

Real-world physics has a system of laws and constants. Scientists have devised numerous physical units such as kilograms, meters, and newtons to measure physical systems. UIKit has a similar set of laws, constants, and units.

In the real world, the standard unit of distance is the meter. In UIKit physics, the unit of distance is the point *(p)*. The mass of an object is its square area times its density. So a 100p × 100p view has half the mass of a 200p × 100p view if both have the same density. UIKit does not give a name to this mass unit. In this chapter, I refer to it as a *UIKit kilogram*.

Unsurprisingly, the unit of time in UIKit physics is the second *(s)*.

In the real world, the standard unit of force is the newton, which is the force required to accelerate one kilogram of mass by 1 m/s^2. The UIKit newton is the force required to accelerate a UIKit kilogram by 100 p/s^2.

Finally, the UIKit gravitational force *(g)* is 1000 p/s^2. There is no gravity by default. It is a behavior that you must add items to, just like any other behavior. See the "Gravity" section later in this chapter.

If you remove all behaviors from an item, the item stops moving. For example, if an item is falling under the influence of gravity, and then you remove the gravity, you might expect that the item would continue moving in the same direction. Instead, the item immediately stops falling. Unless there is a behavior actively modifying the item, there is no reason for the item to do anything. The dynamic animator does not automatically cause anything to move in the absence of a behavior. See the "Dynamic Item" section later in this chapter for techniques for making items move more realistically.

Built-in Behaviors

iOS provides several useful behaviors that can handle the majority of simple needs. The following sections cover all the built-in behaviors: snap, attachment, gravity, collision, push, and dynamic item.

Snap

The snap behavior (`UISnapBehavior`) is one of the simplest behaviors. It causes an item to move to a given location and stay there. It also aligns the item so that its rotation is zero. In other words, it makes the item vertical. When it is applied to an item, the item "snaps" to the new location, potentially overshooting slightly. The following code demonstrates how to use it with an item:

ViewController.m (Dynamic)

```
UISnapBehavior *snap = [[UISnapBehavior alloc] initWithItem:self.box1
                                              snapToPoint:point];
snap.damping = 0.25;
[self.dynamicAnimator addBehavior:snap];
```

This code creates a behavior object that snaps a view to a point. If the view is already at that point, nothing generally happens. If the view is elsewhere, it causes the view to move to the point. The `damping` property is the only configuration available. Higher values make the movement slower and reduce the amount of overshoot when the view arrives at the snap point. The default is 0.5, so the preceding code is a little more "dynamic" than usual.

After the snap behavior is created, `damping` is the only thing you can modify. You cannot change the item or the point. If you want to snap to a new point, you need to remove this behavior and add a new one.

The snap behavior acts somewhat like a tight spring. If two snap behaviors are controlling the same view, the final location of that view is between the two snap points. Objects subject to normal gravity (<1) are very close to their snap point. Objects subject to high gravity (5+) have noticeable "sag." (See the "Gravity" section later in this chapter.) Snap does not work reliably when combined with attachment.

Attachment

The attachment behavior is like the snap behavior but is somewhat more flexible. Whereas snap is always attached to the center of an item, attachments can be offset from the center. You can also attach an item to another item rather than to a specific point. In this example, you attach one view to the background and attach another view to the first view.

ViewController.m (Dynamic)

```
UIAttachmentBehavior *
attach1 = [[UIAttachmentBehavior alloc] initWithItem:self.box1
                                         offsetFromCenter:UIOffsetMake(25, 25)
                                         attachedToAnchor:self.box1.center];
[self.dynamicAnimator addBehavior:attach1];

UIAttachmentBehavior *
attach2 = [[UIAttachmentBehavior alloc] initWithItem:self.box2
                                         attachedToItem:self.box1];
[self.dynamicAnimator addBehavior:attach2];
```

A single attachment can often be unstable. For instance, if you try to re-create `UISnapBehavior` with a single `UIAttachmentBehavior`, you find that the item tends to vibrate more than is probably desired. It also spins around its anchor point because `UIAttachmentBehavior` doesn't prevent rotation. A common solution is to use multiple attachments. For example, you can create a four-point attachment as shown here:

```
CGRect bounds = item.bounds;
CGFloat width = CGRectGetWidth(bounds);
CGFloat height = CGRectGetHeight(bounds);

CGFloat offsetWidth = width/2;
CGFloat offsetHeight = height/2;
UIOffset offsetUL = UIOffsetMake(-offsetWidth, -offsetHeight);
UIOffset offsetUR = UIOffsetMake( offsetWidth, -offsetHeight);
UIOffset offsetLL = UIOffsetMake(-offsetWidth,  offsetHeight);
UIOffset offsetLR = UIOffsetMake( offsetWidth,  offsetHeight);

CGFloat anchorWidth = width/2;
CGFloat anchorHeight = height/2;
CGPoint anchorUL = CGPointMake(center.x - anchorWidth,
                               center.y - anchorHeight);
CGPoint anchorUR = CGPointMake(center.x + anchorWidth,
                               center.y - anchorHeight);
CGPoint anchorLL = CGPointMake(center.x - anchorWidth,
                               center.y + anchorHeight);
```

```
CGPoint anchorLR = CGPointMake(center.x + anchorWidth,
                               center.y + anchorHeight);

[dynamicAnimator addBehavior:[[UIAttachmentBehavior alloc] initWithItem:item
                                               offsetFromCenter:offsetUL
                                               attachedToAnchor:anchorUL]];
[dynamicAnimator addBehavior:[[UIAttachmentBehavior alloc] initWithItem:item
                                               offsetFromCenter:offsetUR
                                               attachedToAnchor:anchorUR]];
[dynamicAnimator addBehavior:[[UIAttachmentBehavior alloc] initWithItem:item
                                               offsetFromCenter:offsetLL
                                               attachedToAnchor:anchorLL]];
[dynamicAnimator addBehavior:[[UIAttachmentBehavior alloc] initWithItem:item
                                               offsetFromCenter:offsetLR
                                               attachedToAnchor:anchorLR]];
```

Getting attachments to work well often requires this kind of tedious, repetitive code. When you develop a behavior that works the way you like, it is a good idea to refactor it into a reusable UIDynamicBehavior subclass.

Push

The push behavior applies a force to an object. As discussed in the section "UIKit 'Physics,'" force is measured in UIKit newtons, and mass is function of square area times density. A UIKit newton is the force required to accelerate a UIKit kilogram by 100 p/s^2.

The same push behavior influences different sized items differently. For a given density, larger items accelerate slower than smaller items.

Pushes can be instantaneous or continuous. An instantaneous push applies a second's worth of acceleration instantaneously. So if an item is one UIKit kilogram, and you apply a one UIKit newton instantaneous force to it, the final velocity is 100 p/s. A continuous push applies acceleration over time. If you apply a one UIKit newton continuous force to a one UIKit kilogram item, after one second the velocity is 100 p/s, and after two seconds the velocity is 200 p/s. The item will continue to accelerate until the force is removed.

The default force vector is nil. If you do not assign one, no force is applied. The following code demonstrates how to create a force vector that points to the lower right with a magnitude $\sqrt{2}$. This force will continue until removed, so the box will accelerate over time.

ViewController.m (Dynamic)

```
UIPushBehavior *
push = [[UIPushBehavior alloc] initWithItems:@[self.box1]
                                        mode:UIPushBehaviorModeContinuous];
push.pushDirection = CGVectorMake(1, 1);
[self.dynamicAnimator addBehavior:push];
```

After an instantaneous push behavior fires, it immediately becomes inactive. You can re-apply the force by setting active to YES.

Gravity

If you want an item to be subject to gravity, you must attach it to a gravity behavior. Just like gravity in the real world, UIKit gravity accelerates all objects along its vector at the same rate. Standard UIKit gravity is (0.0, 1.0), which is a downward vector of magnitude 1.0. It accelerates all items at 1000 p/s^2.

There is no "ground" in UIKit. Items will fall forever if nothing stops them.

A dynamic animator can have only a single gravity behavior. If you add and remove items, you may need to keep track of the gravity behavior in a property so that you can update its collection of items.

Collision

Collision behaviors cause items to interact with other items and with boundaries. When items collide, they behave according to their relative momentum and elasticity, just as in the real world. Items can also interact with boundaries. A boundary is effectively a path containing infinite mass, so boundaries never move when struck by colliding objects.

The most common configuration is a boundary defined by the reference view. This boundary can be configured using `setTranslatesReferenceBoundsIntoBoundaryWithInsets:`. You can also create a boundary using an arbitrary path.

After you create a collision behavior, you can add or remove objects from it, and you can add and remove boundaries. Boundaries are tracked by an identifier, which makes them easy to find and remove.

Collision behaviors also take a delegate, which you can use to track when objects collide. The delegate is commonly used to play sounds but can also be used to add or remove behaviors. For example, by adding an attachment behavior, you can cause items to stick to each other when they collide.

Dynamic Item

`UIDynamicItemBehavior` is unlike other behaviors. It assigns physical properties to items. For example, you can use a dynamic item behavior to set the density or elasticity of an item.

Two dynamic item properties seem similar but are distinct: friction and resistance. Friction influences items as they slide against other items in a straight line. Resistance influences items any time they move in a straight line. An item with low resistance but high friction still moves easily when pulled by an attachment. Angular resistance influences items any time they rotate. You can use the `allowsRotation` property to prevent rotation entirely.

Dynamic item behavior can also directly add velocity to an item. This is similar to push behaviors, but its value is given in absolute points per second rather than UIKit newtons.

Items attached to a dynamic item behavior tend to continue moving when other behaviors are removed. For example, if you apply a push behavior to an object with no other behaviors and then remove the push behavior, the item immediately stops. However, if the item has a dynamic item behavior, even when you remove the push behavior, the item continues moving at its current velocity. If you want items to behave "naturally" in complex systems, it is often useful to attach an empty dynamic item behavior to them.

> Don't confuse `UIDynamicItemBehavior` **with its superclass,** `UIDynamicBehavior`. **Note the word** `Item` **in the class name.**

Behavior Hierarchies

A custom `UIDynamicBehavior` can have child behaviors, which is useful for grouping related behaviors. For example, you may want to apply gravity and collisions to every item in the system. You can package them into a single dynamic behavior, using `addChildBehavior:` in a custom `UIDynamicBehavior` subclass. You generally should not use `addChildBehavior:` in any of the built-in behaviors.

Although this technique is useful for creating groups, it can become confusing and difficult to debug if you create complex hierarchies, so keep it simple.

Custom Actions

During every animation frame, the animator calls the `action` block of each behavior. You can add any functionality you want here, but because it is called frequently, performance is critical.

A common use for custom actions is to modify the behavior when some condition is met. For example, you can check the time to make sure that the animation hasn't run too long, or you can modify the behavior based on the location of the objects.

The WWDC 2013 examples suggest using `-[UIDynamicAnimator elapsedTime]` in the action block to change the animation after it has run for a period of time. This approach should be used carefully. The `elapsedTime` property counts the total amount of time the animator has actually run. It does not include any time that the animator is paused (generally because the system is at rest). It also never resets. So if you add and remove behaviors to an animator, the absolute `elapsedTime` value may be misleading. You may need to check the current `elapsedTime` when the behavior is added. You can determine this by overriding the behavior's `willMoveToAnimator:`.

Putting It Together: A Tear-Off View

Now that you have the basics, you can put everything together into a complex behavior. In this example, you create an application that displays a "tear-off" shape widget. The user drags the shape from its initial location. If the user lets go before dragging very far, the shape snaps back to its initial location. If the user drags further, the shape "tears off" and a new copy snaps back to the initial location. The user can use this widget to create many copies of the shape to play with. Additionally, the shapes have collision and gravity behaviors.

Dragging a View

You start with the top-level view controller that just creates the shape and lets the user drag the shape around the screen:

ViewController.m (TearOff)

```objc
#import "ViewController.h"
#import "DraggableView.h"

const CGFloat kShapeDimension = 100.0;

@interface ViewController ()
@property (nonatomic) UIDynamicAnimator *animator;
@end

@implementation ViewController
- (void)viewDidLoad {
  [super viewDidLoad];
  self.animator =
    [[UIDynamicAnimator alloc] initWithReferenceView:self.view];

  CGRect frame = CGRectMake(0, 0,
                            kShapeDimension, kShapeDimension);
  DraggableView *
  dragView = [[DraggableView alloc] initWithFrame:frame
                                         animator:self.animator];
  dragView.center = CGPointMake(self.view.center.x / 4,
                                self.view.center.y / 4);
  dragView.alpha = 0.5;
  [self.view addSubview:dragView];
}
@end
```

`ViewController` owns the `UIDynamicAnimator` and defines the reference view. It then creates a `DraggableView` and places it on the screen. The following code creates the drag view:

DraggableView.m (TearOff)

```objc
#import "DraggableView.h"

@interface DraggableView ()
@property (nonatomic) UISnapBehavior *snapBehavior;
@property (nonatomic) UIDynamicAnimator *dynamicAnimator;
@property (nonatomic) UIGestureRecognizer *gestureRecognizer;
@end

@implementation DraggableView
- (instancetype)initWithFrame:(CGRect)frame
                     animator:(UIDynamicAnimator *)animator {
  self = [super initWithFrame:frame];
  if (self) {
    _dynamicAnimator = animator;
    self.backgroundColor = [UIColor darkGrayColor];
```

```
      self.layer.borderWidth = 2;
      self.gestureRecognizer = [[UIPanGestureRecognizer alloc]
                                initWithTarget:self
                                action:@selector(handlePan:)];
      [self addGestureRecognizer:self.gestureRecognizer];
    }
    return self;
}
- (void)handlePan:(UIPanGestureRecognizer *)g {
    if (g.state == UIGestureRecognizerStateEnded ||
        g.state == UIGestureRecognizerStateCancelled) {
      [self stopDragging];
    }
    else {
      [self dragToPoint:[g locationInView:self.superview]];
    }
}

- (void)dragToPoint:(CGPoint)point {
    [self.dynamicAnimator removeBehavior:self.snapBehavior];
    self.snapBehavior = [[UISnapBehavior alloc] initWithItem:self
                                                snapToPoint:point];
    self.snapBehavior.damping = .25;
    [self.dynamicAnimator addBehavior:self.snapBehavior];
}
- (void)stopDragging {
    [self.dynamicAnimator removeBehavior:self.snapBehavior];
    self.snapBehavior = nil;
}
@end
```

As the user drags the item around the screen with the `UIPanGestureRecognizer`, this object continually moves its new snap behavior to the new location. That's all that is required to create a view that the user can drag around the screen. The low `damping` setting allows the view to slide around easily and somewhat sloppily, giving it a slightly playful feeling.

Tearing Off the View

Now you add a new behavior to make the view "stick" in place, but tear off a copy if dragged sufficiently. In the view controller, you just add the new behavior as shown here:

ViewController.m (TearOff)

```
- (void)viewDidLoad {
    ...
    [self.view addSubview:dragView];

    TearOffBehavior *tearOffBehavior = [[TearOffBehavior alloc]
                            initWithDraggableView:dragView
                            anchor:dragView.center
                            handler:^(DraggableView *tornView,
                                  DraggableView *newPinView) {
```

(continued)

```
                                         tornView.alpha = 1;
                                     }];
        [self.animator addBehavior:tearOffBehavior];
    }
    @end
```

TearOffBehavior is defined as shown here. It is a custom behavior with a child snap behavior and a custom action (self.action), as described in the "Custom Actions" section. During each animation frame, its action checks whether the view has been dragged too far from its initial position. If so, it creates a copy of the view and attaches a new TearOffBehavior to it. It uses an active property so that the custom action doesn't fire again until the view snaps back into place. When the tear-off action fires, it also calls the handler block so that the view controller can perform other operations if desired. This code is worth studying to understand complex behaviors.

TearOffBehavior.h (TearOff)

```
@class DraggableView;

typedef void(^TearOffHandler)(DraggableView *tornView,
                              DraggableView *newPinView);
@interface TearOffBehavior : UIDynamicBehavior
@property(nonatomic) BOOL active;

- (instancetype) initWithDraggableView:(DraggableView *)view
                                anchor:(CGPoint)anchor
                               handler:(TearOffHandler)handler;

@end
```

TearOffBehavior.m (TearOff)

```
@implementation TearOffBehavior

- (instancetype)initWithDraggableView:(DraggableView *)view
                               anchor:(CGPoint)anchor
                              handler:(TearOffHandler)handler {
    self = [super init];
    if (self) {
      _active = YES;
      [self addChildBehavior:[[UISnapBehavior alloc] initWithItem:view
                                                      snapToPoint:anchor]];
      CGFloat distance = MIN(CGRectGetWidth(view.bounds),
                             CGRectGetHeight(view.bounds));

      TearOffBehavior * __weak weakself = self;
      self.action = ^{
        TearOffBehavior *strongself = weakself;
        if (! PointsAreWithinDistance(view.center, anchor, distance)) {
          if (strongself.active) {
            DraggableView *newView = [view copy];
            [view.superview addSubview:newView];
            TearOffBehavior *newTearOff = [[[strongself class] alloc]
                                          initWithDraggableView:newView
                                          anchor:anchor
                                          handler:handler];
```

```
            newTearOff.active = NO;
            [strongself.dynamicAnimator addBehavior:newTearOff];
            handler(view, newView);
            [strongself.dynamicAnimator removeBehavior:strongself];
          }
        }
        else {
          strongself.active = YES;
        }
      };
    }
    return self;
}

BOOL PointsAreWithinDistance(CGPoint p1,
                             CGPoint p2,
                             CGFloat distance) {
  CGFloat dx = p1.x - p2.x;
  CGFloat dy = p1.y - p2.y;
  CGFloat currentDistance = hypotf(dx, dy);
  return (currentDistance < distance);
}
@end
```

Note that `TearOffBehavior` copies `DraggableView`, so you implement `copyWithZone:`.

DraggableView.m (TearOff)

```
- (instancetype)copyWithZone:(NSZone *)zone {
  DraggableView *newView = [[[self class] alloc]
                            initWithFrame:CGRectZero
                            animator:self.dynamicAnimator];
  newView.bounds = self.bounds;
  newView.center = self.center;
  newView.transform = self.transform;
  newView.alpha = self.alpha;
  return newView;
}
```

Adding Extra Effects

The preceding example allows you to drag off shapes, but they immediately freeze in place when you release them. It would be nice to make them a little more dynamic, so you can add gravity and collision behaviors. All you have to do is add new items to these behaviors as they are created. To make things a little easier, you put these into a custom behavior, as shown here:

DefaultBehavior.h (TearOff)

```
@interface DefaultBehavior : UIDynamicBehavior
- (void)addItem:(id<UIDynamicItem>)item;
- (void)removeItem:(id<UIDynamicItem>)item;
@end
```

DefaultBehavior.m (TearOff)

```objc
#import "DefaultBehavior.h"

@implementation DefaultBehavior

- (instancetype)init {
  self = [super init];
  if (self) {
    UICollisionBehavior *collisionBehavior = [UICollisionBehavior new];
    collisionBehavior.translatesReferenceBoundsIntoBoundary = YES;
    [self addChildBehavior:collisionBehavior];

    [self addChildBehavior:[UIGravityBehavior new]];
  }
  return self;
}

- (void)addItem:(id<UIDynamicItem>)item {
  for (id behavior in self.childBehaviors) {
    [behavior addItem:item];
  }
}

- (void)removeItem:(id<UIDynamicItem>)item {
  for (id behavior in self.childBehaviors) {
    [behavior removeItem:item];
  }
}
@end
```

ViewController.m (TearOff)

```objc
@interface ViewController ()
@property (nonatomic) UIDynamicAnimator *animator;
@property (nonatomic) DefaultBehavior *defaultBehavior;
@end
...
  [self.view addSubview:dragView];

  DefaultBehavior *defaultBehavior = [DefaultBehavior new];
  [self.animator addBehavior:defaultBehavior];
  self.defaultBehavior = defaultBehavior;

  TearOffBehavior *tearOffBehavior = [[TearOffBehavior alloc]
                                initWithDraggableView:dragView
                                anchor:dragView.center
                                handler:^(DraggableView *tornView,
                                          DraggableView *newPinView) {
                                  tornView.alpha = 1;
                                  [defaultBehavior addItem:tornView];
                                }];
```

Multiple Dynamic Animators

Sometimes putting all the behaviors in a single dynamic animator is inconvenient. For example, if you have groups of objects that are animated independently, it may be more convenient to create a separate animator for each group.

The key point is that a dynamic animator is just an object that over time will adjust the location and rotation of its items by following the rules provided in its behaviors. It is not inherently part of a view hierarchy. It doesn't even care if its items are views. So there is no problem having multiple dynamic animators in the same view hierarchy, as long as you don't try to animate the same view with multiple animators. Animating the same view with multiple animators is undefined behavior because the animators have no way to coordinate.

Continuing the previous TearOff example, you can now add an explosion behavior. If an item is double-tapped, it explodes. The exploding pieces interact with each other but not with other views, so this is a good application of a separate dynamic animator.

ViewController.m (TearOff)

```objc
- (void)viewDidLoad {
  ...
  TearOffBehavior *tearOffBehavior = [[TearOffBehavior alloc]
                            initWithDraggableView:dragView
                            anchor:dragView.center
                            handler:^(DraggableView *tornView,
                                    DraggableView *newPinView) {
                              tornView.alpha = 1;
                              [defaultBehavior addItem:tornView];

                              // Double-tap to trash
                              UITapGestureRecognizer *
                              t = [[UITapGestureRecognizer alloc]
                                    initWithTarget:self
                                    action:@selector(trash:)];
                              t.numberOfTapsRequired = 2;
                              [tornView addGestureRecognizer:t];
                            }];
  [self.animator addBehavior:tearOffBehavior];
}

- (void)trash:(UIGestureRecognizer *)g {
  UIView *view = g.view;

  // Calculate the new views. (See sample code.)
  NSArray *subviews = [self sliceView:view
                          intoRows:kSliceCount
                           columns:kSliceCount];

  // Create a new animator
  UIDynamicAnimator *
  trashAnimator = [[UIDynamicAnimator alloc]
                  initWithReferenceView:self.view];
```

(continued)

```objc
// Create a new default behavior
DefaultBehavior *defaultBehavior = [DefaultBehavior new];

for (UIView *subview in subviews) {
    // Add the new "exploded" view to the hierarchy
    [self.view addSubview:subview];
    [defaultBehavior addItem:subview];

    UIPushBehavior *
    push = [[UIPushBehavior alloc]
            initWithItems:@[subview]
            mode:UIPushBehaviorModeInstantaneous];
    [push setPushDirection:CGVectorMake((float)rand()/RAND_MAX - .5,
                                        (float)rand()/RAND_MAX - .5)];
    [trashAnimator addBehavior:push];

    // Fade out the pieces as they fly around.
    // At the end, remove them. Referencing trashAnimator here
    // also allows ARC to keep it around without an ivar.
    [UIView animateWithDuration:1
                     animations:^{
                         subview.alpha = 0;
                     }
                     completion:^(BOOL didComplete){
                         [subview removeFromSuperview];
                         [trashAnimator removeBehavior:push];
                     }];
}

// Remove the old view
[self.defaultBehavior removeItem:view];
[view removeFromSuperview];
}
```

Interacting with UICollectionView

A particularly interesting use of UIKit Dynamics is influencing `UICollectionView` layouts. You can use these layouts to create all kinds of exciting and engaging interfaces.

As a quick reminder, `UICollectionView` relies on `UICollectionViewLayout` to generate a `UICollectionViewLayoutAttributes` object for each item. This layout attributes object defines the center and transform, among other things. That matches exactly with `<UIDynamicItem>`, which also defines a center and transform. This means that a dynamic animator can modify layout attributes over time according to behaviors.

You will almost always implement this layout by subclassing `UICollectionViewLayout` and modifying the layout attributes in `layoutAttributesForElementsInRect:`. In the following example, you create a simple layout that allows the user to drag items around the collection view. The collection view controller uses a press gesture recognizer to track dragging. This code implements the recognizer action. It is available in the CollectionDrag sample code.

DragViewController (CollectionDrag)

```
- (IBAction)handleLongPress:(UIGestureRecognizer *)g {
  DragLayout *dragLayout = (DragLayout *)self.collectionViewLayout;
  CGPoint location = [g locationInView:self.collectionView];

  // Find the indexPath and cell being dragged
  NSIndexPath *indexPath = [self.collectionView
                              indexPathForItemAtPoint:location];
  UICollectionViewCell *cell = [self.collectionView
                                  cellForItemAtIndexPath:indexPath];

  UIGestureRecognizerState state = g.state;
  if (state == UIGestureRecognizerStateBegan) {
    // Change the color and start dragging
    [UIView animateWithDuration:0.25
                     animations:^{
                        cell.backgroundColor = [UIColor redColor];
                     }];
    [dragLayout startDraggingIndexPath:indexPath fromPoint:location];
  }

  else if (state == UIGestureRecognizerStateEnded ||
           state == UIGestureRecognizerStateCancelled) {
    // Change the color and stop dragging
    [UIView animateWithDuration:0.25
                     animations:^{
                        cell.backgroundColor = [UIColor lightGrayColor];
                     }];
    [dragLayout stopDragging];
  }

  else {
    // Drag
    [dragLayout updateDragLocation:location];
  }
}
```

The layout itself is also very simple. It just uses an attachment behavior to move the item center. When the drag ends, it attaches the item back to its original location so that it animates smoothly.

DragLayout.h (CollectionDrag)

```
@interface DragLayout : UICollectionViewFlowLayout
- (void)startDraggingIndexPath:(NSIndexPath *)indexPath
                     fromPoint:(CGPoint)p;
- (void)updateDragLocation:(CGPoint)point;
- (void)stopDragging;
@end
```

DragLayout.m (CollectionDrag)

```objc
@interface DragLayout ()
@property (nonatomic) NSIndexPath *indexPath;
@property (nonatomic) UIDynamicAnimator *animator;
@property (nonatomic) UIAttachmentBehavior *behavior;
@end

@implementation DragLayout

- (void)startDraggingIndexPath:(NSIndexPath *)indexPath
                     fromPoint:(CGPoint)p {
  self.indexPath = indexPath;
  self.animator = [[UIDynamicAnimator alloc]
                  initWithCollectionViewLayout:self];

  UICollectionViewLayoutAttributes *attributes = [super
   layoutAttributesForItemAtIndexPath:self.indexPath];
  // Raise the item above its peers
  attributes.zIndex += 1;

  self.behavior = [[UIAttachmentBehavior alloc] initWithItem:attributes
                                         attachedToAnchor:p];
  self.behavior.length = 0;
  self.behavior.frequency = 10;
  [self.animator addBehavior:self.behavior];

  UIDynamicItemBehavior *dynamicItem = [[UIDynamicItemBehavior alloc]
                                       initWithItems:@[attributes]];
  dynamicItem.resistance = 10;
  [self.animator addBehavior:dynamicItem];

  [self updateDragLocation:p];
}

- (void)updateDragLocation:(CGPoint)p {
  self.behavior.anchorPoint = p;
}

- (void)stopDragging {
  // Move back to the original location (super)
  UICollectionViewLayoutAttributes *
  attributes = [super layoutAttributesForItemAtIndexPath:self.indexPath];
  [self updateDragLocation:attributes.center];
  self.indexPath = nil;
  self.behavior = nil;
}

- (NSArray *)layoutAttributesForElementsInRect:(CGRect)rect {
  // Find all the attributes, and replace the one for our indexPath
  NSArray *existingAttributes = [super
                           layoutAttributesForElementsInRect:rect];
  NSMutableArray *allAttributes = [NSMutableArray new];
  for (UICollectionViewLayoutAttributes *a in existingAttributes) {
```

```
      if (![a.indexPath isEqual:self.indexPath]) {
        [allAttributes addObject:a];
      }
    }

    [allAttributes addObjectsFromArray:[self.animator itemsInRect:rect]];
    return allAttributes;
}
@end
```

That's all there is to it. Using this technique, you can apply a wide variety of behaviors to collection view layouts. Just replace the attributes you want in `layoutAttributesForElementsInRect:`.

Summary

UIKit Physics is one of the most exciting additions in iOS 7. At times, it can be challenging to use well, but with some care and a focus on simplicity, it is possible to get effects that were unavailable in previous versions. Just remember, animators apply behaviors to items over time. Items can be views, collection view layout attributes, or any other object that has a center, bounds, and a transform. When you build up a few simple behaviors, UIKit can be almost physics.

Further Reading

Apple Documentation

The following documents are available in the iOS Developer Library at `developer.apple.com` or through the Xcode Documentation and API Reference.

UIDynamicAnimator Class Reference

UIDynamicBehavior Class Reference

UIDynamicItem Protocol Reference

WWDC Sessions

The following session videos are available at `developer.apple.com`.

WWDC 2013, "Session 206: Getting Started with UIKit Dynamics"

WWDC 2013, "Session 221: Advanced Techniques with UIKit Dynamics"

Chapter 20

Fantastic Custom Transitions

In Chapter 2, you learned about the new user interface paradigm in iOS 7. Two of the most important changes to the UI are to the superfluous use of animations and the way real-world physics are simulated in almost every aspect of the user interface.

Adding custom transitions (especially interactive custom transitions) to your view controllers is an important step to make your app look as though it was built for iOS 7.

Interactive custom transitions aren't new, however. They were present in iOS (but not on the iOS SDK) from at least iOS 3.2, the first version of iOS that ran on iPads. For example, the page flip animation isn't just a transition from one page to another. It's an interactive transition—a transition that follows your finger. In contrast, the "page curl" transition in the maps app (prior to iOS 7) is a normal transition. It's not interactive; the transition doesn't follow your fingers.

In Chapter 4, you learned how to add custom transitions using storyboards. In this chapter, you learn how to build custom transitions the iOS 7 way, without storyboard segues (you may or may not use storyboards and that's up to you), and more importantly, you learn how to build the interactive custom transitions introduced in iOS 7. Interactive custom transitions are, in my opinion, an important tool to push the limits of your app and make it shine in the App Store.

Custom Transitions in iOS 7

Apple has minimized the number of screen transitions that happen in most of its built-in apps. When you open the Calendar app and tap on a date, the Date view is animated from the Month view with a custom transition, as illustrated in Figure 20-1.

Along similar lines, when you create a new event, changing the start and end time of the event is animated using a custom transition. Note that this was the navigation controller's default push navigation animation in previous versions of iOS. Another example of a custom transition is apparent on the Photos application's Photos tab. The uber-long camera roll is now replaced with a collection view that uses a custom transition to navigate between Years, Collections, Moments, and Single Photo views. iOS 7's new UI paradigm emphasizes letting users know where they are, instead of letting users lose their way among the myriad pushed view controllers. In most cases, this is done using a custom transition.

When you design an app for iOS 7, you should carefully consider whether using a custom transition will help users know where they are within your app. The iOS 7 SDK adds APIs that enable you do this without much difficulty.

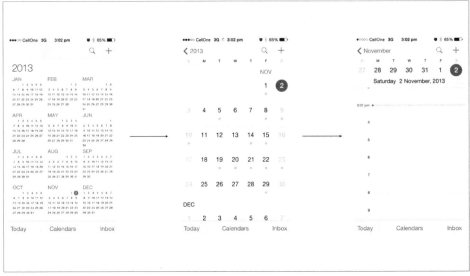

Figure 20-1 The built-in calendar app uses a custom transition when transitioning from one screen to another

The iOS 7 SDK supports two kinds of custom transitions: Custom View Controller Transitions and Interactive View Controller Transitions. Custom View Controller Transitions were possible previously using storyboards and custom segues. Interactive View Controller Transitions enable a gesture (usually a pan gesture) to control the amount of transition (from start to the end). So as users pan or swipe a finger, the transition occurs from one view controller to another.

When the transition is a function of time, it's usually a Custom View Controller Transition, and if it is a function of a parameter of a gesture recognizer or any such event, it is usually an Interactive View Controller Transition.

For example, you can think of a navigation controller's push transition (as of iOS 6) as an example of a custom view controller transition, and you can think of a `UIPageViewController` page transition as an example of an interactive view controller transition. When you use `UIPageViewController` to page between views, the transition is not timed. The page transition follows your finger movement, so this is an Interactive View Controller Transition. The `UINavigationController` transition (on iOS 6) occurs over a period of time, and transitions like these are Custom View Controller Transitions.

The iOS 7 SDK allows you to customize almost any kind of transition—view controller presentations and dismissals, the `UINavigationController` push and pop transitions, the `UITabBarController` transitions (`UITabBarController` does not animate view controllers by default), or even a Collection view's layout change transitions.

Transition Coordinators

All view controllers in iOS 7 have a property called `transitionCoordinator`, which is an object that conforms to the `UIViewControllerTransitionCoordinator`. A transition coordinator is created for all transitions, custom or otherwise. That means, when you are performing a default push or pop navigation transition, you can animate something else in your view. The methods in this protocol are

```
-animateAlongsideTransition:completion:
-animateAlongsideTransitionInView:animation:completion:
-notifyWhenInteractionEndsUsingBlock:
```

The first and second methods are useful when you want to perform some custom animation in your view controller along with the navigation controller animation. Note that in iOS 6 you were doing this by manually hard-coding your animation duration to 0.35, the default navigation controller animation duration. With this new API, you no longer have to use this hack.

With iOS 7, navigation controller transitions (the interactive type) can be canceled. That means, on your second view, `viewWillAppear` gets called, but `viewDidAppear` might not be called. In a classical scenario, a user starts a transition from view 1 to view 2 and decides to cancel it halfway. View 2's view controller gets a `viewWillAppear`, and view 1's view controller gets a `viewWillDisappear`. But because the transition was not completed, `viewDidAppear` and `viewDidDisappear` were never called. If you write code assuming that `viewDidAppear` will always be called after a `viewWillAppear`, you might have to rethink your logic. The third method in the `transitionCoordinator` can be of help in these cases. You get a notification in a block when an interactive transition ends. This block is called regardless of whether the transition ended successfully or was canceled. You can check the state from the context passed and perform any additional layout in this block instead of writing them in your `viewWillAppear`/`viewDidAppear` combo.

Collection View and Layout Transitions

Collection views now have additional methods to support transitions when they are pushed and popped in a navigation controller. When you push a collection view into a navigation stack, you can make the transition happen through a layout change instead of the default push mechanism. The iOS 7 Calendar app and Photos app use this technique to transition from one collection view controller to another.

iOS 7 introduces a `useLayoutToLayoutNavigationTransitions` property on `UICollectionViewController`. When you set this property to `YES`, before pushing your collection view controller into a navigation controller, the push transition uses the `setCollectionViewLayout:animated:` method to transition the collection view layout.

Do note that this method requires both collection views to have the same data (and this is true for the built-in photos and calendar app).

You were able to simulate this animation even in iOS 6, but doing so involved hacks to make the navigation controller believe a new controller had in fact been pushed. This hack is no longer needed with iOS 7. All you need to do is set `useLayoutToLayoutNavigationTransitions`, and the system takes care of the rest. A demo application, `CollectionViewCustomTransitionDemo` that illustrates collection view layout transition is available from the book's website.

Custom View Controller Transitions Using Storyboards and Custom Segues

Storyboards and segues were introduced in iOS 5, so you are probably already familiar with them. If not, turn to Chapter 4 for more information.

Storyboards are yet another way to create interfaces. With nib files, you created views. With storyboards, you create views and specify the navigation (called *segues*) between those interfaces. By writing the custom subclass `UIStoryboardSegue`, you can customize the transition between one view controller to another, as you learned in Chapter 4. With iOS 7, Apple expanded custom transition support greatly, so you can now write a custom transition to transition between any two view controllers. Transitions can also be "interactively driven" based on a gesture.

Custom View Controller Transition: The iOS 7 Way

Custom View Controller Transitions are slightly easier to use than Interactive View Controller Transitions. In this section, you create the same transition effect created in the preceding section using the iOS 7 SDK.

You do Custom View Controller Transitions in iOS 7 by implementing a couple of protocols: namely, `UIViewControllerTransitioningDelegate` and `UIViewControllerAnimatedTransitioning`.

Chapter 4 covers custom transitions using custom segues in a storyboard.

Implementing a Custom View Controller Transition is easy. You use the same `pushViewController:animated:` or `presentViewController:animated:completion:` method (depending on whether it is a push transition or modal presentation transition) and set the `transitioningDelegate` of your view controller to an animator object. The animator object is any object that conforms to the `UIViewControllerAnimatedTransitioning` protocol. If your custom transition effect is unique to a single view controller, the view controller can implement the `UIViewControllerAnimatedTransitioning` protocol. Otherwise, you need to create a separate class (inheriting from `NSObject`) implementing the `UIViewControllerAnimatedTransitioning` protocol. In this example, for the sake of simplicity, you implement the protocol in the view controller.

To implement the protocol, you should implement two of the four optional methods.

The other two methods are used for implementing Interactive View Controller Transitions. You look at them in the next section.

UIViewControllerTransitioningDelegate Methods (SCTMasterViewController.m)

```
- (id <UIViewControllerAnimatedTransitioning>)
  animationControllerForPresentedController:(UIViewController*) presented
  presentingController:(UIViewController *)presenting
  sourceController:(UIViewController *)source {

    return self;
}
```

```
-(id <UIViewControllerAnimatedTransitioning>)
animationControllerForDismissedController:(UIViewController*) dismissed {

    return self;
}
```

The protocol method implementation is responsible for retuning the animation object. In this case, the animation is handled by the view controller itself; hence, it is returning `self`.

The next step is to implement the `UIViewControllerAnimatedTransitioning` protocol. This protocol has two required methods, shown in the following code:.

UIViewControllerAnimatedTransitioning Protocol Methods

```
-  (NSTimeInterval)transitionDuration:(id
   <UIViewControllerContextTransitioning>)transitionContext {

    return 1.0f;
}

-(void)animateTransition:(id
 <UIViewControllerContextTransitioning>)transitionContext {
   UIViewController *src = [transitionContext
 viewControllerForKey:UITransitionContextFromViewControllerKey];
   UIViewController *dest = [transitionContext
viewControllerForKey:UITransitionContextToViewControllerKey];

// the real animation code goes here

[transitionContext completeTransition:YES];

 }
```

The first method tells the system how long the animation will take, and the second method performs the actual animation. If you recall your custom segue implementation, you overrode a method called `perform`. The `animateTransition:` method will contain code similar to your `perform` method. The `animateTransition:` method passes a transitionContext from which you can get pointers to the from and to view controllers. This is shown in the preceding code segment. When you have the pointer to the view controllers, you can use any of your favorite methods (View based, Quartz based, UIKit Dynamics based) to animate them. After the animation is complete, call the `completeTransition:` method in the `transitionContext` to let the system know that the transition is complete.

The code fragments explained here show the necessary protocols and methods that you should implement but don't actually show the animation code. The complete code for the Custom View Controller Transition is available from the book's website.

Interactive Custom Transitions Using the iOS 7 SDK

You learned about Custom View Controller Transitions earlier in this chapter. Interactive transitions are like custom transitions, but the transition, instead of being a function of time, is a function of a gesture recognizer or a similar event. Interactive transitions also have special support for collection views through the `UICollectionViewTransitionLayout` class. Here, I show you how to perform an interactive custom transition.

Interactive transitions are "driven" by an event. It could be a motion event or a gesture, more commonly a gesture. The `UIScreenEdgePanGestureRecognizer` is an example of an interactive transition. In iOS 7, all views can be "popped" by panning the view from the screen's edge to the right. The transition follows your finger movement.

Implementing a custom interactive transition is fairly complicated. However, Apple has made this process much simpler than you might think. The most complicated part is to calculate the intermediate steps and animate your UI. Fortunately, the SDK automatically takes care of animating the intermediate steps. All you need to do in your animation handler is to use one of the UIView animation methods. As of when this chapter was written, you cannot use a layer-based animation (QuartzCore animation) with interactive transitions. But the following new UIView animation methods should obviate the need for layer-based animations:

```
animateKeyframesWithDuration:delay:options:animations:completion:
animateWithDuration:delay:usingSpringWithDamping:

initialSpringVelocity:options:animations:completion:
```

Interaction Controller

To implement an interactive transition, you do the same animation and tell the interaction controller how much of the animation has been completed. It's your responsibility to determine the completion percentage, and the system takes care of the rest. You do this by calculating the amount of the gesture/motion event or whatever is driving the interaction. For example, the amount of pan or the pinch distance/velocity can be a parameter to compute the completion percentage. Later in this chapter, you implement a custom pop view controller animation in which the user pinches the screen to close the view.

For implementing interactive transitions, you use the same `pushViewController:animated:` or `present ViewController:animated:completion:` method (depending on whether it is a push transition or a modal presentation transition) and set the `transitioningDelegate` of your view controller to an animator object.

For custom transitions, the animator object conformed to the `UIViewControllerAnimatedTransitioning` protocol, and you implemented two of the four optional methods. The `UIViewControllerAnimatedTransitioning` protocol asks you for an animation controller and an interaction controller. You implemented the first two methods in the previous section to return an animation controller. For interactive transitions, you should implement all four methods. That is, you should return an animation controller and an interaction controller.

UIViewControllerTransitioningDelegate Methods

```
- (id <UIViewControllerAnimatedTransitioning>)
  animationControllerForPresentedController:(UIViewController*) presented
  presentingController:(UIViewController *)presenting
  sourceController:(UIViewController *)source {
```

```
        return self;
}

-(id <UIViewControllerAnimatedTransitioning>)
animationControllerForDismissedController:(UIViewController*) dismissed {

        return self;
}

- (id <UIViewControllerInteractiveTransitioning>)
interactionControllerForPresentation:(id
 <UIViewControllerAnimatedTransitioning>)animator {

    return self.animator;
}

- (id <UIViewControllerInteractiveTransitioning>)
interactionControllerForDismissal:(id
 <UIViewControllerAnimatedTransitioning>)animator {

    return self.animator;
}
```

The animation controller you use remains the same for the most part. But instead of implementing the `animateTransition:` method, you implement the `startInteractiveTransition:` method. An important consideration here is that you can have only one animation block within this method. This animation block should be UIView based and not layer based. CATransition- or CALayer-based animations are not supported in interactive transitions.

The interaction controller is an object that implements the `UIViewControllerInteractiveTransitioning` protocol. An interaction controller that you use for interactive transitions should be a subclass of `UIPercentDriven InteractiveTransition`.

Now look at the code in the animator. The animator class is responsible for calculating the completion percentage. It's also responsible for updating the system of "how much" transition has been completed. There are methods in the super class `UIPercentDrivenInteractiveTransition` to accomplish this task.

```
- (void)updateInteractiveTransition:(CGFloat)percentComplete;
- (void)cancelInteractiveTransition;
- (void)finishInteractiveTransition;
```

Now, based on your gesture, all you need to do is calculate a percentage and call the update method. If the gesture was completed or canceled, call the respective methods accordingly. Next, you need to create the animator.

Creating the Animator

```
self.animator = [[SCTPercentDrivenAnimator alloc]
  initWithViewController:self];
UIPinchGestureRecognizer *gr = [[UIPinchGestureRecognizer alloc]
  initWithTarget:self.animator action:@selector(pinchGestureAction:)];
[self.view addGestureRecognizer:gr];
```

The animator is also the target of your pinch gesture recognizer. In this gesture recognizer handler, you compute the completion percentage and call the necessary super class methods.

Gesture Recognizer Handler

```
-(void) pinchGestureAction:(UIPinchGestureRecognizer*) gestureRecognizer {

   CGFloat scale = gestureRecognizer.scale;
   if(gestureRecognizer.state == UIGestureRecognizerStateBegan) {

      self.startScale = scale;
      [self.controller dismissViewControllerAnimated:YES completion:nil];
   }
   if(gestureRecognizer.state == UIGestureRecognizerStateChanged) {
      CGFloat completePercent = 1.0 - (scale/self.startScale);
      [self updateInteractiveTransition:completePercent];
   }
   if(gestureRecognizer.state == UIGestureRecognizerStateEnded) {
      [self finishInteractiveTransition];
   }

   if(gestureRecognizer.state == UIGestureRecognizerStateCancelled) {
      [self cancelInteractiveTransition];
   }
}
```

This gesture recognizer handler computes the completion percentage using a simple calculation. The animator remembers the original scale, and the completion percentage is proportional to the new pinch scale. So as the user pinches, the completion percentage increases.

When you do this, the system automatically takes care of updating the intermediate animation states for you. The complete code is available from the book's website.

In the preceding example, you wrote a `UIPercentDrivenInteractiveTransition` for a view controller. You can also use the same example with collection view controllers. In fact, by implementing the protocol `UIViewControllerInteractiveTransitioning`, `UIViewControllerAnimatedTransitioning` methods in your collection view controller, you can interactively drive the collection view layout transitions.

Summary

In this chapter, you learned an important technique introduced in iOS 7. Custom transitions are pervasive throughout Apple's built-in apps, and I imagine that most of the high-quality apps on the App Store, including Facebook, Twitter, or foursquare, will exploit this feature in their apps.

Custom transitions are not just eye candy. You can use them to help your users understand where they are within your application. Using them wisely will help you create apps that shine on the App Store.

Further Reading

Apple Documentation

The following document is available in the iOS Developer Library at developer.apple.com or through the Xcode Documentation and API Reference.

What's New in iOS 7

WWDC Sessions

The following session videos are available at developer.apple.com.

WWDC 2013, "Session 226: Implementing Engaging UI on iOS"

WWDC 2013, "Session 218: Custom Transitions Using View Controllers"

WWDC 2013, "Session 201: Building User Interfaces for iOS 7"

WWDC 2013, "Session 208: What's New in iOS User Interface Design"

WWDC 2013, "Session 225: Best Practices for Great iOS UI Design"

Chapter 21

Fancy Text Layout

Beautiful text is at the heart of the new iOS 7 look and feel. More than ever before, getting your fonts right and making your layout perfect are key to differentiating your app. Luckily, iOS 7 also brings powerful new text tools, including Dynamic Type and Text Kit. Advanced layout that once required low-level Core Text calls can now be handled directly in UIKit. Making your text beautiful has gotten much easier. Of course, that means your users will be demanding beautiful text.

In this chapter, you learn the key concepts behind rich text, as well as its central data structure: `NSAttributedString`. You learn enough typography to understand the frameworks. In particular, you learn the differences between characters and glyphs, fonts and decorations, and how they relate to paragraph styles and layout.

With a strong foundation in place, you learn how to choose fonts correctly to match your users' desires and how to perform advanced layout with Text Kit. Finally, you get a quick tour of the low-level Core Text calls you need for complete layout control.

Understanding Rich Text

Before you can really understand rich text, you need some background in typography. Typographers have studied how to best display printed text for centuries, and the iOS text rendering system is based heavily on techniques that were worked out long before the era of digital computers. Although there are important differences between digital typography and physical printing, much of the terminology is still rooted in its history. Companies that design and distribute typefaces, such as Adobe and Bitstream, are still called *foundries* after their metal-casting forebears.

At the core of typography is the conversion of *characters* into *glyphs*, arranging glyphs into lines, and arranging lines into paragraphs. But first, what are glyphs and characters?

Characters versus Glyphs

As a developer, you are likely very familiar with characters. An `NSString` is a collection of characters, each representable by a unique Unicode value. The letter a is represented by the Unicode character "LATIN SMALL LETTER A," which has the value 97 (U+0061). The Chinese character 我 is represented by "Unicode Han Character 'our, us, i, me, my, we,'" which has the value 25105 (U+6211). Any time that the letter *A* or the character 我 appears in a string, the numeric value is the same. Each character in Figure 21-1 has a unique value.

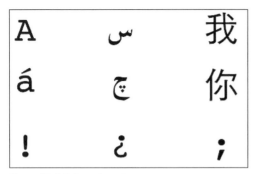

Figure 21-1 Unique characters

But now consider Figure 21-2. Each letter *a* has the same character value (U+0061), but they have dramatically different shapes. These shapes are called glyphs.

Figure 21-2 Unique glyphs

A font is primarily a collection of glyphs that map to characters. This is not a one-to-one mapping, however. In many cases, a font may have several glyphs that could represent a given character. In Arabic, glyphs depend on where a character appears in a word. Figure 21-3 shows various forms of the Arabic letter HEH, depending on whether the character stands alone, begins the word, is in the middle of the word, or ends the word.

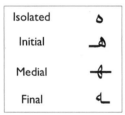

Figure 21-3 Forms of HEH

For a given string, the number of characters and the number of glyphs may be different as well. The most common cause of this difference is ligatures, where multiple characters are combined into a single glyph. In some languages, this is required. In Arabic, the characters LAM and ALIF are often combined as a spelling requirement. This creates the ligature shown in Figure 21-4.

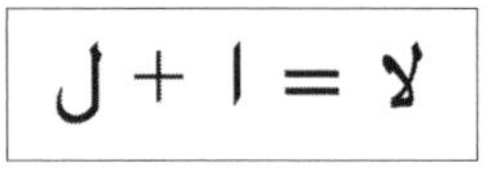

$$ ل + ا = لا $$

Figure 21-4 LAM + ALIF

Some ligatures are font-specific and are designed to improve readability. The most common in English is the *f-i* ligature shown in Figure 21-5. Notice how the dot over the *i* is missing. This is an intentional feature of the font. The two characters are drawn as a single glyph.

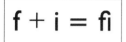

Figure 21-5 f + i

Diacritics ("accent marks") may cause there to be more or fewer glyphs than characters, depending on whether the font provides that particular letter-diacritic combination, and whether the letter is encoded with a "combining diacritic" in the string. In some fonts, long series of characters can be a single glyph. The font Zapfino has a single, enormous glyph for its own name. The details go on and on, and you do not need to understand them all in most cases. You just need to understand that characters and glyphs are different things, and there are often a different number of them for a given string in a given font. iOS has methods for converting between the two when required, and you should let it do the work for you.

Understanding Fonts

A *font* is a collection of glyphs, along with rules for how to choose and position them. Fonts are scalable, so Baskerville 18pt and Baskerville 28pt are just two sizes of the same font. The bold and italic versions of Baskerville, however, are completely different fonts. Note the differences in Figure 21-6.

Baskerville 18pt   *Baskerville Italic 18pt*

Baskerville 28pt   **Baskerville Bold 28pt**

Figure 21-6 Variations of Baskerville

Note how the shapes of the letters in Baskerville are different from the shapes in Baskerville italic. They're not just slanted. In particular, notice the dramatic difference in the shape of the letter *a*. Baskerville Bold may look like Baskerville with thicker lines, but there are several subtle differences. In particular, pay attention to the widths of thin strokes versus heavy strokes. They are not increased equally.

The key lesson here is that when you want to "bold a font," what you really want is to find a new font that is a bold variation of your current font. It's possible that there is no such font available. There might be multiple fonts available. For example, some fonts have "bold," "semi-bold," and "extra-bold" variations. So applying bold and italic is really an exercise in font selection.

On the other hand, underline and strikethrough are *decorations*. A decoration modifies an existing font. When you request underlining, the glyphs don't change and the font doesn't change. The system just draws an extra line.

Fonts also provide metrics and rules (called "hints") for choosing glyphs, adjusting them, and laying them out. These rules can be very complex, and can include conditional logic, loops, and variables, which are processed by a virtual machine provided by the font rendering engine.

Paragraph Styles

Finally, there are *paragraph styles*. They don't modify the font or glyph selection; they just modify glyph positioning. A paragraph style might include justification, indentation, or direction, which indicates how the glyphs should be laid out horizontally. A paragraph style might also include line spacing or paragraph spacing, which indicates how the glyphs should be laid out vertically.

In the preceding sections, you learned how the text layout engine uses a font to convert characters into glyphs. It then uses font metric information and hints, along with paragraph styles, to position the glyphs into lines and paragraphs. With this in mind, you next learn how you can control this process in iOS.

Attributed Strings

An *attributed string* combines a string of characters with metadata about ranges within that string. The most common metadata is rich text style information such as fonts, colors, and paragraph styles, but attributed strings can hold any key-value pairs. Figure 21-7 shows some rich text along with the continuous attribute ranges.

Figure 21-7 Rich text with continuous attribute ranges

This string of contiguous characters with identical attributes has six ranges: a default range, a bolded range, another default range, a small blue range, a normal-sized blue range in the Papyrus font, and a final default range. Each of these ranges is called a *run*.

> In Foundation and UIKit, *run* sometimes means a range of characters that have the same effective value for some attribute. In Core Text, *run* specifically means a range of characters with identical attributes. The distinction is occasionally important but is usually obvious from context.

The following code sample demonstrates one way to create the attributed string in Figure 21-7:

PTLViewController.m (BeBold)

```
NSString *string = @"Be Bold! And a little color wouldn't hurt either.";

NSDictionary *attrs = @{
                        NSFontAttributeName: [UIFont systemFontOfSize:36]
                        };

NSMutableAttributedString *
as = [[NSMutableAttributedString alloc] initWithString:string
                                             attributes:attrs];
```

```
[as addAttribute:NSFontAttributeName
        value:[UIFont boldSystemFontOfSize:36]
        range:[string rangeOfString:@"Bold!"]];

[as addAttribute:NSForegroundColorAttributeName
        value:[UIColor blueColor]
        range:[string rangeOfString:@"little color"]];

[as addAttribute:NSFontAttributeName
        value:[UIFont systemFontOfSize:18]
        range:[string rangeOfString:@"little"]];
[as addAttribute:NSFontAttributeName
        value:[UIFont fontWithName:@"Papyrus" size:36]
        range:[string rangeOfString:@"color"]];
```

In this example, you create a mutable attributed string with `-initWithString:attributes:` and then add attributes to ranges using `-addAttribute:value:range:`. This is a fairly common way to programmatically create an attributed string because there is no built-in way to create an attributed string with varying attributes. The other common way to create an attributed string is to read an RTF file from disk with `-initWithFileURL:options:documentAttributes:error:`. Later in this chapter, you learn about third-party tools to more easily create attributed strings programmatically.

Selecting Fonts with Font Descriptors

The primary tool for finding fonts is a *font descriptor*. `UIFontDescriptor` provides a variety of methods for querying the system for fonts and is the preferred object to serialize font information. A common use of font descriptors is to find related fonts. In the following example, you toggle italics for the attributed string created in the preceding section:

PTLViewController.m (BeBold)

```
- (IBAction)toggleItalic:(id)sender {
  NSMutableAttributedString *as = [self.label.attributedText mutableCopy];

  [as enumerateAttribute:NSFontAttributeName
              inRange:NSMakeRange(0, as.length)
              options:0
           usingBlock:^(id value, NSRange range, BOOL *stop)
  {
    UIFont *font = value;
    UIFontDescriptor *descriptor = font.fontDescriptor;
    UIFontDescriptorSymbolicTraits
    traits = descriptor.symbolicTraits ^ UIFontDescriptorTraitItalic;

    UIFontDescriptor *toggledDescriptor = [descriptor
                          fontDescriptorWithSymbolicTraits:traits];

    UIFont *italicFont = [UIFont fontWithDescriptor:toggledDescriptor
                                        size:0];
```

(continued)

```
        [as addAttribute:NSFontAttributeName value:italicFont range:range];
    }];

    self.label.attributedText = as;
}
```

The `NSAttributedString` method `enumeraterAttribute:inRange:options:usingBlock:` is a useful way to loop over runs in an enumerated string. The block is called once each time the given attribute changes. In this block, you retrieve the font descriptor from the font, and you toggle the "italic" symbolic trait. Then you find a new font for that font descriptor and apply it to the range.

Fonts have "real" attributes such as weight, width, and slant. They take specific values, generally from −1.0 to 1.0, where 0.0 is considered "normal" by the font designer. Fonts also have "symbolic" attributes like bold and italic that are more generic. A given font may come in a variety of weights with names such as Light, Normal, Demi-bold, and Black. These names are not standardized. Requesting the symbolic trait "bold" returns the variation that the font designer considered appropriate for that use.

Not all fonts have every kind of variation. For instance, the Papyrus font does not have an italic variation. When you use `fontDescriptorWithSymbolicTraits:`, it finds the best match for the given descriptor. In this case, the closest match for the descriptor "an italic variation of Papyrus" is Papyrus itself, and so `italicFont` is the same as `font`.

Assigning Paragraph Styles

Most styles apply specifically to the range you attach them to. Fonts, colors, strikethrough, underline, and similar attributes affect individual characters. Paragraph styles, on the other hand, include margins, justification, hyphenation, line breaks, and similar attributes that impact layout. What does it mean to change the margin in the middle of a line?

In iOS, paragraphs are separated by newlines, and the only paragraph style that matters is the one applied to the first character of a paragraph. Paragraph styles applied to other ranges are ignored.

Because paragraph styles have special scoping rules, they are bundled into a single `NSParagraphStyle` object and applied together as a single attribute with the `NSParagraphStyleAttributeName` key. In the following example, you set margins and alignment on two paragraphs:

```
// Apply a basic style to the entire document
NSMutableParagraphStyle *wholeDocStyle = [[NSMutableParagraphStyle new];
  [wholeDocStyle setParagraphSpacing:34.0];
  [wholeDocStyle setFirstLineHeadIndent:10.0];
  [wholeDocStyle setAlignment:NSTextAlignmentJustified];

NSDictionary *attributes = @{
                             NSParagraphStyleAttributeName: wholeDocStyle
                            };
```

```
NSMutableAttributedString *
pas = [[NSMutableAttributedString alloc] initWithString:paragraphs
                                    attributes:attributes];

// Find the second paragraph by looking for \n
NSUInteger
secondParagraphStart = NSMaxRange([pas.string rangeOfString:@"\n"]);

// Add a head and tail ident
NSMutableParagraphStyle *
secondParagraphStyle = [[pas attribute:NSParagraphStyleAttributeName
                           atIndex:secondParagraphStart
                        effectiveRange:NULL] mutableCopy];
secondParagraphStyle.headIndent += 50.0;
secondParagraphStyle.firstLineHeadIndent += 50.0;
secondParagraphStyle.tailIndent -= 50.0;

// Apply the first character of the paragraph.
 [pas addAttribute:NSParagraphStyleAttributeName
            value:secondParagraphStyle
            range:NSMakeRange(secondParagraphStart, 1)];
```

iOS 7 adds a new NSMutableAttributedString method, fixAttributesInRange:. This method cleans up various common inconsistencies in styles. In particular, it applies paragraph styles to the entire paragraph, based on the style of the first character in the paragraph. Keeping attributes normalized this way can improve performance by reducing the number of ranges within the attributed string. It also makes it easier to figure out how the attributed string will be displayed.

HTML

In iOS 7, attributed strings include basic support for HTML and CSS. For example, the following code reads an HTML file into an NSAttributedString:

```
NSURL *URL = ...
NSError *error;
NSDictionary *options = @{NSDocumentTypeDocumentAttribute:
                          NSHTMLTextDocumentType};
NSAttributedString *
  as = [[NSAttributedString alloc] initWithFileURL:URL
                                     options:options
                              documentAttributes:NULL
                                       error:&error];
```

Attributed strings can handle external URL references, such as external CSS and images (which are converted into attachments). Because loading an URL may require network access, you should avoid using this method on the main thread. Errors in the HTML are ignored, much like a web browser. You do not receive any errors for failed network connections either.

Local bundle resources can be read using `file:` URLs. For example:

```
<img src="file:myimage.png">
```

> `NSAttributedString` uses `NSURLConnection` to read external resources. This means you can use `NSURLProtocol` to intercept and modify these connections. For example, you could redirect certain requests to the Documents directory. For more information on using `NSURLProtocol` to rewrite requests, see "Drop-in offline caching for UIWebView (and NSURLProtocol)" (`http://robnapier.net/blog/offline-uiwebview-nsurlprotocol-588`).

Attributed strings are not a replacement for a web view. There are many things in HTML they cannot represent. In particular, they cannot handle most layout features such as tables or CSS float attributes.

You can also generate HTML from attributed strings. For example:

```
NSDictionary *attributes = @{NSDocumentTypeDocumentAttribute:
                                NSHTMLTextDocumentType};
NSData *data = [attrString dataFromRange:NSMakeRange(0, attrString.length)
                    documentAttributes:attributes
                                error:NULL];
NSString *string = [[NSString alloc] initWithData:data
                                encoding:NSUTF8StringEncoding];
```

HTML created this way tends to be fairly complicated, with many style definitions and more `<p>` and `<span>` tags than you might expect. If you convert HTML into an attributed string and then convert it back to HTML, you should expect the result to be radically different from the input. Some loss in accuracy is likely to occur as well. Attachments are lost when converting from attributed strings to HTML, so any images are discarded, even if the original HTML referred to an external URL.

Easier Attributed Strings

Creating long attributed strings programmatically is a bit of a headache. Keeping track of ranges within the original string is tedious and error-prone. Although iOS 7 adds HTML support, it is still somewhat overcomplicated for simple formatting needs. Luckily, a few third-party tools are available to simplify attributed strings.

One of the most convenient helper classes is `MTStringParser` from Mysterious Trousers (see the link in the "Further Reading" section). With it, you can create bundles of attributes that are applied with an HTML-like tag language. Here is an example from its documentation:

```
[[MTStringParser sharedParser] addStyleWithTagName:@"red"
                                font:[UIFont systemFontOfSize:12]
                                color:[UIColor redColor]];

NSAttributedString *string = [[MTStringParser sharedParser]
        attributedStringFromMarkup:@"This is a <red>red section</red>"];
```

Although `MTStringParser` cannot handle full HTML, this is a benefit in most cases. You can create simpler tags such as `<red>` rather than `<font color="red">`. The resulting `NSAttributedString` also is much simpler.

Even so, sometimes you may need something even more flexible. For that, I recommend `DTCoreText` from Cocoanetics (see the link in the "Further Reading" section at the end of this chapter). `DTCoreText` is a full-featured text rendering and editing system built on top of Core Text with powerful HTML support and portability back to iOS 4.2. That said, many of the features of `DTCoreText` are now available in iOS 7, and the attributed strings generated by `DTCoreText` are not completely compatible with UIKit text controls or Text Kit. If support for older versions of iOS are important to you, and you have complex text requirements, `DTCoreText` may be a good choice.

Dynamic Type

iOS 7 adds an important new user preference: Dynamic Type. Users are now able to modify the overall size of text on their device through the General⇨Text Size setting panel, shown in Figure 21-8. This setting is quite powerful. It doesn't just scale fonts. It makes numerous small changes to ensure that the selected fonts are highly legible at all supported sizes and that the fonts work well together.

Figure 21-8 Text Size settings panel

What this means to you, as a developer, is that you should generally not try to control the size of text. Instead of using `systemFontWithSize:`, you should use the new font selector, `preferredFontForTextStyle:`. This method selects a font based on its semantic meaning rather than a specific size. iOS defines six styles:

Headline, Body, Subheadline, Footnote, Caption1, and Caption2. In many ways, this is similar to HTML and CSS. Rather than specify a specific font size, you should specify that a given block of text is a "first-level headline" and let the system select the right font based on the user's preferences.

To determine a user's text size preference, use `[UIApplication preferredContentSizeCategory]`. It returns a constant from either the list of sizes (ranging from extra-small to extra-extra-extra-large) or a constant from the list of accessibility sizes (ranging from medium to extra-extra-extra-large). The accessibility sizes are larger and may include additional readability features, such as automatic bolding. If the preferred size changes while you are running—for instance, if the user double-taps the Home button to switch to Settings—your app receives `UIContentSizeCategoryDidChangeNotification`.

To get the most out of Dynamic Type, you should go beyond just resizing the text. First, you should test your application at various text sizes and under various accessibility options. See the General⇨Accessibility settings panel, as shown in Figure 21-9.

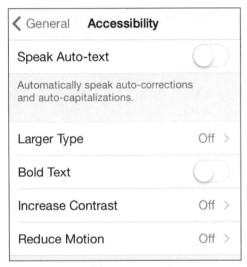

Figure 21-9 Accessibility settings panel

After verifying that your application behaves reasonably for all Dynamic Type sizes, you should consider how to make your layout work best in various sizes, including accessibility sizes. For an excellent example, see the Mail application. As you increase the size of the type, the cells also get taller to accommodate a reasonable amount of preview. This is the kind of attention to detail that you should strive for.

Getting your layout correct with Dynamic Type can be very challenging. Getting a beautiful layout was hard enough with just one text size. Now there are at least a dozen different sizes, plus bold accessibility variants. If you are not using Auto Layout yet, Dynamic Type is yet another reason that you really need to.

Text Kit

Text Kit is a new high-level text layout engine in iOS 7. It is built on top of Core Text and is closely related to Cocoa Text from OS X. It is not as powerful as Cocoa Text, however. With Cocoa Text, creating a multicolumn text editor is fairly simple. Implementing such an editor in Text Kit is almost as hard as in Core Text. That said, Text Kit is a nice addition and does simplify many common text layout problems that formerly required Core Text.

`UITextView` is now built on top of Text Kit rather than WebKit. This removes many subtle rendering differences between `UILabel`, `UITextField`, and `UITextView`. In iOS 7, you should expect these controls to be much more consistent with each other.

Text Kit Components

Text Kit's layout system is based on three important pieces:

- `NSTextStorage`—This subclass of `NSMutableAttributedString` holds the text to be managed.
- `NSTextContainer`—This component represents an area to be filled with text. This area is usually a column or page but can include exclusion areas. In most cases, the area is rectangular, but it does not have to be.
- `NSLayoutManager`— This component coordinates layout across the text containers.

`UITextView` relies on Text Kit but isn't really part of the layout system. Each `UITextView` has an `NSTextContainer` that it uses to handle layout.

Each `NSTextStorage` can have multiple `NSLayoutManager` objects, and each `NSLayoutManager` can have multiple `NSTextContainer` objects. So a given collection of text can be laid out multiple ways simultaneously, and each layout can involve multiple areas.

The DuplicateLayout project provides a simple example of how these pieces work together. In this example, a `UITextView` and a `LayoutView` provide two layouts for the same text. In `awakeFromNib`, `LayoutView` creates a text container and layout manager. It then attaches its layout manager to the text view's storage.

LayoutView.m (DuplicateLayout)

```
- (void)awakeFromNib {
    // Create a text container half as wide as our bounds
    CGSize size = self.bounds.size;
    size.width /= 2;
    self.textContainer = [[NSTextContainer alloc] initWithSize:size];
    self.layoutManager = [NSLayoutManager new];
    self.layoutManager.delegate = self;
    [self.layoutManager addTextContainer:self.textContainer];

    [self.textView.textStorage addLayoutManager:self.layoutManager];
}
```

The important point here is that `LayoutView` and `UITextView` have the same storage, but they have different layout managers. So as you edit `UITextView`, the changes are automatically visible to the `LayoutView`. Still, `LayoutView` needs to know when layout changes have occurred so that it can redraw itself. This capability is provided by the delegate method shown here:

```
- (void)layoutManagerDidInvalidateLayout:(NSLayoutManager *)sender {
    [self setNeedsDisplay];
}
```

Every time the text storage changes, all its layout managers update their layout and inform their delegates. `LayoutView` just needs to redraw itself, as shown in `drawRect:`.

```
- (void)drawRect:(CGRect)rect {
  NSLayoutManager *lm = self.layoutManager;
  NSRange range = [lm glyphRangeForTextContainer:self.textContainer];
  CGPoint point = CGPointZero;
  [lm drawBackgroundForGlyphRange:range atPoint:point];
  [lm drawGlyphsForGlyphRange:range atPoint:point];
}
```

First, `drawRect` asks the layout manager for the range of glyphs within its text container. It then asks the layout manager to draw the background and the glyphs. That's all there is to it.

Multicontainer Layout

A layout manager can support multiple containers. In this example, you modify the DuplicateLayout project to split its text into two boxes:

LayoutView.m (DoubleLayout)

```
- (void)awakeFromNib {
  // Create two text containers
  CGSize size = CGSizeMake(CGRectGetWidth(self.bounds),
                           CGRectGetMidY(self.bounds) * .75);
  self.textContainer1 = [[NSTextContainer alloc] initWithSize:size];
  self.textContainer2 = [[NSTextContainer alloc] initWithSize:size];
  self.layoutManager = [NSLayoutManager new];
  self.layoutManager.delegate = self;
  [self.layoutManager addTextContainer:self.textContainer1];
  [self.layoutManager addTextContainer:self.textContainer2];

  [self.textView.textStorage addLayoutManager:self.layoutManager];
}

- (void)layoutManagerDidInvalidateLayout:(NSLayoutManager *)sender {
  [self setNeedsDisplay];
}
```

```
- (void)drawRect:(CGRect)rect {
    [self drawTextForTextContainer:self.textContainer1
                           atPoint:CGPointZero];

    CGPoint box2Corner = CGPointMake(CGRectGetMinX(self.bounds),
                                     CGRectGetMidY(self.bounds));
    [self drawTextForTextContainer:self.textContainer2
                           atPoint:box2Corner];
}

- (void)drawTextForTextContainer:(NSTextContainer *)textContainer
                         atPoint:(CGPoint)point {

    // Draw a line around the container
    CGRect box = {
      .origin = point,
      .size = textContainer.size
    };
    UIRectFrame(box);

    NSLayoutManager *lm = self.layoutManager;
    NSRange range = [lm glyphRangeForTextContainer:textContainer];
    [lm drawBackgroundForGlyphRange:range atPoint:point];
    [lm drawGlyphsForGlyphRange:range atPoint:point];
}
```

Very little has changed. Instead of adding just one text container, you add two, and `NSLayoutManager` handles the rest. There is a key lesson here, however. Text containers have a size, but no origin. The layout manager does not need to know where a text container will be positioned onscreen to be able to fill it with text. Only the size and order of the text containers are important for layout.

Because you can easily add additional text containers to a layout manager, and because `UITextView` uses a layout manager, you may think it would be easy to create a multicolumn, editable `UITextView`. Unfortunately, this is not the case. If a `UITextView` is assigned multiple text containers, it becomes static and cannot respond to user interaction such as editing or selection. This is a known issue and is currently "as designed."

For static views, such as the examples so far in this chapter, using a `UITextView` with a custom `NSTextContainer` does require slightly less code. I have not given examples in that form because I have found it to be very buggy, fragile, and underdocumented. Unless you do things in just the right way, it crashes or behaves erratically. Things that work well with Cocoa Text on OS X cannot be easily ported to Text Kit. For now, I have found it easier to create custom views rather than use `UITextView`.

Exclusion Paths

Text containers are responsible for deciding where text may be placed. A key feature of Text Kit is the capability to include areas to exclude from layout, called exclusion paths. Creating a "hole" in your layout is as simple as creating an array of `UIBezierPath` objects and assigning it to the text container's `exclusionPaths` property, as shown in this example:

ViewController.m (Exclusion)

```
self.textView.text = string;
self.textView.textAlignment = NSTextAlignmentJustified;

CGRect bounds = self.view.bounds;
CGFloat width = CGRectGetWidth(bounds);
CGFloat height = CGRectGetHeight(bounds);
CGRect rect = CGRectInset(bounds,
                          width/4,
                          height/4);
UIBezierPath *exclusionPath = [UIBezierPath bezierPathWithRoundedRect:rect
                                          cornerRadius:width/10];
self.textView.textContainer.exclusionPaths = @[exclusionPath];
```

This code draws text in the shape shown in Figure 21-10.

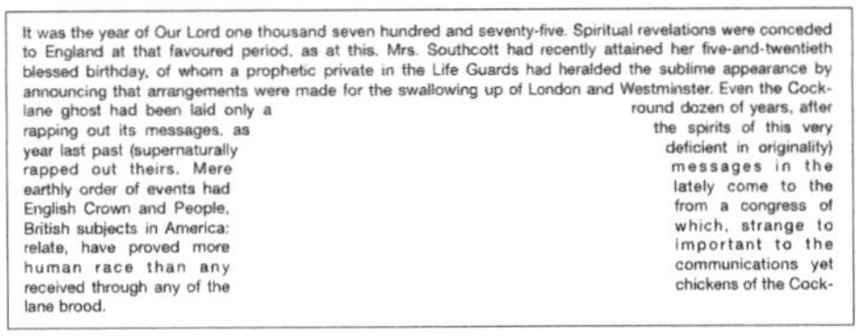

Figure 21-10 Text drawn with an exclusion path

Note how the text flows left to right, skipping over excluded areas. You can include an arbitrary number of paths to exclude, and as shown here, the paths do not have to be rectangular.

In some cases, you can create "inverted" exclusion paths using the even-odd fill rule. This essentially creates "inclusion paths." For example, the following code lays out text in a rounded rectangle. It creates a path that covers the entire view and then appends the path you actually want to fill. When you use the even-odd fill rule, only areas that are covered by an odd number of paths are considered part of the exclusion:

```
UIBezierPath *path = [UIBezierPath bezierPathWithRect:self.view.bounds];
[path appendPath:[UIBezierPath bezierPathWithRoundedRect:CGRectMake(100, 0,
                                                      400, 400)
                                          cornerRadius:100]];
[path setUsesEvenOddFillRule:YES];
textContainer.exclusionPaths = @[ inclusionPath ];
```

At the time of writing, Text Kit does not correctly handle some kinds of exclusion paths. In particular, if your exclusion paths would force some lines to be empty, the entire layout may fail. For example, if you attempt to lay out text in a circle this way, the top of the circle may be too small to include any text, and `NSLayoutManager` will silently fail. This limitation impacts all uses of `NSTextContainer`. Specifically, if `lineFragmentRectForProposedRect:atIndex:writingDirection:remainingRect:` ever returns an empty `CGRect`, the entire layout will fail.

Subclassing the Text Container

In the majority of cases, carefully designed exclusion paths can get you the shape you want, but in some cases, you need to go deeper. For some kinds of layout, it may be convenient to subclass `NSTextContainer`.

The text container's primary responsibility is to answer the following question for the layout manager: Given a rectangle, what part of that rectangle is available for text? This question is answered by the method `lineFragmentRectForProposedRect:atIndex:writingDirection:remainingRect:`. You should keep the following points in mind to implement this correctly:

- The returned rectangle must be contained within the proposed rectangle. If your calculations inject rounding errors, particularly if you ever use `floorf` or `ceilf`, it is possible to accidentally create a rectangle slightly larger than the one proposed. You can use `NSRectIntersection` to make sure your final rectangle is contained in the proposed rectangle.

- It is possible that the proposed rectangle includes multiple text areas. For instance, if there is an exclusion path, there could be text on the left and on the right of the exclusion path. In this case, you should update `remainingRect:` to include the remainder of the rectangle, after you've removed the first text area.

- You must always return a nonempty rectangle from this method. This limitation appears to be a bug in Text Kit and may be resolved in future releases of iOS 7. It makes certain layouts very difficult. It also means that the rectangle you return in `remainingRect` must include more text areas. If there are no more text areas, you must set `remainingRect` to an empty rectangle. Making sure that this is true unfortunately can make the code much more complicated.

In the CircleLayout example, you override the text container to lay out text in a circle within the text view. Notice that the line fragment is calculated twice here. First, the method calls the `super` implementation to apply any exclusion paths. Then the method recalculates the circle-clipped rectangle for the entire line, even if only a portion of the line is requested by the `proposedRect`. This keeps the calculation simpler. Finally, it returns the intersection of these two results. This way of thinking is very important in designing these kinds of layouts. Apply each constraint independently if you can and then intersect them at the end.

CircleTextContainer.m (CircleLayout)

```
- (CGRect)lineFragmentRectForProposedRect:(CGRect)proposedRect
                              atIndex:(NSUInteger)characterIndex
                writingDirection:(NSWritingDirection)baseWritingDirection
                      remainingRect:(CGRect *)remainingRect {

    CGRect rect = [super lineFragmentRectForProposedRect:proposedRect
                                    atIndex:characterIndex
                        writingDirection:baseWritingDirection
                            remainingRect:remainingRect];

    CGSize size = [self size];
    CGFloat radius = fmin(size.width, size.height) / 2.0;
    CGFloat ypos = fabs((proposedRect.origin.y +
                    proposedRect.size.height / 2.0) - radius);
    CGFloat width = (ypos < radius) ? 2.0 * sqrt(radius * radius
                                        - ypos * ypos) : 0.0;
    CGRect circleRect = CGRectMake(radius - width / 2.0,
                            proposedRect.origin.y,
                            width,
                            proposedRect.size.height);

    return CGRectIntersection(rect, circleRect);
}
```

Subclassing the Text Storage

`NSTextStorage` is a subclass of `NSMutableAttributedString`. You can subclass it to automatically apply or manage styles. There are a few tricks to deal with, however. In this example, you create a text storage that automatically applies formats to certain words. It is available in the `ScribbleLayout` sample project and is based on the `TKDInteractiveTextColoringTextStorage` example from WWDC 2013: "Session 210, Introducing Text Kit."

`NSTextStorage` is an unusual class. It is a "semiconcrete" subclass of `NSMutableAttributedString`. This means that it does not actually implement the required primitive methods of `NSMuttableAttributedString`. To subclass it yourself, you must implement the following methods:

- `string;`
- `attributesAtIndex:effectiveRange:`
- `replaceCharactresInRange:withString:`
- `setAttributes:range:`

The easiest way to implement these methods is to have an `NSMutableAttributedString` property and use that as a backing storage. In practice, this means that `NSTextStorage` is generally both IS-A `NSMutableAttributedString` and HAS-A `NSMutableAttributedString`. This class is very unusual.

The mutation methods need to be wrapped in calls to `beginEditing` and `endEditing`, and also must call `edited:range:changeInLength:`. In practice, this means that most subclasses of `NSTextStorage` include the following boilerplate code:

PTLScribbleTextStorage.m (ScribbleLayout)

```objc
@interface PTLScribbleTextStorage ()
@property (nonatomic, readwrite) NSMutableAttributedString *backingStore;
@end

@implementation PTLScribbleTextStorage

- (id)init {
  self = [super init];
  if (self) {
    _backingStore = [NSMutableAttributedString new];
  }
  return self;
}

- (NSString *)string {
  return [self.backingStore string];
}

- (NSDictionary *)attributesAtIndex:(NSUInteger)location
                     effectiveRange:(NSRangePointer)range {
  return [self.backingStore attributesAtIndex:location
                               effectiveRange:range];
}

- (void)replaceCharactersInRange:(NSRange)range
                      withString:(NSString *)str {
  [self beginEditing];
  [self.backingStore replaceCharactersInRange:range withString:str];
  [self edited:NSTextStorageEditedCharacters|
NSTextStorageEditedAttributes
         range:range
changeInLength:str.length - range.length];
  [self endEditing];
}

- (void)setAttributes:(NSDictionary *)attrs
                range:(NSRange)range {
  [self beginEditing];
  [self.backingStore setAttributes:attrs range:range];
  [self edited:NSTextStorageEditedAttributes
         range:range
changeInLength:0];
  [self endEditing];
}
```

When you have this template, you can add your special handling, generally in `processEditing:`. In this example, you add a `tokens` dictionary that maps key words to automatically applied attributes:

PTLScribbleTextStorage.h (ScribbleLayout)

```
NSString * const PTLDefaultTokenName;

NSString * const PTLRedactStyleAttributeName;
NSString * const PTLHighlightColorAttributeName;

@interface PTLScribbleTextStorage : NSTextStorage
@property (nonatomic, readwrite, copy) NSDictionary *tokens;
@end
```

PTLScribbleTextStorage.m (ScribbleLayout)

```
#import "PTLScribbleTextStorage.h"

NSString * const PTLDefaultTokenName = @"PTLDefaultTokenName";
NSString * const PTLRedactStyleAttributeName =
  @"PTLRedactStyleAttributeName";
NSString * const PTLHighlightColorAttributeName =
  @"PTLHighlightColorAttributeName";

@interface PTLScribbleTextStorage ()
@property (nonatomic, readwrite) NSMutableAttributedString *backingStore;
@property (nonatomic, readwrite) BOOL dynamicTextNeedsUpdate;
@end

@implementation PTLScribbleTextStorage

...

- (void)replaceCharactersInRange:(NSRange)range
                      withString:(NSString *)str {
  [self beginEditing];
  [self.backingStore replaceCharactersInRange:range withString:str];
  [self edited:NSTextStorageEditedCharacters|NSTextStorageEditedAttributes
          range:range
changeInLength:str.length - range.length];
  self.dynamicTextNeedsUpdate = YES;
  [self endEditing];
}

- (void)performReplacementsForCharacterChangeInRange:(NSRange)changedRange {
  NSString *string = [self.backingStore string];
  NSRange startLine = NSMakeRange(changedRange.location, 0);
  NSRange endLine = NSMakeRange(NSMaxRange(changedRange), 0);
  NSRange
  extendedRange = NSUnionRange(changedRange,
                               [string
                                lineRangeForRange:startLine]);
  extendedRange = NSUnionRange(extendedRange,
                               [string
```

```
                                        lineRangeForRange:endLine]);
    [self applyTokenAttributesToRange:extendedRange];
}

- (void)processEditing {
  if(self.dynamicTextNeedsUpdate) {
    self.dynamicTextNeedsUpdate = NO;
    [self performReplacementsForCharacterChangeInRange:[self editedRange]];
  }
  [super processEditing];
}

- (void)applyTokenAttributesToRange:(NSRange)searchRange {
  NSDictionary *defaultAttributes = self.tokens[PTLDefaultTokenName];

  NSString *string = [self.backingStore string];
  [string enumerateSubstringsInRange:searchRange
                             options:NSStringEnumerationByWords
                          usingBlock:^(NSString *substring,
                                       NSRange substringRange,
                                       NSRange enclosingRange,
                                       BOOL *stop) {
                    NSDictionary *
                    attributesForToken = self.tokens[substring];

                    if(!attributesForToken)
                      attributesForToken = defaultAttributes;

                    if(attributesForToken)
                      [self setAttributes:attributesForToken
                                    range:substringRange];
                  }];
}
```

Because the text in this text storage is modified, it is automatically reformatted according to the definitions in the tokens dictionary. Here is how it is used. This example makes *France* appear in blue and *England* appear in red.

ViewController.m (ScribbleLayout)

```
// Create the text storage
PTLScribbleTextStorage *text = [[PTLScribbleTextStorage alloc] init];

text.tokens = @{ @"France" : @{ NSForegroundColorAttributeName :
                                  [UIColor blueColor] },
                 @"England" : @{ NSForegroundColorAttributeName :
                                  [UIColor redColor] },

                 PTLDefaultTokenName : @{
                     NSParagraphStyleAttributeName: style,
                     NSFontAttributeName:
                         [UIFont
                         preferredFontForTextStyle:UIFontTextStyleCaption2]
                 } };
```

(continued)

```
[text setAttributedString:attributedString];

// Create the layout manager
NSLayoutManager *layoutManager = [NSLayoutManager new];
[text addLayoutManager:layoutManager];

// Create the text container
CGRect textViewFrame = CGRectMake(30, 40, 708, 400);
NSTextContainer *
textContainer = [[NSTextContainer alloc] initWithSize:textViewFrame.size];
[layoutManager addTextContainer:textContainer];

// Create the text view
UITextView *textView = [[UITextView alloc] initWithFrame:textViewFrame
                                           textContainer:textContainer];

[self.view addSubview:textView];
```

Subclassing the Layout Manager

Subclassing the text storage is useful for applying attributes, but it can't change how those attributes are drawn. By subclassing the layout manager, you can change the appearance of various attributes and create entirely new drawing styles. This capability is particularly useful for specialized underlining or implementing other text decorations.

If you only want to tweak layout, you may implement a layout manager delegate rather than subclassing NSLayoutManager. For instance, delegate methods such as layoutManager: shouldBreakLineByHyphenatingBeforeCharacterAtIndex: allow you to take fine-grained control of hyphenation and similar layout issues.

The following example builds on ScribbleLayout by adding two new attributes: redacting and highlighting. Redacted text is completely replaced by a crossed-out box. The glyphs are never drawn. Highlighted text has an oversized yellow background and is circled. These are the kinds of styles that you cannot create easily using a traditional NSAttributedString.

To implement the redacting attribute, you override drawGlyphsForGlyphRange:atPoint:. This is the primitive method for drawing glyphs. If the attributes include a true value for PTLRedactStyleAttributeName, draw a crossed-out box. Otherwise, use the default glyph drawing implementation.

PTLScribbleLayoutManager.m (ScribbleLayout)

```
- (void)drawGlyphsForGlyphRange:(NSRange)glyphsToShow
                        atPoint:(CGPoint)origin {
    // Determine the character range so you can check the attributes
    NSRange characterRange = [self characterRangeForGlyphRange:glyphsToShow
                                              actualGlyphRange:NULL];

    // Enumerate each time PTLRedactStyleAttributeName changes
    [self.textStorage enumerateAttribute:PTLRedactStyleAttributeName
                                 inRange:characterRange
                                 options:0
```

```objc
                     usingBlock:^(id value,
                         NSRange attributeCharacterRange,
                         BOOL *stop) {
                 [self redactCharacterRange:attributeCharacterRange
                                      ifTrue:value
                                     atPoint:origin];
                 }];
}

- (void)redactCharacterRange:(NSRange)characterRange
                      ifTrue:(NSNumber *)value
                     atPoint:(CGPoint)origin {

  // Switch back to glyph ranges, since we're drawing
  NSRange glyphRange = [self glyphRangeForCharacterRange:characterRange
                                     actualCharacterRange:NULL];
  if ([value boolValue]) {

    // Prepare the context. origin is in view coordinates.
    // The methods below will return text container coordinates,
    // so apply a translation
    CGContextRef context = UIGraphicsGetCurrentContext();
    CGContextSaveGState(context);
    CGContextTranslateCTM(context, origin.x, origin.y);
    [[UIColor blackColor] setStroke];
    // Enumerate contiguous rectangles that enclose the redacted glyphs
    NSTextContainer *
    container = [self textContainerForGlyphAtIndex:glyphRange.location
                                      effectiveRange:NULL];

    [self enumerateEnclosingRectsForGlyphRange:glyphRange
                     withinSelectedGlyphRange:NSMakeRange(NSNotFound, 0)
                             inTextContainer:container
                                  usingBlock:^(CGRect rect, BOOL *stop){
                                    [self drawRedactionInRect:rect];
                                  }];
    CGContextRestoreGState(context);
  }
  else {
    // Wasn't redacted. Use default behavior.
    [super drawGlyphsForGlyphRange:glyphRange atPoint:origin];
  }
}

- (void)drawRedactionInRect:(CGRect)rect {
  // Draw a box with an X through it.
  // You could draw anything here.
  UIBezierPath *path = [UIBezierPath bezierPathWithRect:rect];
  CGFloat minX = CGRectGetMinX(rect);
  CGFloat minY = CGRectGetMinY(rect);
  CGFloat maxX = CGRectGetMaxX(rect);
  CGFloat maxY = CGRectGetMaxY(rect);
```

(continued)

```
    [path moveToPoint:    CGPointMake(minX, minY)];
    [path addLineToPoint:CGPointMake(maxX, maxY)];
    [path moveToPoint:    CGPointMake(maxX, minY)];
    [path addLineToPoint:CGPointMake(minX, maxY)];
    [path stroke];
}
```

Notice that there is a lot of switching between character ranges and glyph ranges in this code. It is very important to use good variable and method names to keep from getting confused.

The layout manager supports several drawing phases. The previous example customized glyph drawing. There are also drawing phases for the background, underline, and strikethrough. The highlighting attribute is best implemented as part of the background, so you override `drawBackgroundForGlyphRange:atPoint:`. The background is drawn first, then the glyphs, and then the underlines and strikethroughs.

This example is similar to the redacting example, but there are a couple of differences. Highlighting takes a color value rather than a Boolean. It also always calls `[super drawBackgroundForGlyphRange:atPoint:]` to merge highlighting with any other background decoration. Both of these are just design decisions. You could also decide not to call `super` if you wanted to replace rather than extend background decorations.

PTLScribbleLayoutManager.m (ScribbleLayout)

```
- (void)drawBackgroundForGlyphRange:(NSRange)glyphsToShow
                            atPoint:(CGPoint)origin {
  [super drawBackgroundForGlyphRange:glyphsToShow atPoint:origin];

  CGContextRef context = UIGraphicsGetCurrentContext();
  NSRange characterRange = [self characterRangeForGlyphRange:glyphsToShow
                                            actualGlyphRange:NULL];
  [self.textStorage enumerateAttribute:PTLHighlightColorAttributeName
                              inRange:characterRange
                              options:0
                            usingBlock:^(id value,
                                        NSRange highlightedCharacterRange,
                                        BOOL *stop) {
                      [self highlightCharacterRange:highlightedCharacterRange
                                            color:value
                                          atPoint:origin
                                        inContext:context];
                    }];
}

- (void)highlightCharacterRange:(NSRange)highlightedCharacterRange
                        color:(UIColor *)color
                      atPoint:(CGPoint)origin
                    inContext:(CGContextRef)context {
  if (color) {
    CGContextSaveGState(context);
    [color setFill];
    CGContextTranslateCTM(context, origin.x, origin.y);
```

```
        NSRange highlightedGlyphRange = [self
                glyphRangeForCharacterRange:highlightedCharacterRange
                    actualCharacterRange:NULL];
        NSTextContainer *container = [self
                textContainerForGlyphAtIndex:highlightedGlyphRange.location
                        effectiveRange:NULL];

        [self enumerateEnclosingRectsForGlyphRange:highlightedGlyphRange
                    withinSelectedGlyphRange:NSMakeRange(NSNotFound, 0)
                            inTextContainer:container
                                usingBlock:^(CGRect rect, BOOL *stop){
                                    [self drawHighlightInRect:rect];
                                }];
        CGContextRestoreGState(context);
    }
}

- (void)drawHighlightInRect:(CGRect)rect {
    CGRect highlightRect = CGRectInset(rect, -3, -3);
    UIRectFill(highlightRect);
    [[UIBezierPath bezierPathWithOvalInRect:highlightRect] stroke];
}
```

Per-Glyph Layout

Using Text Kit, you can alter glyph locations on a glyph-by-glyph basis. In this example, you learn how to draw text on any Bézier curve, and the techniques are applicable to drawing on any path. You also preserve kerning and ligatures. The result is shown in Figure 21-11. This example is available from the downloads for this chapter, in CurvyTextView.m in the CurvyText project.

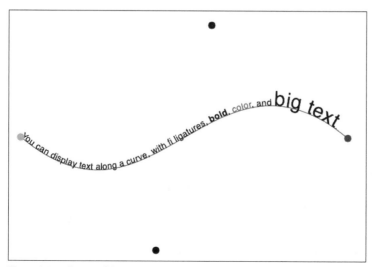

Figure 21-11 Output of CurvyTextView

Although `CGPath` can represent a Bézier curve and Core Graphics can draw it, iOS doesn't have any functions that allow you to calculate the points along the curve. You need these points, provided by `Bezier()`, and the slope along the curve, provided by `BezierPrime()`:

CurvyTextView.m (CurvyText)

```
static double Bezier(double t, double P0, double P1, double P2,
                     double P3) {
  return
               (1-t) * (1-t) * (1-t)         * P0
    + 3 *               (1-t) * (1-t) *     t * P1
    + 3 *                       (1-t) *   t*t * P2
    +                                   t*t*t * P3;
}

static double BezierPrime(double t, double P0, double P1,
                          double P2, double P3) {
  return
    -   3 *  (1-t) * (1-t)  * P0
    + (3 *  (1-t) * (1-t)  * P1)  - (6 * t *  (1-t)  * P1)
    - (3 *             t*t  * P2)  + (6 * t *  (1-t)  * P2)
    +   3 * t*t * P3;
}
```

`P0` is the starting point, drawn in green by `CurvyTextView`. `P1` and `P2` are the control points, drawn in black. `P3` is the endpoint, drawn in red. You call these functions twice, once for the x coordinate and once for the y coordinate. To get a point and angle along the curve, you pass a number between 0 and 1 to `pointForOffset:` and `angleForOffset:`.

```
- (CGPoint)pointForOffset:(double)t {
  double x = Bezier(t, _P0.x, _P1.x, _P2.x, _P3.x);
  double y = Bezier(t, _P0.y, _P1.y, _P2.y, _P3.y);
  return CGPointMake(x, y);
}
- (double)angleForOffset:(double)t {
  double dx = BezierPrime(t, _P0.x, _P1.x, _P2.x, _P3.x);
  double dy = BezierPrime(t, _P0.y, _P1.y, _P2.y, _P3.y);
  return atan2(dy, dx);
}
```

> These methods are called so many times that I've made an exception to the rule to always use accessors. This is the major hotspot of this program, and optimizations to speed it up or call it less frequently are worthwhile. You can improve the performance of this code in many ways, but this approach is fast enough for an example. See `robnapier.net/bezier` for more details on Bézier calculations.

With these two functions to define your path, you can now lay out the text. The following method draws an attributed string into the current context along this path:

```
- (void)drawText {
    if ([self.attributedString length] == 0) { return; }

    NSLayoutManager *layoutManager = self.layoutManager;

    CGContextRef context = UIGraphicsGetCurrentContext();
    NSRange glyphRange;
    CGRect lineRect = [layoutManager lineFragmentRectForGlyphAtIndex:0
                                          effectiveRange:&glyphRange];
    double offset = 0;
    CGPoint lastGlyphPoint = self.P0;
    CGFloat lastX = 0;
    for (NSUInteger glyphIndex = glyphRange.location;
         glyphIndex < NSMaxRange(glyphRange);
         ++glyphIndex) {
      CGContextSaveGState(context);

      CGPoint location = [layoutManager locationForGlyphAtIndex:glyphIndex];

      CGFloat distance = location.x - lastX;  // Assume single line
      offset = [self offsetAtDistance:distance
                            fromPoint:lastGlyphPoint
                            andOffset:offset];
      CGPoint glyphPoint = [self pointForOffset:offset];
      double angle = [self angleForOffset:offset];

      lastGlyphPoint = glyphPoint;
      lastX = location.x;

      CGContextTranslateCTM(context, glyphPoint.x, glyphPoint.y);
      CGContextRotateCTM(context, angle);

      [layoutManager drawGlyphsForGlyphRange:NSMakeRange(glyphIndex, 1)
                  atPoint:CGPointMake(-(lineRect.origin.x + location.x),
                                      -(lineRect.origin.y + location.y))];
      CGContextRestoreGState(context);
    }
}
```

The translation at the end of `drawText` is particularly important. All the glyphs are drawn in container coordinates. You use a transform to move the glyph into the correct position. This method's basic concept is to ask the layout manager for the correct location of the glyph using `locationForGlyphAtIndex:`, adjust the transform based on the curve calculation, and then tell the layout manager to draw the glyph using `drawGlyphsForGlyphRange:atPoint:`. Most custom drawing code has a similar structure. If rotation is not required, you could simply modify the `location` directly instead of applying a transform.

To maintain proper spacing, you need to find the point along the curve that is the same distance as the original glyph. This is not trivial for a Bézier curve. Offsets are not linear, and it's almost certain that the offset 0.25 will not be a quarter of the way along the curve. A simple solution is to repeatedly increment the offset and calculate a new point on the curve until the distance to that point is at least equal to your advance. The larger the increment you choose, the more characters tend to spread out. The smaller the increment you choose, the longer it takes to calculate. My experience is that values between 1/1000 (0.001) and 1/10,000 (0.0001) work well. Although 1/1000 has visible errors when compared to 1/10,000, the speed improvement is generally worth it. You could try to optimize this with a binary search, but that approach can fail if the loop wraps back on itself or crosses itself. Here is a simple implementation of the search algorithm:

```
- (double)offsetAtDistance:(double)aDistance
                 fromPoint:(CGPoint)aPoint
                    offset:(double)anOffset {
  const double kStep = 0.001; // 0.0001 - 0.001 work well
  double newDistance = 0;
  double newOffset = anOffset + kStep;
  while (newDistance <= aDistance && newOffset < 1.0) {
    newOffset += kStep;
    newDistance = Distance(aPoint,
                           [self pointForOffset:newOffset]);
  }
  return newOffset;
```

For more information on finding lengths on a curve, search the Web for "arc length parameterization."

With these tools, you can typeset rich text along any path you can calculate.

Core Text

Core Text is the low-level text layout and font-handling engine in iOS. It's extremely fast and powerful. Although most of its functionality is now available through Text Kit, it is still useful when performance is critical or if you need to support older versions of iOS.

Core Text is a C-based API that uses Core Foundation naming and memory management. If you're not familiar with Core Foundation patterns, see Chapter 10.

Core Text uses attributed strings, just as Text Kit does. In most cases, Core Text is quite forgiving about the kind of attributed string you pass. For instance, you can use the keys `NSForegroundColorAttributedName` or `kCTForegroundColorAttributeName`, and they will work the same. You can generally use `UIColor` or `CGColor` interchangeably, even though they are not toll-free bridged. Similarly, you can use `UIFont` or `CTFont` and get the same results. The SimpleLayout project in the sample code for this chapter shows how to create a `CFMutableAttributedString`, but I don't go into detail about that here because it's similar to `NSMutableAttributedString`, and you can use the two interchangeably.

Simple Layout with CTFramesetter

After you have an attributed string, you generally lay out the text using `CTFramesetter`. A *framesetter* is responsible for creating frames of text. A `CTFrame` (frame) is an area enclosed by a `CGPath` containing one or more lines of text. When you generate a frame, you draw it into a graphics context using `CTFrameDraw`. In the next example, you draw an attributed string into the current view using `drawRect:`.

First, you need to flip the view context. Core Text was originally designed on the Mac, and it performs all calculations in Mac coordinates. The origin is in the lower-left corner—lower-left origin (LLO)—and the y coordinates run from bottom to top as in a mathematical graph. `CTFramesetter` doesn't work properly unless you invert the coordinate space, as shown in the following code:

CoreTextLabel.m (SimpleLayout)

```
- (id)initWithFrame:(CGRect)frame {
  if ((self = [super initWithFrame:frame])) {
    CGAffineTransform
    transform = CGAffineTransformMakeScale(1, -1);
    CGAffineTransformTranslate(transform,
                               0, -self.bounds.size.height);
    self.transform = transform;
    self.backgroundColor = [UIColor whiteColor];
  }
  return self;
}
```

Before drawing the text, you need to set the text transform, or matrix. The text matrix is not part of the graphics state and is not always initialized the way you expect. It isn't included in the state saved by `CGContextSaveGState`. If you're going to draw text, always call `CGContextSetTextMatrix` in `drawRect:`.

```
- (void)drawRect:(CGRect)rect {
  CGContextRef context = UIGraphicsGetCurrentContext();
  CGContextSetTextMatrix(context, CGAffineTransformIdentity);

  // Create a path to fill. In this case, use the whole view
  CGPathRef path = CGPathCreateWithRect(self.bounds, NULL);

  CFAttributedStringRef
  attrString = (__bridge CFTypeRef)self.attributedString;

  // Create the framesetter using the attributed string
  CTFramesetterRef framesetter =
  CTFramesetterCreateWithAttributedString(attrString);

  // Create a single frame using the entire string (CFRange(0,0))
  // that fits inside of path.
  CTFrameRef
  frame = CTFramesetterCreateFrame(framesetter,
                                   CFRangeMake(0, 0),
                                   path,
                                   NULL);

  // Draw the frame into the current context
  CTFrameDraw(frame, context);
```

```
        CFRelease(frame);
        CFRelease(framesetter);
        CGPathRelease(path);
    }
```

There's no guarantee that all the text will fit within the frame. CTFramesetterCreateFrame simply lays out text within the path until it runs out of space or runs out of text.

Creating Frames for Noncontiguous Paths

Since at least iOS 4.2, CTFramesetterCreateFrame has accepted nonrectangular and noncontiguous frames. The *Core Text Programming Guide* has not been updated since before the release of iPhoneOS 3.2 and is occasionally ambiguous on this point. Because CTFramePathFillRule was added in iOS 4.2, Core Text has explicitly supported complex paths that cross themselves, including paths with embedded holes.

CTFramesetter always typesets the text from top to bottom (or right to left for vertical layouts such as for Japanese). This approach works well for contiguous paths, but can be a problem for noncontiguous paths such as for multicolumn text. For example, you can define a series of columns this way:

ColumnView.m (Columns)

```
- (CGRect *)copyColumnRects {
    CGRect bounds = CGRectInset([self bounds], 20.0, 20.0);

    int column;
    CGRect* columnRects = (CGRect*)calloc(kColumnCount,
                                          sizeof(*columnRects));

    // Start by setting the first column to cover the entire view.
    columnRects[0] = bounds;
    // Divide the columns equally across the frame's width.
    CGFloat columnWidth = CGRectGetWidth(bounds) / kColumnCount;
    for (column = 0; column < kColumnCount - 1; column++) {
        CGRectDivide(columnRects[column], &columnRects[column],
                &columnRects[column + 1], columnWidth, CGRectMinXEdge);
    }

    // Inset all columns by a few pixels of margin.
    for (column = 0; column < kColumnCount; column++) {
        columnRects[column] = CGRectInset(columnRects[column],
                                          10.0, 10.0);
    }
    return columnRects;
}
```

You have two choices for how to combine these rectangles. First, you can create a single path that contains all of them, like this:

```
CGRect *columnRects = [self copyColumnRects];

// Create a single path that contains all columns
CGMutablePathRef path = CGPathCreateMutable();
for (int column = 0; column < kColumnCount; column++) {
  CGPathAddRect(path, NULL, columnRects[column]);
}
free(columnRects);
```

This code typesets the text as shown in Figure 21-12.

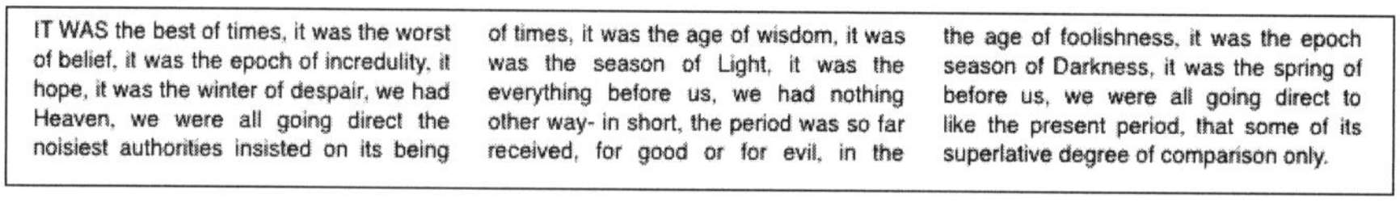

IT WAS the best of times, it was the worst of belief, it was the epoch of incredulity, it hope, it was the winter of despair, we had Heaven, we were all going direct the noisiest authorities insisted on its being   of times, it was the age of wisdom, it was was the season of Light, it was the everything before us, we had nothing other way- in short, the period was so far received, for good or for evil, in the   the age of foolishness, it was the epoch season of Darkness, it was the spring of before us, we were all going direct to like the present period, that some of its superlative degree of comparison only.

Figure 21-12 Column layout using a single path

Most of the time what you see in Figure 21-12 isn't what you want. Instead, you need to typeset the first column, then the second column, and finally the third. To do so, you need to create three paths and add them to a CFMutableArray called paths:

```
CGRect *columnRects = [self copyColumnRects];
// Create an array of layout paths, one for each column.
for (int column = 0; column < kColumnCount; column++) {
  CGPathRef path = CGPathCreateWithRect(columnRects[column], NULL);
  CFArrayAppendValue(paths, path);
  CGPathRelease(path);
}
free(columnRects);
```

You then iterate over this array, typesetting the text that hasn't been drawn yet:

```
CFIndex pathCount = CFArrayGetCount(paths);
CFIndex charIndex = 0;
for (CFIndex pathIndex = 0; pathIndex < pathCount; ++pathIndex) {
  CGPathRef path = CFArrayGetValueAtIndex(paths, pathIndex);

  CTFrameRef
  frame = CTFramesetterCreateFrame(framesetter,
                                   CFRangeMake(charIndex, 0),
                                   path,
                                   NULL);
  CTFrameDraw(frame, context);
  CFRange frameRange = CTFrameGetVisibleStringRange(frame);
  charIndex += frameRange.length;
  CFRelease(frame);
}
```

The call to `CTFrameGetVisibleStringRange` returns the range of characters within the attributed string included in this frame. That range lets you know where to start the next frame. The zero-length range passed to `CTFramesetterCreateFrame` indicates that the framesetter should typeset as much of the attributed string as will fit.

Using these techniques, you can typeset text into any shape you can draw with `CGPath`, as long as the text fits into lines.

Typesetters, Lines, Runs, and Glyphs

The framesetter is responsible for combining typeset lines into frames that can be drawn. The typesetter is responsible for choosing and positioning the glyphs in those lines. `CTFramesetter` automates this process, so you usually don't need to deal with the underlying typesetter (`CTTypesetter`). You generally use the framesetter or move further down the stack to lines, runs, and glyphs.

The typesetter is responsible for choosing the glyphs for a given attributed string and for collecting them into runs. A *run* (`CTRun`) is a series of glyphs with the same attributes and direction (such as left-to-right or right-to-left). Attributes include font, color, shadow, and paragraph style. You cannot directly create `CTRun` objects, but you can draw them into a context with `CTRunDraw`. Each glyph is positioned in the run, taking into account individual glyph size and kerning. *Kerning* is small adjustments to the spacing between glyphs to make text more readable. For example, the letters *V* and *A* are often kerned very close together.

The typesetter combines runs into lines. A line (`CTLine`) is a series of runs oriented either horizontally or vertically (for languages such as Japanese). `CTLine` is the lowest-level typesetting object that you can directly create from an attributed string. This is convenient for drawing small blocks of rich text. You can directly draw a line into a context using `CTLineDraw`.

Generally in Core Text, you work with either a `CTFramesetter` for large blocks or a `CTLine` for small labels. From any level in the hierarchy, you can fetch the lower-level objects. For example, given a `CTFramesetter`, you create a `CTFrame`, and from that, you can fetch its array of `CTLine` objects. Each line includes an array of `CTRun` objects, and within each run is a series of glyphs, along with positioning information and attributes. Behind the scenes is the `CTTypesetter` doing most of the work, but you seldom interact with it directly.

> To see an example of all these pieces working together, see the CurvyText example from the iOS 6 edition of this book. It implements the same functionality as the CurvyText example earlier in this chapter (see `https://github.com/iosptl/ios6ptl/tree/master/ch26/CurvyText`).

Summary

Apple provides a variety of powerful text layout tools, from `UILabel` to Text Kit to Core Text. In this chapter, you looked at the major options and how to choose among them. Most of all, you should have a good understanding of how to use Text Kit to create beautiful text layout in even the most complex applications.

Further Reading

Apple Documentation

The following documents are available in the iOS Developer Library at `developer.apple.com` or through the Xcode Documentation and API Reference.

Text Programming Guide for iOS: "Drawing and Managing Text"

Core Text Programming Guide

Quartz 2D Programming Guide: "Text"

String Programming Guide: "Drawing Strings"

NSAttributedString UIKit Additions Reference

WWDC Sessions

The following session videos are available at `developer.apple.com`.

WWDC 2013, "Session 210: Introducing Text Kit"

WWDC 2013, "Session 220: Advanced Text Layouts and Effects with Text Kit"

WWDC 2013, "Session 223: Using Fonts with Text Kit"

WWDC 2012, "Session 222: Introduction to Attributed Strings for iOS"

WWDC 2012, "Session 226: Core Text and Fonts"

WWDC 2012, "Session 230: Advanced Attributed Strings for iOS"

Other Resources

Clegg, Jay. *Jay's Projects.* "Warping Text to a Bézier Curves." Useful background on techniques for laying out text along a curve. The article is in C# and GDI+, but the math is useful on any platform.

`planetclegg.com/projects/WarpingTextToSplines.html`

Drobnik, Oliver. *DTCoreText.* Powerful framework for drawing attributed strings in iOS and converting between attributed strings and HTML.

`github.com/Cocoanetics/DTCoreText`

Kirk, Adam. *MTStringAttributes.* Useful wrapper to simplify creation of `NSAttributeString`.
`https://github.com/mysterioustrousers/MTStringAttributes`

Kosmaczewski, Adrian. *CoreTextWrapper.* Wrapper to simplify multicolumn layout using Core Text.

`github.com/akosma/CoreTextWrapper`

## Chapter 22

Cocoa's Biggest Trick: Key-Value Coding and Observing

There is no magic in Cocoa. It's just C. But one particular trick borders on "magic," and that's key-value observing (KVO). This chapter explores how and when to use KVO, as well as its nonmagical cousin, key-value coding (KVC).

Key-value coding is a mechanism that allows you to access an object's properties by name rather than by calling explicit accessors. This way, you can determine property bindings at runtime rather than at compile time. For instance, you can request the value of the property named by the string variable `someProperty` using `[object valueForKey:someProperty]`. You can set the value of the property named by `someProperty` using `[object setValue:someValue forKey:someProperty]`. This indirection allows you to determine the specific properties to access at runtime rather than at compile time, allowing more flexible and reusable objects. To get this flexibility, you need to name methods in specific ways. This naming convention is called key-value coding, and this chapter covers the rules for creating indirect getters and setters and how to access items in collections and manage KVC with nonobjects. You also find out how to implement advanced KVC techniques such as Higher Order Messaging and collection operators.

If your objects follow the KVC naming rules, you can also make use of *key-value observing*. KVO is a mechanism for notifying objects of changes in the properties of other objects. Cocoa has several observer mechanisms including delegation and `NSNotification`, but KVO has lower overhead. The observed object does not have to include any special code to notify observers, and if there are no observers, KVO has no runtime cost. The KVO system adds the notification code only when the class is actually observed. This makes it very attractive for situations in which performance is at a premium. In this chapter, you find out how to use KVO with properties and collections, and you also learn the trick Cocoa uses to make KVO so transparent.

You can find all the code samples in the online files for this chapter in the projects `KVC`, `KVC-Collection`, and `KVO`.

Key-Value Coding

Key-value coding is a standard part of Cocoa that allows your properties to be accessed by name *(key)* rather than by calling an explicit accessor. KVC allows other parts of the system to ask for "the property named `foo`" rather than calling `foo` directly. This system permits dynamic access by parts of the system that don't know your keys at compile time. This dynamic access is particularly important for nib file loading and Core Data in iOS. On the Mac, KVC is a fundamental part of the AppleScript interface.

The following code listings demonstrate how KVC works with an example of a cell that can display any object using `valueForKeyPath:`.

KVCTableViewCell.h (KVC)

```objc
@interface KVCTableViewCell : UITableViewCell
- (id)initWithReuseIdentifier:(NSString*)identifier;

// Object to display.
@property (nonatomic, readwrite, strong) id object;

// Name of property of object to display
@property (nonatomic, readwrite, copy) NSString *property;
@end
```

KVCTableViewCell.m (KVC)

```objc
- (BOOL)isReady {
  // Only display something if configured
  return (self.object && [self.property length] > 0);
}

- (void)update {
  NSString *text;
  if (self.isReady) {
    // Ask the target for the value of its property that has the
    // name given in self.property. Then convert that into a human
    // readable string
    id value = [self.object valueForKeyPath:self.property];
    text = [value description];
  }
  else {
    text = @"";
  }
  self.textLabel.text = text;
}

- (id)initWithReuseIdentifier:(NSString *)identifier {
  return [self initWithStyle:UITableViewCellStyleDefault
            reuseIdentifier:identifier];
}

- (void)setObject:(id)anObject {
  _object = anObject;
  [self update];
}
- (void)setProperty:(NSString *)aProperty {
  _property = aProperty;
  [self update];
}
```

KVCTableViewController.m (KVC)

```
- (NSInteger)tableView:(UITableView *)tableView
  numberOfRowsInSection:(NSInteger)section {
    return 100;
}
- (UITableViewCell *)tableView:(UITableView *)tableView
        cellForRowAtIndexPath:(NSIndexPath *)indexPath {

    static NSString *CellIdentifier = @"KVCTableViewCell";

    KVCTableViewCell *cell = [tableView
        dequeueReusableCellWithIdentifier:CellIdentifier];

    if (cell == nil) {
      cell = [[KVCTableViewCell alloc]
          initWithReuseIdentifier:CellIdentifier];
      // You want the "intValue" of the row's NSNumber.
      // The property will be the same for every row, so you set it
      // here in the cell construction section.
      cell.property = @"intValue";
    }

    // Each row's object is an NSNumber representing that integer
    // Since each row has a different object (NSNumber), you set
    // it here, in the cell configuration section.
    cell.object = @(indexPath.row);

    return cell;
}
```

This example is quite simple, displaying 100 rows of integers, but imagine if KVCTableViewCell had animation effects or special selection behaviors. You could apply those to arbitrary objects without the object or the cell needing to know anything about the other. That's the ultimate goal of a good model-view-controller (MVC) design, which is the heart of Cocoa's architecture.

The update method of KVCTableViewCell demonstrates valueForKeyPath:, which is the main KVC method you use in this example. Here is the important section:

```
id value = [self.object valueForKeyPath:self.property];
text = [value description];
```

In this example, self.property is the string @"intValue" and self.target is an NSNumber object representing the row index. So the first line is effectively the same as this code:

```
id value = [NSNumber numberWithInt:[self.object intValue]];
```

The call to numberWithInt: is automatically inserted by valueForKeyPath:, which automatically converts number types (int, float, and so on) into NSNumber objects and all other nonobject types (structs, pointers) into NSValue objects.

Although this example utilizes an `NSNumber`, the key take-away is that `target` could be any object and `property` could be the name of any property of `target`.

Setting Values with KVC

KVC can also modify writable properties using `setValue:forKey:`. For example, the following two lines are roughly identical:

```
cell.property = @"intValue";
[cell setValue:@"intValue" forKey:@"property"];
```

Both of these lines call `setProperty:` as long as `property` is an object. See the later section "KVC and Nonobjects" for a discussion on how to handle `nil` and nonobject properties.

Methods that modify properties are generally called *mutators* in the Apple documentation.

Traversing Properties with Key Paths

You may have noticed that KVC methods have `key` and `keyPath` versions. For instance, there are `valueForKey:` and `valueForKeyPath:`. The difference between a key and a key path is that a key path can have nested relationships, separated by a period. The `valueForKeyPath:` method traverses the relationships. For instance, the following two lines are roughly identical:

```
[[self department] name];
[self valueForKeyPath:@"department.name"];
```

On the other hand, `valueForKey:@"department.name"` would try to retrieve the property `department.name`, which in most cases would throw an exception.

The `keyPath` version is more flexible, whereas the `key` version is slightly faster. If the key is passed to me, I generally use `valueForKeyPath:` to provide the most flexibility to my caller. If the key is hard-coded, I generally use `valueForKey:`.

KVC and Collections

Object properties can be one-to-one or one-to-many. One-to-many properties are either ordered (arrays) or unordered (sets).

Immutable ordered (`NSArray`) and unordered (`NSSet`) collection properties can be fetched normally using `valueForKey:`. If you have an `NSArray` property called `items`, then `valueForKey:@"items"` returns it as you would expect. But there are more flexible ways of managing acess to this property. The sample code for this section is in the KVC-Collection project.

Consider the model object, `TwoTimesArray`.

TwoTimesArray.h (KVC-Collection)

```
@interface TwoTimesArray: NSObject
-  (void)incrementCount;
-  (NSUInteger)countOfNumbers;
-  (id)objectInNumbersAtIndex:(NSUInteger)index;
@end
```

TwoTimesArray.m (KVC-Collection)

```
#import "TimesTwoArray.h"

@interface TimesTwoArray ()
@property (nonatomic, readwrite, assign) NSUInteger count;
@end

@implementation TimesTwoArray

-  (NSUInteger)countOfNumbers {
   return self.count;
}

-  (id)objectInNumbersAtIndex:(NSUInteger)index {
   return @(index * 2);
}

-  (void)incrementCount {
   self.count++;
}

@end
```

Given this object, you can request the value for `numbers`:

```
NSArray *numbers = [twoTimesArray valueForKey:@"numbers"];
```

Even though there is no `numbers` property, the KVC system will create a proxy object that behaves like an array. The reason is that `TwoTimesArray` implements `countOfNumbers` and `objectInNumbersAtIndex:`. These are very specially named methods. When `valueForKey:` looks for `items`, it searches for the following methods:

- `getNumbers` or `numbers` or `isNumbers`—These methods are searched in order, and the first one found is used to return the value.

- `countOfNumbers` and either `objectInNumbersAtIndex:` or `numbersAtIndexes`—This is the combination you use in this example. KVC generates a proxy array that's discussed shortly.

- `countOfNumbers` and `enumeratorOfNumbers` and `memberOfNumbers`—This combination causes KVC to return a proxy set.

- An instance variable named _numbers, _isNumbers, numbers, or isNumbers—KVC will directly access the ivar. You generally want to avoid this behavior. Direct instance variable access breaks encapsulation and makes the code more fragile. You can prevent this behavior by overriding +accessInstanceVariablesDirectly and returning NO.

In this example, valueForKey: automatically generates and returns a proxy NSKeyValueArray. It is a subclass of NSArray, and you can use it like any other array, but calls to count, objectAtIndex:, and related methods are forwarded to the appropriate KVC methods. The proxy caches its requests, making it very efficient. See the *Key-Value Coding Programming Guide* in the iOS Developer Library for the full set of methods you can implement for this form.

In this example, the property is numbers, so KVC looks for countOfNumbers. If the property were boxes instead, KVC would look for countOfBoxes. KVC requires that you name your methods in a standard way so that it can construct these method names. This is why getters must begin with a lowercase letter.

For mutable collection properties, you have two options. You can use the mutator (property-changing) methods such as the following (again, see the *Key-Value Coding Programming Guide* for the full list):

```
- (void) insertObject: (id) object inNumbersAtIndex: (NSUInteger) index;
- (void) removeObject: (id) object inNumbersAtIndex: (NSUInteger) index;
```

Or you can return a special proxy object by calling mutableArrayValueForKey: or mutableSetValueForKey:. Modifying this object automatically calls the appropriate KVC methods on your object.

KVC and Dictionaries

Dictionaries are just a special kind of nested relationship. For most keys, calling valueForKey: is the same as calling objectForKey: (the exception is if the key begins with @, which is used to refer to the NSDictionary itself, if needed). This is a convenient way to handle nested dictionaries because you can use valueForKeyPath: to access arbitrary layers.

KVC and Nonobjects

Not every method returns an object, but valueForKey: always returns an id. Nonobject return values are wrapped in an NSValue or NSNumber. These two classes can handle just about anything from numbers and Booleans to pointers and structures. Although valueForKey: automatically wraps scalar values into objects, you cannot pass nonobjects to setValue:forKey:. You must wrap scalars in NSValue or NSNumber.

Setting a nonobject property to nil presents a special case. Whether or not doing so is permissible depends on the situation, so KVC does not guess. If you call setValue:forKey: with a value of nil, the key is passed to setNilValueForKey:. You need to override this method to do the right thing if you want to handle setting nil for a nonobject property. Its default behavior is to throw an exception.

Higher Order Messaging with KVC

valueForKey: is filled with useful special cases, such as the fact that it's overridden for collections like NSArray and NSSet. Rather than operating on the collection, valueForKey: is passed to each member of the collection. The results are added to the returned collection. Consequently, you can easily construct collections from other collections such as

```
NSArray *array = @[ @"foo", @"bar", @"baz" ];
NSArray *capitals = [array valueForKey:@"capitalizedString"];
```

This code passes the method capitalizedString to each item in the NSArray and returns a new NSArray with the results. Passing messages (capitalizedString) as parameters is called Higher Order Messaging. Multiple messages can be passed using key paths:

```
NSArray *array = @[ @"foo", @"bar", @"baz" ];
NSArray *capitalLengths =
[array valueForKeyPath:@"capitalizedString.length"];
```

The preceding code calls capitalizedString on each element of array, then calls length, and finally wraps the return into an NSNumber object. The results are collected into a new array called capitalLengths.

Collection Operators

KVC also provides a few complex functions. It can, for instance, sum or average a list of numbers automatically. Consider this:

```
NSArray *array = @[ @"foo", @"bar", @"baz" ];
NSUInteger totalLength = [[array valueForKeyPath:@"@sum.length"] intValue];
```

@sum is an operator that sums the indicated property (length). Note that this example can be hundreds of times slower than the equivalent loop:

```
NSArray *array = @[ @"foo", @"bar", @"baz" ];
NSUInteger totalLength = 0;
for (NSString *string in array) {
  totalLength += [string length];
}
```

The performance issues are generally significant when you are dealing with arrays of thousands or tens of thousands of elements. Beyond @sum, you can find many other operators in the *Key-Value Coding Programming Guide* in the iOS Developer Library. The operations are particularly valuable when working with Core Data and can be faster than the equivalent loop because they can be optimized into database queries. You cannot create your own operations, however.

Key-Value Observing

Key-value observing is a mechanism for transparently notifying observers of changes in object properties. At the beginning of the "Key-Value Coding" section, you built a table view cell that could display any object. In that example, the data was static. If you changed the data, the cell wouldn't update. You improve that now. You can make the cell automatically update whenever its object changes. You need a changeable object, so use the current date and time. You use key-value observing to get a callback every time a property you care about changes.

KVO has a lot of similarities to `NSNotificationCenter`. You start observing using `addObserver:forKeyPath:options:context:`. To stop observing, you use `removeObserver:forKeyPath:context:`. The callback is always `observeValueForKeyPath:ofObject:change:context:`. Here are the modifications required to create 1,000 rows of date cells that automatically update every second:

KVCTableViewCell.m (KVO)

```
- (void) removeObservation {
  if (self.isReady) {
    [self.object removeObserver:self
                  forKeyPath:self.property];
  }
}

- (void) addObservation {
  if (self.isReady) {
    [self.object addObserver:self forKeyPath:self.property
                  options:0
                  context:(void*)self];
  }
}

- (void) observeValueForKeyPath:(NSString *)keyPath
                  ofObject:(id)object
                    change:(NSDictionary *)change
                   context:(void *)context {

  if ((__bridge id)context == self) {
    // Our notification, not our superclass_s
      [self update];
  }
  else {
    [super observeValueForKeyPath:keyPath ofObject:object
                      change:change context:context];
  }
}

- (void) dealloc {
  if (_object && [_property length] > 0) {
    [_object removeObserver:self
               forKeyPath:_property
                  context:(void *)self];
  }
}
```

```objc
- (void)setObject:(id)anObject {
  [self removeObservation];
  _object = anObject;
  [self addObservation];
  [self update];
}

- (void)setProperty:(NSString *)aProperty {
  [self removeObservation];
  _property = aProperty;
  [self addObservation];
  [self update];
}
```

KVCTableViewController.m (KVO)

```objc
#import "RNTimer.h"
@interface KVCTableViewController ()
@property (readwrite, retain) RNTimer *timer;
@property (readwrite, retain) NSDate *now;
@end

@implementation KVCTableViewController

- (void)updateNow {
  self.now = [NSDate date];
}

- (void)viewDidLoad {
  [self updateNow];

  __weak id weakSelf = self;
  self.timer =
      [RNTimer repeatingTimerWithTimeInterval:1
                                        block:^{
                                          [weakSelf updateNow];
                                        }];
}
- (void)viewDidUnload {
  self.timer = nil;
  self.now = nil;
}

...

- (UITableViewCell *)tableView:(UITableView *)tableView
        cellForRowAtIndexPath:(NSIndexPath *)indexPath {

  static NSString *CellIdentifier = @"KVCTableViewCell";

  id cell = [tableView
    dequeueReusableCellWithIdentifier:CellIdentifier];
```

(continued)

```
   if (cell == nil) {
     cell = [[[KVCTableViewCell alloc]
        initWithReuseIdentifier:CellIdentifier] autorelease];
     [cell setProperty:@"now"];
     [cell setObject:self];
   }

   return cell;
 }
 @end
```

In `KVCTableViewCell`, you observe the requested property on your target in `addObservation`. When you register for KVO, you pass `self` as the `context` pointer so that in the callback you can determine whether this was your observation. Because a class has only one KVO callback, you may receive a callback for a property that your superclass registered for. If so, you need to pass it along to `super`. Unfortunately, you can't always pass to `super` because `NSObject` will throw an exception. So you use a unique `context` to identify your observations. For more details about this topic, see the "KVO Tradeoffs" section later in this chapter.

In `RootViewController`, you create a property `now` and ask the cell to observe it. Once a second, you update it. Observers are notified and the cells update. This approach is quite efficient because, at any given time, you have only about one screen of cells due to cell reuse.

The real power of KVO is seen in `[KVCTableViewController updateNow]`:

```
 - (void)updateNow {
     self.now = [NSDate date];
 }
```

The only thing you have to do is update your data. You don't have to worry that someone might be observing you, and if no one is observing you, you don't pay any overhead as you would for `NSNotificationCenter`. The incredible simplicity on the part of the model class is the real benefit of KVO. As long as you use accessors to modify your properties, all the observation mechanism is handled automatically, with no cost when you don't need it. All the complexity is moved into the observer rather than the observed. It's no wonder that KVO is popular in low-level Apple frameworks.

KVO and Collections

Observing collections often causes confusion. The point to remember is that observing a collection is not the same as observing the objects in it. If a collection contains Adam, Bob, and Carol, adding Denise changes the collection. Changes to Adam do not change the collection. If you want to observe changes to the objects in a collection, you must observe those objects, not the collection. Generally, you do that by observing the collection and then observing objects as they're added and stopping when they're removed.

How Is KVO Implemented?

Key-value observing notifications rely on two `NSObject` methods: `willChangeValueForKey:` and `didChangeValueForKey:`. Before an observed property change is made, something must call `willChangeValueForKey:`. This records the old value. After the change is made, something must call `didChangeValueForKey:`, which calls `observeValueForKeyPath:ofObject:change:context:`. You can do this by hand, but that's fairly uncommon. Generally, you do so only if you're trying to control when the callbacks are made. Most of the time, the change notifications are called automatically.

There is very little magic in Objective-C. Even message dispatching, which can seem mysterious at first, is actually pretty straightforward. (Message dispatching is covered in Chapter 24.) However, KVO borders on magic. Somehow when you call `setNow:`, an extra call is made to `willChangeValueForKey:`, `didChangeValueForKey:`, and `observeValueForKeyPath:ofObject:change:context:` in the middle. You might think that the reason is you synthesized `setNow:`, and occasionally you'll see people write code like this:

```
- (void)setNow:(NSDate *)aDate {
    [self willChangeValueForKey:@"now"]; // Unnecessary
    _now = aDate;
    [self didChangeValueForKey:@"now"]; // Unnecessary
}
```

This approach is redundant and will call the KVO methods twice. KVO always calls `willChangeValue-ForKey:` before an accessor and `didChangeValueForKey:` afterward. How? The answer is class swizzling. Swizzling is discussed further in Chapter 24, but when you first call `addObserver:forKeyPath:option s:context:` on an object, the framework creates a new KVO subclass of the class and converts the observed object to that new subclass. In that special KVO subclass, Cocoa creates setters for the observed properties that work effectively like this:

```
- (void)setNow:(NSDate *)aDate {
    [self willChangeValueForKey:@"now"];
    [super setValue:aDate forKey:@"now"];
    [self didChangeValueForKey:@"now"];
}
```

This subclassing and method injection are done at runtime, not compile time. That's why it's so important that you name things correctly. KVO can create the required overrides only if you use the KVC naming convention.

It's difficult to detect the KVO class swizzling. It overrides `class` to return the original class. But occasionally you'll see references to `NSKVONotifying_MYClass` instead of `MYClass`.

KVO Tradeoffs

KVC is powerful technology, but other than possibly being slower than direct method calls, it's generally a good thing. The one major downside is that you lose compile-time checks of your property names. Always code following KVC naming conventions, whether or not you use KVC directly. Doing so will save you a lot of grief when you want to instantiate objects from a nib file, which requires KVC. It also makes your code readable by other Objective-C developers, who expect certain names to mean certain things. For the most part, this means naming your getters and setters `property` and `setProperty:`, respectively.

KVO, on the other hand, is a mixed bag. It can be useful, and it can cause trouble. It's implemented in a highly magical way, and some of its usage is quite awkward. There's only one callback method, which means you often wind up with a lot of unrelated code in that method. The callback method passes you a change dictionary that is somewhat tedious to use.

Because `removeObserver:forKeyPath:context:` will crash if you're not an observer for that key path, you must keep track of exactly which properties you're observing. KVO has no equivalent to `NSNotificationCenter removeObserver:`, which conveniently cleans up all observations you might have.

KVO creates subtle code-path surprises. When you call `postNotification:`, you know that some other code may run. You can search your code for the notification name and generally find all the things that might happen. It can be surprising that just setting one of your own properties can cause code in another part of the program to execute. It can be difficult to search the code to discover this interaction. KVO bugs in general are difficult to solve because so much of the activity "just happens," without any visible code causing it.

So KVO's greatest strength is also its greatest danger. It can sometimes dramatically reduce the amount of common code you write. In particular, it can get rid of the frequent problem of hand-building all your setters just so you can call an `updateSelf` method. In this way, KVO can reduce bugs caused by incorrectly cut-and-pasted code. But it can also inject really confusing bugs, and with the introduction of Automatic Reference Counting, handwritten setters are even easier to write correctly.

> Several third parties have attempted to improve KVO. In particular, KVO is an obvious candidate for a block-based interface. In the "Further Reading" section at the end of this chapter, you'll find links to some examples by very accomplished developers. Still, I'm very nervous about using a third-party solution here. KVO is complicated and magical. Blocks are complicated and magical. Combining the two without a large testing group seems extremely dangerous. Personally, I'll wait for Apple to improve this interface, but if you're interested in other options, see the "Further Reading" section.

My recommendation is to use KVO sparingly, simply, and only in places where doing so is a real benefit. For situations in which you need a very large number of observations (a few hundred or more), its performance scales much better than `NSNotification`. It gives you the advantages of `NSNotification` without modifying the observed class. And it sometimes requires less code, although you need to include all the special-case code you may need to work around subtle KVO problems. In the `KVCTableViewCell` example, hand-coding `setProperty:` and `setTarget:` makes the file about 15 lines shorter than the equivalent KVO solution.

Avoid KVO in situations in which you have complex interdependencies or a complicated class hierarchy. Simple solutions with delegates and `NSNotification` are often better than excessively clever solutions using KVO.

On the other hand, Apple relies on KVO in performance-critical frameworks. It's the primary way to deal with `CALayer` and `NSOperation`. It has the advantage of zero-overhead observation. If there are no observers of a given instance, KVO costs nothing because there is no KVO code. Delegate methods and `NSNotification` still have to do work even if there are no observers. For low-level, performance-critical objects, KVO is an obvious win. Use it wisely. And hope Apple improves the API.

Summary

In this chapter you have learned two of the most powerful techniques in Objective-C: KVC and KVO. These techniques provide a level of runtime flexibility that is difficult to achieve in other languages. Writing your code to conform to KVC is a critical part of a Cocoa program, whether or not you call `valueForKey:` directly. KVO can be challenging to use well but is a powerful tool when you need high-performance observations. As a Cocoa developer, you need to keep KVC and KVO in mind when designing your classes. Following a few simple naming rules will make all the difference.

Further Reading

Apple Documentation

The following documents are available in the iOS Developer Library at `developer.apple.com` or through the Xcode Documentation and API Reference.

Key-Value Coding Programming Guide

Key-Value Observing Programming Guide

NSKeyValueCoding Protocol Reference

NSKeyValueObserving Protocol Reference

Other Resources

Ash, Mike. "Key-Value Observing Done Right." One of the earliest non-Apple investigations into how to manage KVO.

```
http://www.mikeash.com/pyblog/key-value-observing-done-right.html
```

Matuschak, Andy. "KVO+Blocks: Block Callbacks for Cocoa Observers." This is a very promising category for adding block support to KVO. In my opinion, it hasn't gotten the kind of testing time it needs to trust it in a complex project, but I definitely like the approach. It's the foundation of several other wrappers. Be sure to read the update link at the bottom.

```
http://blog.andymatuschak.org/post/156229939/kvo-blocks-block-callbacks-
for-cocoa-observers
```

Waldowski, Zachary. *BlocksKit*. An extensive set of blocks-based enhancements, including a KVO observations wrapper. Although I haven't used it extensively enough to recommend it, this package has the most users and active development. It is the one I'd most likely use. Its KVO wrapper is based on Andy Matuschak's code, among others.

```
https://github.com/pandamonia/BlocksKit
```

Wight, Jonathan. "KVO-Notification-Manager." Another interesting example, based on Mike Ash's work.

```
https://github.com/schwa/KVO-Notification-Manager
```

Chapter 23

Beyond Queues: Advanced GCD

Grand Central Dispatch (GCD) is one of the most fundamental technologies in iOS. When it was first introduced in OS X 10.6, it revolutionized multithreaded programming. The majority of GCD is focused on queues, which are covered at length in Chapter 9. In this chapter, you learn the other parts of GCD, including semaphores, dispatch sources, Dispatch Data, and Dispatch I/O.

Semaphores

Semaphores manage concurrent access to resources. A semaphore has a value that can be atomically incremented or decremented. An attempt to decrement a semaphore below zero is blocked until some other caller (on another thread) increments the semaphore. This tool is very useful and flexible for multithreaded programming. Typically, a semaphore is initialized to the maximum number of callers who may simultaneously access a resource. Often this value is 1, but it can be a higher value, particularly to put limits on parallelization. A semaphore is often initialized to zero in producer-consumer patterns.

GCD provides a fairly typical semaphore implementation. Semaphores can be initialized to any value and then can be incremented or decremented by one. There is no mechanism to read a semaphore's current value. Like other GCD objects since iOS 5, semaphores are Automatic Reference Counting managed, so you do not need to manually manage them with `dispatch_retain` or `dispatch_release` under ARC. Unlike most GCD objects, semaphores do not rely on dispatch queues. They can be directly manipulated on any thread.

The following example demonstrates how to use semaphores to manage a limited pool of work "slots." In this example, your user can start an unlimited number of jobs by calling `runProcess:`, as shown here. The full code is available in the ProducerConsumer sample project on this book's website.

ViewController.m (ProducerConsumer)

```
@interface ViewController ()
...
@property (nonatomic) dispatch_semaphore_t semaphore;
....
@end

- (void)viewDidLoad {
  ...
  self.semaphore = dispatch_semaphore_create([self.progressViews count]);
  ...
}
```

(continued)

```
- (IBAction)runProcess:(UIButton *)button {
    // Make sure we're on the main queue
    RNAssertMainQueue();

    // Update the UI to display the number of pending jobs
    [self adjustPendingJobCountBy:1];

    // Dispatch a new work unit to the serial pending queue.
    dispatch_async(self.pendingQueue, ^{
        // Wait for an open slot
        dispatch_semaphore_wait(self.semaphore, DISPATCH_TIME_FOREVER);

        // Fetch an available resource.
        // We're on a serial queue, so we know there is no race condition
        UIProgressView *availableProgressView = [self reserveProgressView];

        // Dispatch actual work to the concurrent work queue
        dispatch_async(self.workQueue, ^{
            // Perform the dummy work
            [self performWorkWithProgressView:availableProgressView];

            // Let go of our resource
            [self releaseProgressView:availableProgressView];

            // Update the UI
            [self adjustPendingJobCountBy:-1];

            // Release our slot so another job can start
            dispatch_semaphore_signal(self.semaphore);
        });
    });
}
```

> **This example is similar to** - [NSOperationQueue maxConcurrentOperationCount]. **If you need this kind of flexibility, it is often better to use an operation queue instead of building your own solution with GCD and semaphores.**

Semaphores are also useful for converting asynchronous operations into synchronous operations. This capability is particularly useful in unit testing asynchronous APIs. For example, if you want to test methods that have a completion block, you can wait for the semaphore after calling the method and then signal the semaphore in the completion block. For example:

SyncSemaphoreTests.m (SyncSemaphore)

```
- (void)testDownload {
    NSURL *URL = [NSURL URLWithString:@"http://iosptl.com"];

    // Block variables to hold results
    __block NSURL *location;
```

```objc
    __block NSError *error;

    // Create a synchronization semaphore
    dispatch_semaphore_t semaphore = dispatch_semaphore_create(0);

    [[[NSURLSession sharedSession] downloadTaskWithURL:URL
                                      completionHandler:
      ^(NSURL *l, NSURLResponse *r, NSError *e) {
        // Unload and test data
        location = l;
        error = e;

        // Signal that the operation has completed
        dispatch_semaphore_signal(semaphore);
      }] resume];

    // Setup the timeout
    double timeoutInSeconds = 2.0;
    dispatch_time_t timeout = dispatch_time(DISPATCH_TIME_NOW,
                              (int64_t)(timeoutInSeconds * NSEC_PER_SEC));

    // Wait for the signal for a limited time
    long timeoutResult = dispatch_semaphore_wait(semaphore, timeout);

    // Test that everything worked
    XCTAssertEqual(timeoutResult, 0L, @"Timed out");
    XCTAssertNil(error, @"Received an error:%@", error);
    XCTAssertNotNil(location, @"Did not get a location");
}
```

Semaphores are low-level tools. They are very powerful, but in many cases, it is better to rethink your design to eliminate the need for them. You should first see whether a higher-level object such as an operation queue is a better choice. Often you can avoid the need for semaphores by adding another dispatch queue, using `dispatch_suspend`, or splitting up your operations in another way. Semaphores aren't bad, but they are locks, and you should look for ways to remove locks when you can. You should also usually rely on the highest-level abstractions you can in the Cocoa framework, and semaphores are near the bottom of the stack. But sometimes, such as when you want to convert an asynchronous task into a synchronous task, a semaphore is just the right tool for the job.

Dispatch Sources

A dispatch source provides an efficient way to handle events. You register an event handler, and when the event occurs, you are notified. If the event occurs multiple times before the system is able to notify you, the events are coalesced into a single event. This functionality is highly useful for low-level, high-performance I/O, but an iOS app developer rarely needs it. Similarly, dispatch sources can respond to UNIX signals, file system changes, changes in other processes, and mach port events. Many of these are useful on the Mac, but again, they don't often apply to iOS developers.

Custom sources, however, can be useful on iOS. They are particularly useful for reporting progress in performance-critical sections, as shown here. First, you create a source like this:

ViewController.m (ProgressReport)

```
dispatch_source_t
source = dispatch_source_create(DISPATCH_SOURCE_TYPE_DATA_ADD,
                                0, 0, dispatch_get_main_queue());
```

Custom sources accumulate the values passed in events. They can either accumulate these values by adding them (`DISPATCH_SOURCE_TYPE_DATA_ADD`) or logically ORing them (`. . . _DATA_OR`). Custom sources also require a queue, which is used to process all handler blocks.

After you create a source, you need to provide a handler. There are registration handlers that are dispatched as soon as the source becomes available, event handlers that are dispatched when events occur, and cancellation handlers that are dispatched when the source is canceled. Custom sources generally need only an event handler, which you create as shown here:

```
__block long totalComplete = 0;
dispatch_source_set_event_handler(source, ^{
  long value = dispatch_source_get_data(source);
  totalComplete += value;
  self.progressView.progress = (CGFloat) totalComplete /100.0f;
});
```

Only one instance of this handler block is dispatched at a time. If another event occurs before this handler completes, the events are accumulated according to the type (`ADD` or `OR`). By coalescing events, the system can work well even under load. When the handler finally runs, it can find out the total data passed by calling `dispatch_source_get_data`. This value is reset after every event handler block is run, which is why this example accumulates the final total in `totalComplete`.

Dispatch sources always start suspended. You must resume them before they dispatch handlers. Forgetting to add this call is very common and a common source of bugs. To resume the queue, call `dispatch_resume`:

```
dispatch_resume(source);
```

After the source is resumed, you can start sending it events using `dispatch_source_merge_data` as shown in this trivial example:

```
dispatch_queue_t
queue = dispatch_get_global_queue(DISPATCH_QUEUE_PRIORITY_DEFAULT,
                                  0);
dispatch_async(queue, ^{
  for (int i = 0; i <= 100; ++i) {
    dispatch_source_merge_data(source, 1);
    usleep(20000);
  }
});
```

This code adds a value of 1 for each iteration. You might also pass the number of records processed or the number of bytes written. You are free to call `dispatch_source_merge_data` from any thread. Note that you cannot pass 0 as the value (the event will not fire) and you cannot pass negative values.

---

iOS 7 introduces a new `NSProgress` class for dealing with this kind of problem. `NSProgress` does not offer coalescing, so it is not as tolerant of high-frequency updates. It also relies heavily on key-value observing (KVO), which can be more difficult to manage than GCD. On the other hand, `NSProgress` has powerful features for managing large, parallel operations that may be subdivided into many child operations.

Dispatch Timer Sources

Another particularly useful dispatch source type is the timer. GCD timers are based on dispatch queues rather than run loops like `NSTimer`, which means that they are much easier to use in multithreaded apps. Because they use blocks rather than selectors, GCD timers don't require a separate method to process the timer. They also make it easier to avoid retain cycles with repeating GCD timers.

`RNTimer` (`https://github.com/rnapier/RNTimer`) implements a simple GCD timer that avoids retain loops (as long as the block doesn't capture `self`) and automatically invalidates when deallocated. It creates a timer dispatch source with `dispatch_source_create` and ties it to the main dispatch queue. This means that the timer will always fire on the main thread. You could, of course, use a different queue for this task if you like. It then sets the timer and the event handler and calls `dispatch_resume` to start the timer. Following is the full code:

RNTimer.m (RNTimer)

```
+ (RNTimer *)repeatingTimerWithTimeInterval:(NSTimeInterval)seconds
                                      block:(dispatch_block_t)block {
  RNTimer *timer = [[self alloc] init];
  timer.block = block;
  timer.source = dispatch_source_create(DISPATCH_SOURCE_TYPE_TIMER,
                                        0, 0,
                                        dispatch_get_main_queue());
  uint64_t nsec = (uint64_t)(seconds * NSEC_PER_SEC);
  dispatch_source_set_timer(timer.source,
                            dispatch_time(DISPATCH_TIME_NOW, nsec),
                            nsec, 0);
  dispatch_source_set_event_handler(timer.source, block);
  dispatch_resume(timer.source);
  return timer;
}

- (void)invalidate {
  if (self.source) {
    dispatch_source_cancel(self.source);
```

(continued)

```
      dispatch_release(self.source);
      self.source = nil;
    }
    self.block = nil;
  }

-  (void)dealloc {
     [self invalidate];
   }
```

Dispatch Once

Most developers encounter dispatch_once as part of the common (and correct) singleton pattern:

```
+  (instancetype)sharedInstance {
     static id sharedInstance;
     static dispatch_once_t onceToken;
     dispatch_once(&onceToken, ^{
       sharedInstance = [self new];
     });
     return sharedInstance;
   }
```

You can also use dispatch_once to initialize nontrivial constants. For example, creating a date formatter can be expensive. Rather than store it in a property, you can keep it in a static variable declared within the method. You could create a simple date formatting method like this one:

```
-  (NSString *)formattedDate {
     static dispatch_once_t onceToken;
     static NSDateFormatter *dateFormatter;
     dispatch_once(&onceToken, ^{
       dateFormatter = [NSDateFormatter new];
       dateFormatter.dateStyle = NSDateFormatterLongStyle;
       dateFormatter.timeStyle = NSDateFormatterNoStyle;
     });

     return [dateFormatter stringFromDate:self.someDate];
   }
```

It is very important to remember that a dispatch_once block executes no more than once during the entire run of the program. You cannot safely use dispatch_once to initialize instance variables or properties. If the program contains more than one instance of the calling class, only one of them executes the dispatch_once block.

Another common mistake is to forget to declare the initializing variable as static. The static storage type indicates that a single instance of that variable exists in the program. In Objective-C, this is essentially a class variable. In the preceding example, if you did not include the static keyword before NSDateFormatter, then dateFormatter would be a local variable, and formattedDate would work only the first time it was called. The token (onceToken) must also be static.

dispatch_once is thread-safe, so even if it is called simultaneously on multiple threads, only one block executes. The other calls to dispatch_once are blocked until the block completes, so you are assured that the full block has completed before the next line of code runs, no matter what thread you are working on.

dispatch_once is designed to be very fast in the "has already run" case. You should have little concern about using it even in methods that are called often. Obviously, you should avoid all function calls (let alone method calls) inside tight loops, but if you are making method calls already, the extra overhead of dispatch_once is fairly trivial.

Queue-Specific Data

Much like associative references discussed in Chapter 3, queue-specific data allows you to attach a piece of data directly to a queue. This approach can sometimes be a useful and extremely fast way to pass information in and out of a queue.

Like associative references, queue-specific data uses a unique address as its key, rather than a string or other identifier. This unique address is usually achieved by passing the address of a static char. Unlike associative references, queue-specific data does not know how to retain and release. You have to pass it a destructor function that it calls when the value is replaced. For memory you've allocated with malloc, the destructor is free. Using queue-specific data with Objective-C objects under ARC is cumbersome, but Core Foundation objects are a bit easier to use, as demonstrated here. In this example, value is released automatically when the queue is destroyed or if another value is set for kMyKey:

```
static char kMyKey;
CFStringRef *value = CFStringCreate...;
dispatch_queue_set_specific(queue,
                            &kMyKey,
                            (void*)value,
                            (dispatch_function_t)CFRelease);
...

dispatch_sync(queue, ^{
  CFStringRef *string = dispatch_get_specific(&kMyKey);
  ...
});
```

One nice thing about queue-specific data is that it respects queue hierarchies. So if the current queue doesn't have the given key assigned, dispatch_get_specific automatically checks the target queue, then that queue's target queue, and on up the chain.

The Myth of the "Current Queue"

In iOS 6 dispatch_get_current_queue was deprecated. It was always a dangerous call and needed to be removed, but Apple didn't provide much guidance on what to use instead. First, you should understand why it is such a dangerous call. Apple unfortunately does not explain this in the documentation. The reason is explained only in the header file (/usr/include/dispatch/queue.h):

Recommended for debugging and logging purposes only:

The code must not make any assumptions about the queue returned, unless it is one of the global queues or a queue the code has itself created.

The code must not assume that synchronous execution onto a queue is safe from deadlock if that queue is not the one returned by dispatch_get_current_queue().

When dispatch_get_current_queue() is called on the main thread, it may or may not return the same value as dispatch_get_main_queue(). Comparing the two is not a valid way to test whether code is executing on the main thread.

Developers have generally ignored this note. One of the most common mistakes developers make is using `dispatch_get_current_queue` to determine whether `dispatch_sync` would deadlock, which is exactly the use case the documentation says is not guaranteed to work.

The problem is that the "current queue" is not well defined. Blocks are only scheduled from the well-known queues (the queues returned by `dispatch_get_main_queue` and `dispatch_get_global_queue`). But that may not have been the queue where you originally scheduled it. The block may have moved through several targeted queues before being scheduled.

This is the point where `dispatch_get_specific` is helpful. Because it honors the queue target hierarchy, you can use it to perform many of the checks that were previously (incorrectly) handled by `dispatch_get_current_queue`. You can attach a "tag" to a queue you care about, and later you can determine whether that tag is part of the current target hierarchy. `RNQueueCreateTagged` creates a new "tagged" queue that can be searched for later:

RNQueue.m (ProducerConsumer)

```
static const char sQueueTagKey;
void RNQueueTag(dispatch_queue_t q) {
  // Make q point to itself by assignment.
  // This doesn't retain, but it can never dangle.
  dispatch_queue_set_specific(q, &sQueueTagKey, (__bridge void*)q, NULL);
}

dispatch_queue_t RNQueueCreateTagged(const char *label,
                                     dispatch_queue_attr_t attr) {
  dispatch_queue_t q = dispatch_queue_create(label, attr);
  RNQueueTag(q);
  return q;
}

// Usage (just like dispatch_queue_create):
// self.pendingQueue = RNQueueCreateTagged("ProducerConsumer.pending",
//                                         DISPATCH_QUEUE_SERIAL);
```

Now that the queue is tagged, you can check whether it is in the current target hierarchy by using `RNQueueCurrentIsTaggedQueue`:

```
dispatch_queue_t RNQueueGetCurrentTagged() {
  return (__bridge dispatch_queue_t)dispatch_get_specific(&sQueueTagKey);
}
```

```
BOOL RNQueueCurrentIsTaggedQueue(dispatch_queue_t q) {
  return (RNQueueGetCurrentTagged() == q);
}
```

And you can easily create an assertion to check this:

RNQueue.h (ProducerConsumer)

```
// Assert that the q is within the current queue hierarchy
#define RNAssertQueue(q) NSAssert(RNQueueCurrentIsTaggedQueue(q),
                                  @"Must run on queue: " #q )
// Usage:
// RNAssertQueue(self.pendingQueue);
```

> Note that this approach can fail if there are multiple tagged queues in the same hierarchy.

This code leads naturally to another common (but incorrect) use for `dispatch_get_current_queue`: safely running a synchronous block. If you dispatch a synchronous block onto the current queue, your program deadlocks. As the documentation notes, however, you cannot trust `dispatch_get_current_queue` for this use. Instead, you can use a tagged queue just as described and shown here:

```
void RNQueueSafeDispatchSync(dispatch_queue_t q, dispatch_block_t block) {
  if (RNQueueCurrentIsTaggedQueue(q)) {
    block();
  }
  else {
    dispatch_sync(q, block);
  }
}
```

Rather than rely on the "current" queue, you might find a better approach is to have the caller pass you the queue for callbacks, as you find in many asynchronous APIs in Cocoa. But for some problems, it is still very useful to determine whether a given queue is part of the current target hierarchy, and `dispatch_get_specific` can help.

Dispatch Data and Dispatch I/O

Dispatch data are immutable memory buffers that can be joined very quickly and split between blocks with minimal copying. This incredibly robust system is the basis for a feature called *dispatch I/O*, which provides significant I/O performance improvements on multicore iOS devices, and particularly on the Mac. However, in many cases, you get all the benefit free by using the higher-level abstractions without taking on the complexity of using dispatch I/O directly. These techniques are used when you need more control over your data.

In this section, I present a simple TCP client implemented with dispatch I/O. This client connects to an HTTP server and performs a GET. It then writes the full result (including HTTP headers) to a file. The full source is available in the DispatchDownload project on this book's website.

First, you need a queue to process dispatch I/O callbacks. Dispatch I/O is highly asynchronous and can handle many things in parallel. As it processes data, it dispatches blocks to the queue of your choice. You are free to use any convenient queue here. If your callbacks are simple, you can even safely use the main queue. If your callbacks are complicated, you may want to create your own queue. It can be serial or concurrent, depending on whether or not you want callbacks to be serialized.

ViewController.m (DispatchDownload)

```
- (void)HTTPDownloadContentsFromHostName:(NSString *)hostName
                                    port:(int)port
                                    path:(NSString *)path {

    dispatch_queue_t queue = dispatch_get_main_queue();
```

Next, you create a socket connection to the server using the low-level `socket` and `connect` calls and use that connection to create a *channel*:

```
    // Code for this method is available in the example project.
    // It just looks up the hostname and creates a socket connection.
    dispatch_fd_t socket = [self connectToHostName:hostName port:port];

    dispatch_io_t
    serverChannel = dispatch_io_create(DISPATCH_IO_STREAM, socket, queue,
                            ^(int error) {
                                NSAssert(!error,
                                        @"Failed socket:%d", error);
                                NSLog(@"Closing connection");
                                close(socket);
                            });
```

The first parameter (`DISPATCH_IO_STREAM`) determines if the channel is streaming or random access. Network sockets are always streaming. The second parameter is a file descriptor. (Sockets are just file handles in iOS.) The last parameter is a cleanup block, which is called when the channel relinquishes control over the file handle. You should never access the file handle directly between when you call `dispatch_io_create` and when the cleanup block executes. The third parameter is the queue for running the cleanup block.

Channels are generic. They don't care whether they are tied to an actual disk file, a pipe, or a network socket. They can work with anything that has a file descriptor.

After you create a channel, you need some data to write. Dispatch I/O works with dispatch objects. They are somewhat similar to `NSData` objects but are optimized to avoid memory copies. They are immutable and thread-safe. Because dispatch data objects are immutable, they can be combined very quickly just by creating a new dispatch data object that points to both. Similarly, creating a subrange just requires a pointer to an existing dispatch data object, an offset, and a length.

The tradeoff for this speed is some convenience. Getting access to the contents of a dispatch data object can be cumbersome. I discuss that issue more later in this example. For now, you can create a new dispatch data object like this:

```
- (dispatch_data_t)requestDataForHostName:(NSString *)hostName
                                     path:(NSString *)path {
   NSString *
   getString = [NSString
                stringWithFormat:@"GET %@ HTTP/1.1\r\nHost: %@\r\n\r\n",
                path, hostName];

   NSData *getData = [getString dataUsingEncoding:NSUTF8StringEncoding];
   return dispatch_data_create(getData.bytes,
                               getData.length,
                               NULL,
                               DISPATCH_DATA_DESTRUCTOR_DEFAULT);
}
```

The last parameter is a block for releasing the data. There are three built-in destructors: . . . DEFAULT, . . . FREE, and . . . MUNMAP. The DEFAULT destructor is special. It causes dispatch_data_create to copy the bytes passed. The dispatch data object then memory manages the buffer itself. The FREE destructor uses free to release memory that was allocated with malloc. The MUNMAP destructor uses munmap to release memory that was memory mapped to a file descriptor with mmap.

You can also pass a destructor block as the fourth parameter. When the dispatch data object is released, the block is placed on the queue passed in the third parameter. You can pass an empty block here if you do not need the dispatch_data_create to release the memory (if it is a static constant, for instance).

In requestDataForHostName:path:, the data is very small, so it is fine to let dispatch data copy it.

Now that you have something to write, you need somewhere to write it. For that, you create another channel. This is a file channel, created with dispatch_io_create_with_path. It takes parameters similar to open, as shown in the following code:

```
NSString *writePath = [self outputFilePathForPath:path];
dispatch_io_t
fileChannel = dispatch_io_create_with_path(DISPATCH_IO_STREAM,
                                           [writePath UTF8String],
                                           O_WRONLY|O_CREAT|O_TRUNC,
                                           S_IRWXU,
                                           queue,
                                           nil);
```

> Note that the queue parameter is required even if there is no cleanup block. If you pass NULL as the queue, the call crashes in iOS 7.0.2 (radar://15160726).

At this point, you have two channels: a server channel and a file channel. To download the file, you need to write a request to the server channel, then read from the server channel, strip off the header, and write the remainder to the file channel.

You start here by queuing a request to the server and returning. This is the end of the HTTPDownloadContents . . . method. The system asynchronously sends the request and calls the write handler. It may call the write handler multiple times to give progress. That is why you must check the `serverWriteDone` parameter to make sure that the write has completed. The code that sends the request is as follows:

```
dispatch_io_write(serverChannel, 0, requestData, queue,
  ^(bool serverWriteDone,
    dispatch_data_t serverWriteData,
    int serverWriteError) {

  NSAssert(!serverWriteError,
           @"Server write error:%d", serverWriteError);
  if (serverWriteDone) {
    [self readFromChannel:serverChannel
           writeToChannel:fileChannel
                    queue:queue];
  }
});
```

When this write is done, the system moves on to reading from the server channel by calling `readFromChannel:writeToChannel:queue:`. This method has to deal with three cases: the time before the HTTP header is completely read, the time after the HTTP header is completely read, and the time when the file is finished downloading.

```
__block dispatch_data_t previousData = dispatch_data_empty;
__block dispatch_data_t headerData;
dispatch_io_read(readChannel, 0, SIZE_MAX, queue,
  ^(bool serverReadDone,
    dispatch_data_t serverReadData,
    int serverReadError) {

  NSAssert(!serverReadError,
           @"Server read error:%d", serverReadError);
  if (serverReadDone) {
    [self handleDoneWithChannels:@[writeChannel,
                                   readChannel];
  }
  else {
    if (! headerData) {
      headerData = [self findHeaderInData:serverReadData
                            previousData:&previousData
                            writeChannel:writeChannel
                                   queue:queue];
      if (headerData) {
        // See the sample code for this simple method
        [self printHeader:headerData];
      }
    }
    else {
      [self writeToChannel:writeChannel
                      data:serverReadData
                     queue:queue];
    }
  }
});
}
```

The `serverReadDone` case is trivial. At this point, you have read everything and written everything. You just need to close the channels like this:

```
- (void)handleDoneWithChannels:(NSArray *)channels {
  NSLog(@"Done Downloading");
  for (dispatch_io_t channel in channels) {
    dispatch_io_close(channel, 0);
  }
}
```

Writing the data after you've found the header is also fairly straightforward. You call `dispatch_io_write` as discussed earlier in `HTTPDownloadContentsFromHostName:`. Almost all the code in this routine is to calculate the number of bytes so you can log it. Without error handling or progress reporting, this method would be one line long.

```
- (void)writeToChannel:(dispatch_io_t)channel
                  data:(dispatch_data_t)writeData
                 queue:(dispatch_queue_t)queue {

  dispatch_io_write(channel, 0, writeData, queue,
    ^(bool done, dispatch_data_t remainingData, int error) {
      NSAssert(!error, @"File write error:%d", error);
      size_t unwrittenDataLength = 0;
      if (remainingData) {
        unwrittenDataLength = dispatch_data_get_size(remainingData);
      }
      NSLog(@"Wrote %zu bytes",
            dispatch_data_get_size(writeData) - unwrittenDataLength);
    });
}
```

The most complicated part of the system is finding the HTTP header and splitting it off from the contents. The header is terminated with two blank lines (`\r\n\r\n`). The problem is that these four bytes may be split across two blocks, so you can't simply search for these bytes in the current block.

You can employ several solutions to this problem. In this example, you use a simple approach, which is to keep track of the last block, attach it to the current block, and then search the resulting memory. This is not the most efficient solution, however. It requires a memory copy. In this particular case, because the header is likely very short, it is unlikely that you need to copy much memory. In the most common case, you will not need to copy any memory because the header terminator will come in the first block. But in other cases, this memory copy would be excessive and you would need to develop a more complicated solution. For instance, you could keep track of just the last few bytes of the previous block and check them together with the first few bytes of the current block.

```
- (dispatch_data_t)findHeaderInData:(dispatch_data_t)newData
                       previousData:(dispatch_data_t *)previousData
                       writeChannel:(dispatch_io_t)writeChannel
                              queue:(dispatch_queue_t)queue {
```

(continued)

```
// Glue the previous data to new data. This is a cheap operation.
*previousData = dispatch_data_create_concat(*previousData,
                                            newData);

// Create a contiguous memory region. This requires a memory copy.
dispatch_data_t mappedData = dispatch_data_create_map(*previousData,
                                                      NULL, NULL);


__block dispatch_data_t headerData;
__block dispatch_data_t bodyData;

// The dispatch_data_apply is unnecessary; we could have gotten
// buffer and size from dispatch_data_create_map, but this shows
// how to use dispatch_data_apply, and there are subtle ARC issues
// with accessing the values through dispatch_data_create_map. We
// have to be careful that mappedData doesn't get released immediately.

dispatch_data_apply(mappedData,
  ^bool(dispatch_data_t region, size_t offset,
        const void *buffer, size_t size) {

    // We know that region is all the data because we just mapped it.
    // Convert it into an NSData for simpler searching.
    NSData *search =
    [[NSData alloc] initWithBytesNoCopy:(void*)buffer
                                 length:size
                            freeWhenDone:NO];
    NSRange r =
    [search rangeOfData:[self headerDelimiter]
                options:0
                  range:NSMakeRange(0,
                                    search.length)];

    // If we found the delimiter, split into header and body
    if (r.location != NSNotFound) {
      headerData = dispatch_data_create_subrange(region,
                                                 0,
                                                 r.location);
      size_t body_offset = NSMaxRange(r);
      size_t body_size = size - body_offset;
      bodyData = dispatch_data_create_subrange(region,
                                               body_offset,
                                               body_size);
    }

    // We only need to process one block
    return false;
  });

if (bodyData) {
```

```
    [self writeToChannel:writeChannel
                     data:bodyData
                    queue:queue];
    }
    return headerData;
}
```

This example covers the main topics you need to work with dispatch data and dispatch I/O. Dispatch I/O supports a few more features to help manage how often your handler block is called. If you configure an interval, your handler is called periodically even if there is not much data. This is helpful for updating progress bars if the network is slow. High and low water marks control how much data can be passed to your handler. Your handler won't be called with less data than the low water mark unless the channel reaches the end or there is an interval configured with `DISPATCH_IO_STRICT_INTERVAL`. Your handler will never be called with more than the high water mark, which is useful if you have fixed-sized buffers.

Dispatch data and dispatch I/O are powerful tools for getting the most performance out of data-intensive operations. Much of this performance, however, is included in higher-level objects such as `NSURLSession`, but when you need to control memory very closely, dispatch data and dispatch I/O are the right tools for the job.

Summary

There is more to Grand Central Dispatch than `dispatch_async`. In this chapter you have seen how to use some of the more advanced features, including semaphores, sources, and dispatch I/O. You've also learned the subtle problems with the concept of "the current queue," and how to use queue-specific data to replace deprecated calls to `dispatch_get_current_queue`. You how have a strong foundation in Grand Central Dispatch and can explore the documentation yourself to learn more.

Further Reading

Apple Documentation

The following documents are available in the iOS Developer Library at `developer.apple.com` or through the Xcode Documentation and API Reference.

File System Programming Guide. "Techniques for Reading and Writing Files." The section "Processing a File Using GCD" includes sample code explaining dispatch I/O channels.

Grand Central Dispatch (GCD) Reference.

Chapter 24
Deep Objective-C

Much of Objective-C is straightforward in practice. There is no multiple inheritance or operator overloading as in C++. All objects have the same memory-management rules, which rely on a simple set of naming conventions. With the addition of Automatic Reference Counting (ARC), you don't even need to worry about memory management in most cases. The Cocoa framework is designed with readability in mind, so most things do exactly what they say they do.

Still, many parts of Objective-C appear mysterious until you dig into them, such as creating new methods and classes at runtime, introspecting the class hierarchy, and message passing. Most of the time, you don't need to understand how this language works, but for some problems, it's very useful to harness the full power of Objective-C. For example, the flexibility of Core Data relies heavily on the dynamic nature of Objective-C.

The heart of this power is the *Objective-C runtime*, provided by `libobjc`. The Objective-C runtime is a collection of functions that provide the dynamic features of Objective-C. It includes such core functions as `objc_msgSend`, which is called every time you use the `[object message]` syntax. It also includes functions to allow you to inspect and modify the class hierarchy at runtime, including creating new classes and methods.

This chapter shows you how to use these features to achieve the same kind of flexibility, power, and speed as Core Data and other Apple frameworks. All code samples in this chapter can be found in the online files for Chapter 24.

> **The 64-bit version of iOS 7 (arm64) includes significant changes to many of the implementation details discussed in this section. Unless otherwise noted, these changes should have no impact on code that uses the published API. The source code for the arm64 runtime was not available at the time of this writing, and internal implementation details are, by definition, undocumented. If you want to know some of the details, the best source is Greg Parker's blog, *Hamster Emporium* (sealiesoftware.com/blog).**

Understanding Classes and Objects

The first thing to understand about Objective-C objects is that they're really C structs. Every Objective-C object has the same layout, as shown in Figure 24-1.

First, notice the pointer to your class definition. Then each of your superclasses' ivars (instance variables) are laid out as struct properties, and then your class's ivars are laid out as struct properties. This structure is called `objc_object`, and a pointer to it is called `id`:

```
typedef struct objc_object {
    Class isa;
} *id;
```

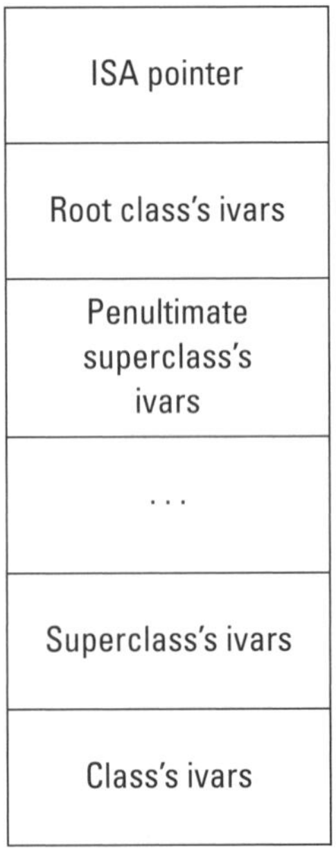

Figure 24-1 Layout of an Objective-C object

The `Class` structure contains a metaclass pointer (more on that in a moment), a superclass pointer, and data about the class. The data of particular interest are the name, ivars, methods, properties, and protocols. Don't worry too much about the internal structure of `Class`. There are public functions to access all the information you need.

> The Objective-C runtime is open source, so you can see exactly how it's implemented. Go to the Apple Open Source site (`www.opensource.apple.com`), and look for the package `objc` in the Mac code. It isn't included in the iOS packages, but the Mac code is identical or very similar. These particular structures are defined in `objc.h` and `objc-runtime-new.h`. You will notice two definitions of many things in these files because of the switch from Objective-C 1.0 to Objective-C 2.0. Look for things marked "new" when there is a conflict.

`Class` is itself much like an object. You can send messages to a `Class` instance—for example, when you call `[Foo alloc]`—so you need a place to store the list of class methods. They are stored in the metaclass, which is the `isa` pointer for a `Class`. It's extremely rare to need to access metaclasses, so I don't dwell on them here; see the "Further Reading" section at the end of this chapter for links to more information. Also see the section "How Message Passing Really Works," later in this chapter, for more information on message passing.

The superclass pointer creates the hierarchy of classes, and the list of ivars, methods, properties, and protocols defines what the class can do. An important point here is that the methods, properties, and protocols are all stored in the writable section of the class definition. They can be changed at runtime, and that's exactly how categories are implemented. Ivars are stored in the read-only section and cannot be modified (because that could impact existing instances). That's why categories cannot add ivars.

In the definition of `objc_object`, shown at the beginning of this section, notice that the `isa` pointer is not `const`. That is not an oversight. The class of an object can be changed at runtime. The superclass pointer of `Class` is also not `const`. The hierarchy can be modified. This topic is covered in more detail in the "ISA Swizzling" section later in this chapter.

> **In iOS 7, the** `objc_object` **structure is more complicated. The** `isa` **pointer is deprecated, and in 64-bit iOS, it is not actually a pointer. You should not rely on the internal implementation details of this structure. See "[objc explain]: Non-pointer isa" in the "Further Reading" section for more details. ISA swizzling, as discussed in this chapter, is still supported in iOS 7.**

Now that you've seen the data structures underlying Objective-C objects, you next look at the kinds of functions you can use to inspect and manipulate them. These functions are written in C, and they use naming conventions somewhat similar to Core Foundation (see Chapter 10). All the functions shown here are public and are documented in the *Objective-C Runtime Reference*. The following is the simplest example:

```
#import <objc/objc-runtime.h>
...
const char *name = class_getName([NSObject class]);
printf("%s\n", name);
```

Runtime methods begin with the name of the thing they act upon, which is almost always also their first parameter. Because this example includes `get` rather than `copy`, you don't own the memory that is returned to you and should not call `free`.

The next example prints a list of the selectors that `NSObject` responds to. The call to `class_copyMethodList` returns a copied buffer that you must dispose of by using `free`.

PrintObjectMethods.m (Runtime)

```
void PrintObjectMethods() {
  unsigned int count = 0;
  Method *methods = class_copyMethodList([NSObject class],
                                         &count);
  for (unsigned int i = 0; i < count; ++i) {
    SEL sel = method_getName(methods[i]);
    const char *name = sel_getName(sel);
    printf("%s\n", name);
  }
  free(methods);
}
```

Because there is no reference counting (automatic or otherwise) in the runtime, there is no equivalent to `retain` or `release`. If you fetch a value with a function that includes the word `copy`, you should call `free` on it. If you use a function that does not include the word `copy`, you must not call `free` on it.

Working with Methods and Properties

The Objective-C runtime defines several important types:

- `Class`—Defines an Objective-C class, as described in the previous section, "Understanding Classes and Objects."

- `Ivar`—Defines an instance variable of an object, including its type and name.

- `Protocol`—Defines a formal protocol.

- `objc_property_t`—Defines a property. Its unusual name is probably to avoid colliding with user types defined in Objective-C 1.0 before properties existed.

- `Method`—Defines an object method or a class method. This provides the name of the method (its *selector*), the number and types of parameters it takes and its return type (collectively its *signature*), and a function pointer to its code (its *implementation*).

- `SEL`—Defines a selector. A selector is a unique identifier for the name of a method.

- `IMP`—Defines a method implementation. It's just a pointer to a function that takes an object, a selector, and a variable list of other parameters (varargs), and returns an object:

```
typedef id (*IMP)(id, SEL, ...);
```

Now you use this knowledge to build your own simplistic *message dispatcher*. A message dispatcher maps selectors to function pointers and calls the referenced function. The heart of the Objective-C runtime is the message dispatcher `objc_msgSend`, which you learn much more about in the "How Message Passing Really Works" section. The sample `myMsgSend` is how `objc_msgSend` might be implemented if it needed to handle only the simplest cases.

The following code is written in C just to prove that the Objective-C runtime is really just C. I added comments to demonstrate the equivalent Objective-C.

MyMsgSend.c (Runtime)

```c
static const void *myMsgSend(id receiver, const char *name) {
  SEL selector = sel_registerName(name);
  IMP methodIMP =
  class_getMethodImplementation(object_getClass(receiver),
                                selector);
  return methodIMP(receiver, selector);
}

void RunMyMsgSend() {
  // NSObject *object = [[NSObject alloc] init];
  Class class = (Class)objc_getClass("NSObject");
  id object = class_createInstance(class, 0);
  myMsgSend(object, "init");

  // id description = [object description];
  id description = (id)myMsgSend(object, "description");

  // const char *cstr = [description UTF8String];
```

```
    const char *cstr = myMsgSend(description, "UTF8String");

    printf("%s\n", cstr);
}
```

You can use this same technique in Objective-C using `methodForSelector:` to avoid the complex message dispatch of `objc_msgSend`. This approach makes sense only if you're going to call the same method thousands of times on an iPhone. On a Mac, you won't see much improvement unless you're calling the same method millions of times. Apple has highly optimized `objc_msgSend`. But for very simple methods called many times, you may be able to improve performance 5% to 10% this way.

The following example demonstrates how to bypass `objc_msgSend` and shows the performance impact:

FastCall.m (Runtime)

```
const NSUInteger kTotalCount = 10000000;

typedef void (*voidIMP)(id, SEL, ...);

void FastCall() {
  NSMutableString *string = [NSMutableString string];
  NSTimeInterval totalTime = 0;
  NSDate *start = nil;
  NSUInteger count = 0;

  // With objc_msgSend
  start = [NSDate date];
  for (count = 0; count < kTotalCount; ++count) {
    [string setString:@"stuff"];
  }

  totalTime = -[start timeIntervalSinceNow];
  printf("w/ objc_msgSend = %f\n", totalTime);

  // Skip objc_msgSend.
  start = [NSDate date];
  SEL selector = @selector(setString:);
  voidIMP
  setStringMethod = (voidIMP)[string methodForSelector:selector];

  for (count = 0; count < kTotalCount; ++count) {
    setStringMethod(string, selector, @"stuff");
  }

  totalTime = -[start timeIntervalSinceNow];
  printf("w/o objc_msgSend  = %f\n", totalTime);
}
```

Be careful with this technique. If you use it incorrectly, it can actually be slower than using normal message dispatch. Because `IMP` returns an `id`, ARC will retain and later release the return value, even though this specific method returns nothing (see http://openradar.appspot.com/10002493 for details). That overhead is more expensive than just using the normal messaging system. In some cases, the extra `retain` can cause a crash. That's why you have to add the extra `voidIMP` type. When you declare that the `setStringMethod` function pointer returns `void`, the compiler skips the `retain`.

The important take-away is that you need to do testing on anything you do to improve performance. Don't assume that bypassing the message dispatcher is going to be faster. In most cases, you can get much better and more reliable performance improvements by simply rewriting your code as a function rather than a method. And in the vast majority of cases, `objc_msgSend` is the least of your performance overhead.

Method Signatures and Invocations

`NSInvocation` is a traditional implementation of the Command pattern. It bundles a target, a selector, a method signature, and all the parameters into an object that can be stored and invoked at a later time. When the invocation is invoked, it sends the message, and the Objective-C runtime finds the correct method implementation to execute.

A *method implementation* (`IMP`) is a function pointer to a C function with the following signature:

```
id function(id self, SEL _cmd, ...)
```

Every method implementation takes two parameters: `self` and `_cmd`. The first parameter is the `self` pointer that you're familiar with. The second parameter, `_cmd`, is the selector that was sent to this object. This reserved symbol in the language is accessed exactly like `self`.

> Although the `IMP` typedef suggests that every Objective-C method returns an `id`, obviously many Objective-C methods return other types such as integers or floating-point numbers, and many Objective-C methods return nothing at all. The actual return type is defined by the message signature, discussed later in this section, not the `IMP` typedef.

`NSInvocation` includes a target and a selector. A target is the object to send the message to, and the selector is the message to send. A selector is roughly the name of a method. I say "roughly" because selectors don't have to map exactly to methods. A selector is just a name, like `initWithBytes:length:encoding:`. A selector isn't bound to any particular class or any particular return value or parameter types. It isn't even specifically a class or instance selector. You can think of a selector as a string. So `- [NSString length]` and `- [NSData length]` have the same selector, even though they map to different methods' implementations.

`NSInvocation` also includes a method signature (`NSMethodSignature`), which encapsulates the return type and the parameter types of a method. An `NSMethodSignature` does not include the name of a method, only the return value and the parameters. Here is how you can create one by hand:

```
NSMethodSignature *sig =
            [NSMethodSignature signatureWithObjCTypes:"@@:*"];
```

This code is the signature for `- [NSString initWithUTF8String:]`. The first character (@) indicates that the return value is an `id`. To the message passing system, all Objective-C objects are the same. It can't tell the difference between an `NSString` and an `NSArray`. The next two characters (@:) indicate that this method takes an id and a SEL. As discussed previously, every Objective-C method takes these as its first two parameters. They're implicitly passed as self and _cmd. Finally, the last character (*) indicates that the first "real" parameter is a character string (char*).

If you do work with type encoding directly, you can use @encode(type) **to get the string that represents that type rather than hard-coding the letter. For example,** @encode(id) **is the string "@".**

You should seldom call signatureWithObjCTypes:. I do it here only to show that it's possible to build a method signature by hand. The way you generally get a method signature is to ask a class or instance for it. Before you do that, you need to consider whether the method is an instance method or a class method. The method −init is an instance method and is marked with a leading hyphen (-). The method +alloc is a class method and is marked with a leading plus (+). You can request instance method signatures from instances and class method signatures from classes by using methodSignatureForSelector:. If you want the instance method signature from a class, you use instanceMethodSignatureForSelector:. The following example demonstrates this for +alloc and −init:

```
SEL initSEL = @selector(init);
SEL allocSEL = @selector(alloc);
NSMethodSignature *initSig, *allocSig;

// Instance method signature from instance
initSig = [@"String" methodSignatureForSelector:initSEL];

// Instance method signature from class
initSig = [NSString instanceMethodSignatureForSelector:initSEL];

// Class method signature from class
allocSig = [NSString methodSignatureForSelector:allocSEL];
```

If you compare initSig and allocSig, you will discover that they are the same. They each take no additional parameters (besides self and _cmd) and return an id. This is all that matters to the message signature.

Now that you have a selector and a signature, you can combine them with a target and parameter values to construct an NSInvocation. An NSInvocation bundles everything needed to pass a message. Here is how you create an invocation of the message [set addObject:stuff] and invoke it:

```
NSMutableSet *set = [NSMutableSet set];
NSString *stuff = @"Stuff";
SEL selector = @selector(addObject:);
NSMethodSignature *sig = [set methodSignatureForSelector:selector];

NSInvocation *invocation =
    [NSInvocation invocationWithMethodSignature:sig];
[invocation setTarget:set];
[invocation setSelector:selector];
// Place the first argument at index 2.
[invocation setArgument:&stuff atIndex:2];
[invocation invoke];
```

Note that the first argument is placed at index 2. As discussed previously, index 0 is the target (`self`), and index 1 is the selector (`_cmd`). `NSInvocation` sets these automatically. Also note that you must pass a pointer to the argument, not the argument itself.

Invocations are extremely flexible, but they're not fast. Creating an invocation is hundreds of times slower than passing a message. Invoking an invocation is efficient, however, and invocations can be reused. They can be dispatched to different targets using `invokeWithTarget:` or `setTarget:`. You can also change their parameters between uses. Much of the cost of creating an invocation is in `methodSignatureForSelector:`, so caching this result can significantly improve performance.

Invocations do not retain their object arguments by default, nor do they make a copy of C string arguments. To store the invocation for later use, you call `retainArguments` on it. This method retains all object arguments and copies all C string arguments. When the invocation is released, it releases the objects and frees its copies of the C strings. Invocations do not provide any handling for pointers other than Objective-C objects and C strings. If you're passing raw pointers to an invocation, you're responsible for managing the memory yourself.

> If you use an invocation to create an `NSTimer`, such as by using `timerWithTimeInterval:invocation:repeats:`, the timer automatically calls `retainArguments` on the invocation.

Invocations are a key part of the Objective-C message dispatching system, discussed later in this chapter. This integration with the message dispatching system makes them central to creating trampolines.

Using Trampolines

A *trampoline* "bounces" a message from one object to another. This technique allows a proxy object to move messages to another thread, cache results, coalesce duplicate messages, or perform any other intermediary processing you'd like. Trampolines generally use `forwardInvocation:` to handle arbitrary messages. If an object does not respond to a selector, before Objective-C throws an error, it creates an `NSInvocation` and passes it to the object's `forwardInvocation:`. You can use this trampoline to forward the message in any way that you'd like.

In this example, you create a trampoline called `RNObserverManager`. Any message sent to the trampoline is forwarded to registered observers that respond to that selector. This provides functionality similar to `NSNotification`, but is easier to use and faster if there are many observers.

The public interface for `RNObserverManager` is as follows:

RNObserverManager.h (ObserverTrampoline)

```
#import <objc/runtime.h>
@interface RNObserverManager: NSObject

- (id)initWithProtocol:(Protocol *)protocol
            observers:(NSSet *)observers;
```

```
- (void) addObserver: (id) observer;
- (void) removeObserver: (id) observer;

@end
```

You initialize this trampoline with a protocol and an initial set of observers. You can then add or remove observers. Any method defined in the protocol is forwarded to all the current observers if they implement it.

Here is the skeleton implementation for `RNObserverManager`, without the trampoline piece. Everything should be fairly obvious.

RNObserverManager.m (ObserverTrampoline)

```
@interface RNObserverManager ()
@property (nonatomic, readonly, strong)
                                 NSMutableSet *observers;
@property (nonatomic, readonly, strong) Protocol *protocol;
@end

@implementation RNObserverManager
- (id) initWithProtocol: (Protocol *) protocol
               observers: (NSSet *) observers {
  if ((self = [super init])) {
    _protocol = protocol;
    _observers = [NSMutableSet setWithSet:observers];
  }
  return self;
}

- (void) addObserver: (id) observer {
  NSAssert([observer conformsToProtocol:self.protocol],
           @"Observer must conform to protocol.");
  [self.observers addObject:observer];
}

- (void) removeObserver: (id) observer {
  [self.observers removeObject:observer];
}
@end
```

Now you override `methodSignatureForSelector:`. The Objective-C message dispatcher uses this method to construct an `NSInvocation` for unknown selectors. You override it to return method signatures for methods defined in `protocol`, using `protocol_getMethodDescription`. You need to get the method signature from the protocol rather than from the observers because the method may be optional, and the observers might not implement it.

```
- (NSMethodSignature *) methodSignatureForSelector: (SEL) sel
{
  // Check the trampoline itself
  NSMethodSignature *
  result = [super methodSignatureForSelector:sel];
  if (result) {
```

(continued)

```
      return result;
   }

   // Look for a required method
   struct objc_method_description desc =
            protocol_getMethodDescription(self.protocol,
                                  sel, YES, YES);
   if (desc.name == NULL) {
      // Couldn_t find it. Maybe it's optional
      desc = protocol_getMethodDescription(self.protocol,
                                  sel, NO, YES);
   }

   if (desc.name == NULL) {
      // Couldn_t find it. Raise NSInvalidArgumentException
      [self doesNotRecognizeSelector: sel];
      return nil;
   }

   return [NSMethodSignature
                     signatureWithObjCTypes:desc.types];
}
```

Finally, you override `forwardInvocation:` to forward the invocation to the observers that respond to the selector:

```
- (void)forwardInvocation:(NSInvocation *)invocation {
   SEL selector = [invocation selector];
   for (id responder in self.observers) {
      if ([responder respondsToSelector:selector]) {
         [invocation setTarget:responder];
         [invocation invoke];
      }
   }
}
```

To use this trampoline, you create an instance, set the observers, and then send messages to it as the following code shows. Variables that hold a trampoline should generally be of type `id` so that you can send any message to it without generating a compiler warning.

```
@protocol MyProtocol <NSObject>
- (void)doSomething;
@end

...
id observerManager = [[RNObserverManager alloc]
                  initWithProtocol:@protocol(MyProtocol)
                        observers:observers];
[observerManager doSomething];
```

Passing a message to this trampoline is similar to posting a notification. You can use this technique to solve a variety of problems. For example, you can create a proxy trampoline that forwards all messages to the main thread as shown here:

RNMainThreadTrampoline.h (ObserverTrampoline)

```
@interface RNMainThreadTrampoline : NSObject
@property (nonatomic, readwrite, strong) id target;
- (id)initWithTarget:(id)aTarget;
@end
```

RNMainThreadTrampoline.m (ObserverTrampoline)

```
@implementation RNMainThreadTrampoline
- (id)initWithTarget:(id)aTarget {
  if ((self = [super init])) {
    _target = aTarget;
  }
  return self;
}
- (NSMethodSignature *)methodSignatureForSelector:(SEL)sel
{
  return [self.target methodSignatureForSelector:sel];
}

- (void)forwardInvocation:(NSInvocation *)invocation {
  [invocation setTarget:self.target];
  [invocation retainArguments];
  [invocation performSelectorOnMainThread:@selector(invoke)
                               withObject:nil
                            waitUntilDone:NO];
}
@end
```

`forwardInvocation:` can transparently coalesce duplicate messages, add logging, forward messages to other machines, and perform a wide variety of other functions.

How Message Passing Really Works

As demonstrated in the "Working with Methods and Properties" section, earlier in this chapter, calling a method in Objective-C eventually translates into calling a method implementation function pointer and passing it an object pointer, a selector, and a set of function parameters. Like the sample `myMsgSend`, every Objective-C message expression is converted into a call to `objc_msgSend` (or a closely related function; I get to that in "The Flavors of `objc_msgSend`," later in this chapter). However, `objc_msgSend` is much more powerful than `myMsgSend`. Here is how it works:

1. Check whether the receiver is `nil`. If so, call the `nil`-handler. This is really obscure, undocumented, unsupported, and difficult to make useful. The default is to do nothing, and I don't go into more detail here. See the "Further Reading" section for more information.

2. In a garbage-collected environment (which iOS doesn't support, but I include for completeness), check for one of the short-circuited selectors (`retain`, `release`, `autorelease`, `retainCount`), and if it matches, return `self`. Yes, that means `retainCount` returns `self` in a garbage-collected environment. You shouldn't have been calling it anyway.

3. Check the class's cache to see if it's already worked out this method implementation. If so, call it.

4. Compare the requested selector to the selectors defined in the class. If the selector is found, call its method implementation.

5. Compare the requested selector to the selectors defined in the superclass, and then its superclass, and so on. If the selector is found, call its method implementation.

6. Call `resolveInstanceMethod:` (or `resolveClassMethod:`). If it returns `YES`, start over. The object is promising that the selector will resolve this time, generally because it has called `class_addMethod`.

7. Call `forwardingTargetForSelector:`. If it returns non-`nil`, send the message to the returned object. Don't return `self` here. That would be an infinite loop.

8. Call `methodSignatureForSelector:`, and if it returns non-`nil`, create an `NSInvocation` and pass it to `forwardInvocation:`.

9. Call `doesNotRecognizeSelector:`. The default implementation of this just throws an exception.

Some changes have been made to how this process works in the 64-bit version of iOS 7. At the time of writing, these changes are not well documented. See *Hamster Emporium* in the "Further Reading" section at the end of this chapter for more information.

Dynamic Implementations

The first interesting thing you can do with message dispatch is provide an implementation at runtime by using `resolveInstanceMethod:` and `resolveClassMethod:`. These methods are a common way `@dynamic` synthesis is handled. When you declare a property to be `@dynamic`, you're promising the compiler that an implementation will be available at runtime, even though the compiler can't find one now. Using `@dynamic` prevents it from automatically synthesizing an ivar.

Here's an example of how to use this implementation to dynamically create getters and setters for properties stored in an `NSMutableDictionary`:

Person.h (Person)

```
@interface Person : NSObject
@property (copy) NSString *givenName;
@property (copy) NSString *surname;
@end
```

Person.m (Person)

```
@interface Person ()
@property (strong) NSMutableDictionary *properties;
@end

@implementation Person
@dynamic givenName, surname;
```

```objc
- (id)init {
  if ((self = [super init])) {
    _properties = [[NSMutableDictionary alloc] init];
  }
  return self;
}

static id propertyIMP(id self, SEL _cmd) {
  return [[self properties] valueForKey:
          NSStringFromSelector(_cmd)];
}

static void setPropertyIMP(id self, SEL _cmd, id aValue) {
  id value = [aValue copy];

  NSMutableString *key =
  [NSStringFromSelector(_cmd) mutableCopy];

  // Delete "set" and ":" and lowercase first letter
  [key deleteCharactersInRange:NSMakeRange(0, 3)];
  [key deleteCharactersInRange:
                        NSMakeRange([key length] - 1, 1)];
  NSString *firstChar = [key substringToIndex:1];
  [key replaceCharactersInRange:NSMakeRange(0, 1)
                  withString:[firstChar lowercaseString]];

  [[self properties] setValue:value forKey:key];
}
+ (BOOL)resolveInstanceMethod:(SEL)aSEL {
  if ([NSStringFromSelector(aSEL) hasPrefix:@"set"]) {
    class_addMethod([self class], aSEL,
                    (IMP)setPropertyIMP, "v@:@");
  }
  else {
    class_addMethod([self class], aSEL,
                    (IMP)propertyIMP, "@@:");
  }
  return YES;
}
@end
```

main.m (Person)

```objc
int main(int argc, char *argv[]) {
  @autoreleasepool {
    Person *person = [[Person alloc] init];
    [person setGivenName:@"Bob"];
    [person setSurname:@"Jones"];

    NSLog(@"%@ %@", [person givenName], [person surname]);
  }
}
```

In this example, you use `propertyIMP` as the generic getter and `setPropertyIMP` as the generic setter. Note how these functions make use of the selector to determine the name of the property. Also note that `resolveInstanceMethod:` assumes that any unrecognized selector is a property setter or getter. In many cases, this is okay. You still get compiler warnings if you pass unknown methods like this:

```
[person addObject:@"Bob"];
```

But if you pass unknown methods this way, you get a slightly surprising result:

```
NSArray *persons = [NSArray arrayWithObject:person];
id object = [persons objectAtIndex:0];
[object addObject:@"Bob"];
```

You get no compiler warning because you can send any message to `id`. And you don't get a runtime error either. You just retrieve the key `addObject:` (including the colon) from the `properties` dictionary and do nothing with it. This kind of bug can be difficult to track down, and you may want to add additional checking in `resolveInstanceMethod:` to guard against it. But the approach is extremely powerful. Although dynamic getters and setters are the most common use of `resolveInstanceMethod:`, you can also use it to dynamically load code in environments that allow dynamic loading. iOS doesn't allow this approach, but on the Mac you can use `resolveInstanceMethod:` to avoid loading entire libraries until the first time one of the library's classes is accessed. This approach can be useful for large but rarely used classes.

Fast Forwarding

The runtime gives you one more fast option before falling back to the standard forwarding system. You can implement `forwardingTargetForSelector:` and return another object to pass the message to. This option is particularly useful for proxy objects or objects that add functionality to another object. The `CacheProxy` example demonstrates an object that caches the getters and setters for another object.

CacheProxy.h (Person)

```
@interface CacheProxy : NSProxy
- (id)initWithObject:(id)anObject
          properties:(NSArray *)properties;
@end

@interface CacheProxy ()
@property (readonly, strong) id object;
@property (readonly, strong)
                    NSMutableDictionary *valueForProperty;
@end
```

`CacheProxy` is a subclass of `NSProxy` rather than `NSObject`. `NSProxy` is a very thin root class designed for classes that forward most of their methods, particularly classes that forward their methods to objects hosted on another machine or on another thread. It's not a subclass of `NSObject`, but it does conform to the `<NSObject>` protocol. The `NSObject` class implements dozens of methods that might be very hard to proxy. For example, methods that require the local run loop, such as `performSelector:withObject:afterDelay:`, might not make sense for a proxied object. `NSProxy` avoids most of these methods.

To implement a subclass of NSProxy, you must override methodSignatureForSelector: and forwardInvocation:. These throw exceptions if they're called otherwise.

First, you need to create the getter and setter implementations, as in the Person example. In this case, if the value is not found in the local cache dictionary, you forward the request to the proxied object, as shown here:

CacheProxy.m (Person)

```objc
@implementation CacheProxy

// setFoo: => foo
static NSString *propertyNameForSetter(SEL selector) {
  NSMutableString *name =
  [NSStringFromSelector(selector) mutableCopy];
  [name deleteCharactersInRange:NSMakeRange(0, 3)];
  [name deleteCharactersInRange:
                      NSMakeRange([name length] - 1, 1)];
  NSString *firstChar = [name substringToIndex:1];
  [name replaceCharactersInRange:NSMakeRange(0, 1)
                withString:[firstChar lowercaseString]];
  return name;
}
// foo => setFoo:
static SEL setterForPropertyName(NSString *property) {
  NSMutableString *name = [property mutableCopy];
  NSString *firstChar = [name substringToIndex:1];
  [name replaceCharactersInRange:NSMakeRange(0, 1)
                withString:[firstChar uppercaseString]];
  [name insertString:@"set" atIndex:0];
  [name appendString:@":"];
  return NSSelectorFromString(name);
}

// Getter implementation
static id propertyIMP(id self, SEL _cmd) {
  NSString *propertyName = NSStringFromSelector(_cmd);
  id value = [[self valueForProperty] valueForKey:propertyName];
  if (value == [NSNull null]) {
    return nil;
  }

  if (value) {
    return value;
  }

  value = [[self object] valueForKey:propertyName];
  [[self valueForProperty] setValue:value
                            forKey:propertyName];
  return value;
}

// Setter implementation
static void setPropertyIMP(id self, SEL _cmd, id aValue) {
```

(continued)

```
    id value = [aValue copy];
    NSString *propertyName = propertyNameForSetter(_cmd);
    [[self valueForProperty] setValue:(value != nil ? value :
                                          [NSNull null])
                             forKey:propertyName];
    [[self object] setValue:value forKey:propertyName];
}
```

Note the use of [NSNull null] to manage nil values. You cannot store nil in an NSDictionary. In the next block of code, you synthesize accessors for the properties requested. All other methods are forwarded to the proxied object.

```
- (id)initWithObject:(id)anObject
          properties:(NSArray *)properties {
    _object = anObject;
    _valueForProperty = [[NSMutableDictionary alloc] init];
    for (NSString *property in properties) {
        // Synthesize a getter
        class_addMethod([self class],
                        NSSelectorFromString(property),
                        (IMP)propertyIMP,
                        "@@:");
        // Synthesize a setter
        class_addMethod([self class],
                        setterForPropertyName(property),
                        (IMP)setPropertyIMP,
                        "v@:@");
    }
    return self;
}
```

The next block of code overrides methods that are implemented by NSProxy. Because NSProxy has default implementations for these methods, they aren't automatically forwarded by forwardingTargetForSelector:

```
- (NSString *)description {
    return [NSString stringWithFormat:@"%@ (%@)",
            [super description], self.object];
}

- (BOOL)isEqual:(id)anObject {
    return [self.object isEqual:anObject];
}

- (NSUInteger)hash {
    return [self.object hash];
}

- (BOOL)respondsToSelector:(SEL)aSelector {
    return [self.object respondsToSelector:aSelector];
}
```

```
-  (BOOL)isKindOfClass:(Class)aClass {
    return [self.object isKindOfClass:aClass];
}
```

Finally, you implement the forwarding methods. Each of them simply passes unknown messages to the proxied object. See the "Method Signatures and Invocations" section, earlier in this chapter, for more details.

Whenever an unknown selector is sent to `CacheProxy`, `objc_msgSend` calls `forwardingTarget ForSelector:`. If it returns an object, then `objc_msgSend` tries to send the selector to that object. This process is called *fast forwarding*. In this example, `CacheProxy` sends all unknown selectors to the proxied object. If the proxied object doesn't appear to respond to that selector, then `objc_msgSend` falls back to normal forwarding by calling `methodSignatureForSelector:` and `forwardInvocation:`. This topic is covered in the next section, "Normal Forwarding." `CacheProxy` forwards these requests to the proxied object as well. The rest of the `CacheProxy` methods are as follows:

```
-  (id)forwardingTargetForSelector:(SEL)selector {
    return self.object;
}

-  (NSMethodSignature *)methodSignatureForSelector:(SEL)sel
{
    return [self.object methodSignatureForSelector:sel];
}
-  (void)forwardInvocation:(NSInvocation *)anInvocation {
    [anInvocation setTarget:self.object];
    [anInvocation invoke];
}
@end
```

Normal Forwarding

After trying everything described in the previous sections, the runtime tries the slowest of the forwarding options: `forwardInvocation:`. This option can be tens to hundreds of times slower than the mechanisms covered in the previous sections, but it's also the most flexible. You are passed an `NSInvocation`, which bundles the target, the selector, the method signature, and the arguments. You may then do whatever you want with it. The most common thing to do is to change the target and `invoke` it, as demonstrated in the `CacheProxy` example.

If you implement `forwardInvocation:`, you also must implement `methodSignatureForSelector:`. That's how the runtime determines the method signature for the `NSInvocation` it passes to you. Often `methodSignatureForSelector:` is implemented by asking the object you're forwarding to.

There is a special limitation of `forwardInvocation:`. It doesn't support *vararg methods*. These methods, such as `arrayWithObjects:`, take a variable number of arguments. The runtime has no way to automatically construct an `NSInvocation` for this kind of method because it has no way to know how many parameters will be passed. Although many vararg methods terminate their parameter list with a `nil`, that is not required or universal (`stringWithFormat:` does not), so determining the length of the parameter list is implementation-dependent. The other forwarding methods, such as Fast Forwarding, do support vararg methods.

Even though `forwardInvocation:` returns nothing itself, the runtime system returns the result of the `NSInvocation` to the original caller. It does so by calling `getReturnValue:` on the `NSInvocation` after `forwardInvocation:` returns. Generally, you call `invoke`, and the `NSInvocation` stores the return value of the called method, but that isn't required. You could call `setReturnValue:` yourself and return. This capability can be handy for caching expensive calls.

Forwarding Failure

Okay, so you've made it through the entire message resolution chain and haven't found a suitable method. What happens now? Technically, `forwardInvocation:` is the last link in the chain. If it does nothing, then nothing happens. You can use it to swallow certain methods if you want to. But the default implementation of `forwardInvocation:` does do something. It calls `doesNotRecognizeSelector:`. The default implementation of that method just raises an `NSInvalidArgumentException`, but you could override this behavior. Doing so is not particularly useful because this method is required to raise `NSInvalidArgumentException` (either directly or by calling `super`), but it's legal.

You can also call `doesNotRecognizeSelector:` yourself in some situations. For example, if you do not want anyone to call your `init`, you could override it like this:

```
- (id)init {
    [self doesNotRecognizeSelector:_cmd];
}
```

This makes calling `init` a runtime error. Personally, I often do it this way instead:

```
- (id)init {
    NSAssert(NO, @"Use -initWithOptions:");
    return nil;
}
```

That way, it crashes when I'm developing, but not in the field. Which form you prefer is somewhat a matter of taste.

You should, of course, call `doesNotRecognizeSelector:` in methods such as `forwardInvocation:` when the method is unknown. Don't just return unless you specifically mean to swallow the error. That can lead to very challenging bugs.

The Flavors of objc_msgSend

In this chapter, I've referred generally to `objc_msgSend`, but there are several related functions: `objc_msgSend_fpret`, `objc_msgSend_stret`, `objc_msgSendSuper`, and `objc_msgSendSuper_stret`. The SendSuper form is obvious. It sends the message to the superclass. The `stret` forms handle most cases when you return a struct. This is for processor-specific reasons related to how arguments are passed and returned in registers versus on the stack. I don't go into all the details here, but if you're interested in this kind of low-level detail, you should read *Hamster Emporium* (see "Further Reading"). Similarly, the `fpret` form handles the case when you return a floating-point value on an Intel processor. It isn't used on the ARM-based processors that iOS runs on, but it is used when you compile for the simulator. There is no `objc_msgSendSuper_fpret` because the floating-point return matters only when the object you're messaging is `nil` (on an Intel processor), and that's not possible when you message `super`.

The point of all this is not, obviously, to address the processor-specific intricacies of message passing. If you're interested in that, read *Hamster Emporium*. The point is that not all message passing is handled by `objc_msgSend`, and you cannot use `objc_msgSend` to handle any arbitrary method call. In particular, you cannot return a "large" struct with `objc_msgSend` on any processor, and you cannot safely return a floating point with `objc_msgSend` on Intel processors (such as when compiling for the simulator). This generally translates into: Be careful when you try to bypass the compiler by calling `objc_msgSend` by hand.

Method Swizzling

In Objective-C, *swizzling* refers to transparently swapping one thing for another. Generally, it means replacing methods at runtime. Using method swizzling, you can modify the behavior of objects that you do not have the code for, including system objects. In practice, swizzling is fairly straightforward, but it can be a little confusing to read. For this example, you add logging every time you add an observer to `NSNotificationCenter`.

> **Since iOS 4.0, Apple has rejected some applications from the App Store for using this technique.**

First, you add a category on `NSObject` to simplify swizzling:

RNSwizzle.h (MethodSwizzle)

```
@interface NSObject (RNSwizzle)
+ (IMP)swizzleSelector:(SEL)origSelector
              withIMP:(IMP)newIMP;
@end
```

RNSwizzle.m (MethodSwizzle)

```
@implementation NSObject (RNSwizzle)
+ (IMP)swizzleSelector:(SEL)origSelector
              withIMP:(IMP)newIMP {
  Class class = [self class];
  Method origMethod = class_getInstanceMethod(class,
                                        origSelector);
  IMP origIMP = method_getImplementation(origMethod);

  if(!class_addMethod(self, origSelector, newIMP,
                  method_getTypeEncoding(origMethod))) {
    method_setImplementation(origMethod, newIMP);
  }

  return origIMP;
}
@end
```

Now, look at this code in more detail. You pass a selector and a function pointer (`IMP`) to this method. What you want to do is to swap the current implementation of that method with the new implementation and return a pointer to the old implementation so you can call it later. You have to consider three cases: The class may implement this method directly, the method may be implemented by one of the superclass hierarchy, or the method may not be implemented at all. The call to class_getInstanceMethod returns an IMP if either the class or one of its superclasses implements the method; otherwise, it returns NULL.

If the method was not implemented at all, or if it's implemented by a superclass, you need to add the method with class_addMethod. This process is identical to overriding the method normally. If class_addMethod fails, you know the class directly implemented the method you're swizzling. You instead need to replace the old implementation with the new implementation using method_setImplementation.

When you're done, you return the original IMP, and it's your caller's problem to make use of it. You do that in a category on the target class, NSNotificationCenter, as shown in the following code:

NSNotificationCenter+RNSwizzle.h (MethodSwizzle)

```
@interface NSNotificationCenter (RNSwizzle)
+ (void)swizzleAddObserver;
@end
```

NSNotificationCenter+RNSwizzle.m (MethodSwizzle)

```
@implementation NSNotificationCenter (RNSwizzle)
typedef void (*voidIMP)(id, SEL, ...);
static voidIMP sOrigAddObserver = NULL;

static void MYAddObserver(id self, SEL _cmd, id observer,
                          SEL selector,
                          NSString *name,
                          id object) {
  NSLog(@"Adding observer: %@", observer);

  // Call the old implementation
  NSAssert(sOrigAddObserver,
           @"Original addObserver: method not found.");
  if (sOrigAddObserver) {
    sOrigAddObserver(self, _cmd, observer, selector, name,
                  object);
  }
}
+ (void)swizzleAddObserver {
  NSAssert(!sOrigAddObserver,
           @"Only call swizzleAddObserver once.");
  SEL sel = @selector(addObserver:selector:name:object:);
  sOrigAddObserver = (void *)[self swizzleSelector:sel
                                withIMP:(IMP)MYAddObserver];
}
@end
```

You call swizzleSelector:withIMP:, passing a function pointer to your new implementation. Notice that this is a function, not a method, but as covered in "How Message Passing Really Works" earlier in this chapter, a method implementation is just a function that accepts an object pointer and a selector. Notice also the `voidIMP` type. See the section "Working with Methods and Properties," earlier in this chapter, for how this interacts with ARC. Without that, ARC tries to retain the nonexistent return value, causing a crash.

You then save the original implementation in a static variable, `sOrigAddObserver`. In the new implementation, you add the functionality you want and then call the original function directly.

Finally, you need to actually perform the swizzle somewhere near the beginning of your program:

```
[NSNotificationCenter swizzleAddObserver];
```

Some people suggest doing the swizzle in a `+load` method in the category. That makes it much more transparent, which is why I don't recommend that approach. Method swizzling can lead to very surprising behaviors. Using `+load` means that just linking the category implementation causes it to be applied. I've personally encountered this situation when bringing old code into a new project. One of the debugging assistants from the old project had this kind of auto-load trick. It wasn't being compiled in the old project; it just happened to be in the sources directory. When I used "add folder" in Xcode, even though I didn't make any other changes to the project, the debug code started running. Suddenly, the new project had massive debug files showing up on customer machines, and it was very difficult to figure out where they were coming from. So my experience is that using `+load` for this task can be dangerous. However, it's very convenient and automatically ensures that it's called only once. Use your best judgment here.

Method swizzling is a powerful technique and can lead to bugs that are hard to track down. It allows you to modify the behaviors of Apple-provided frameworks, but that can make your code much more dependent on implementation details. It always makes the code more difficult to understand. I typically do not recommend using method swizzling for production code except as a last resort, but it's extremely useful for debugging, performance profiling, and exploring Apple's frameworks.

> There are several other method swizzling techniques. The most common is to use `method_exchangeImplementations` **to swap one implementation for another. That approach modifies the selector, which can sometimes break things. It also creates an awkward pseudo-recursive call in the source code that is very misleading to the reader. This is why I recommend using the function pointer approach detailed here. For more information on swizzling techniques, see the "Further Reading" section.**

ISA Swizzling

As discussed in the "Understanding Classes and Objects" section, earlier in this chapter, an object's ISA pointer defines its class. And, as discussed in "How Message Passing Really Works" (also earlier in this chapter), message dispatch is determined at runtime by consulting the list of methods defined on that class. So far, you've learned ways of modifying the list of methods, but it's also possible to modify an object's class (ISA swizzling). The next example demonstrates ISA swizzing to achieve the same `NSNotificationCenter` logging you did in the previous section, "Method Swizzling."

First, you create a normal subclass of `NSNotificationCenter`, which you use to replace the default `NSNotificationCenter`:

MYNotificationCenter.h (ISASwizzle)

```
@interface MYNotificationCenter : NSNotificationCenter
// You MUST NOT define any ivars or synthesized properties here.
@end

@implementation MYNotificationCenter
- (void)addObserver:(id)observer selector:(SEL)aSelector
               name:(NSString *)aName object:(id)anObject
{
  NSLog(@"Adding observer: %@", observer);
  [super addObserver:observer selector:aSelector name:aName
             object:anObject];
}
@end
```

There's nothing really special about this subclass. You could +alloc it normally and use it, but you want to replace the default NSNotificationCenter with your class.

Next, you create a category on NSObject to simplify changing the class:

NSObject+SetClass.h (ISASwizzle)

```
@interface NSObject (SetClass)
- (void)setClass:(Class)aClass;
@end
```

NSObject+SetClass.m (ISASwizzle)

```
@implementation NSObject (SetClass)
- (void)setClass:(Class)aClass {
  NSAssert(
    class_getInstanceSize([self class]) ==
      class_getInstanceSize(aClass),
    @"Classes must be the same size to swizzle.");
  object_setClass(self, aClass);
}
@end
```

Now, you can change the class of the default NSNotificationCenter:

```
    id nc = [NSNotificationCenter defaultCenter];
    [nc setClass:[MYNotificationCenter class]];
```

The most important point to note here is that the size of MYNotificationCenter must be the same as the size of NSNotificationCenter. In other words, you can't declare any ivars or synthesized properties (synthesized properties are just ivars in disguise). Remember, the object you're swizzling has already been allocated. If you added ivars, they would point to offsets beyond the end of that allocated memory. This has a pretty good chance of overwriting the isa pointer of some other object that just happens to be after this object in memory. In all likelihood, when you finally do crash, the other (innocent) object will appear to be the problem. Tracking down this bug is incredibly difficult, which is why I take the trouble of building a category to wrap `object_setClass`. I believe it's worth the effort to include the `NSAssert` ensuring the two classes are the same size.

After you perform the swizzle, the impacted object is identical to a normally created subclass. This means that it's very low risk for classes that are designed to be subclassed. As discussed in Chapter 22, key-value observing (KVO) is implemented with ISA swizzling. This way, the system frameworks can inject notification code into your classes, just as you can inject code into the system frameworks.

Method Swizzling Versus ISA Swizzling

Both method and ISA swizzling are powerful techniques that can cause a lot of problems if used incorrectly. In my experience, ISA swizzling is a better technique and should be used when possible because it impacts only the specific objects you target, rather than all instances of the class. However, sometimes your goal is to impact every instance of the class, so method swizzling is the only option. The following list defines the differences between method swizzling and ISA swizzling:

- **Method Swizzling**

    - Impacts every instance of the class.

    - Is highly transparent. All objects retain their class.

    - Requires unusual implementations of override methods.

- **ISA Swizzling**

    - Impacts only the targeted instance.

    - Changes object class (although this can be hidden by overriding `class`).

    - Allows override methods to be written with standard subclass techniques.

Summary

The Objective-C runtime can be an incredibly powerful tool when you understand it. With it, you can modify classes and instances at runtime, injecting new methods and even whole new classes. When these techniques are used recklessly, they can lead to incredibly difficult bugs, but when the Objective-C runtime is used carefully and in isolation, it is an important part of advanced iOS development.

Further Reading

Apple Documentation

The following document is available in the iOS Developer Library at `developer.apple.com` or through the Xcode Documentation and API Reference.

Objective-C Runtime Programming Guide

Other Resources

Ash, Mike. *NSBlog*. A very insightful blog covering all kinds of low-level topics.

`www.mikeash.com/pyblog`

- Friday Q&A 2009-03-20: "Objective-C Messaging"
- Friday Q&A 2010-01-29: "Method Replacement for Fun and Profit." The method-swizzling approach in this chapter is a refinement of Mike Ash's approach.

bbum. *weblog-o-mat*. bbum is a prolific contributor to Stackoverflow, and his blog has some of my favorite low-level articles, particularly his four-part opcode-by-opcode analysis of `objc_msgSend`.

`friday.com/bbum`

- "Objective-C: Logging Messages to Nil"
- "objc_msgSend() Tour"

CocoaDev, "MethodSwizzling." CocoaDev is an invaluable wiki of all-things-Cocoa. The MethodSwizzling page covers the major implementations out there.

`cocoadev.com/MethodSwizzling`

Parker, Greg. *Hamster Emporium*. Although it doesn't have a lot of posts, this blog provides incredibly useful insights into the Objective-C runtime.

`www.sealiesoftware.com/blog/archive`

- "[objc explain]: Non-pointer isa"
- "[objc explain]: objc_msgSend_vtable"
- "[objc explain]: Classes and metaclasses"
- "[objc explain]: objc_msgSend_fpret"
- "[objc explain]: objc_msgSend_stret"
- "[objc explain]: So you crashed in objc_msgSend()"
- "[objc explain]: Non-fragile ivars"

Index